Cell and Material Interface

Interface

Advances in Tissue Engineering, Biosensor, Implant, and Imaging Technologies

Devices, Circuits, and Systems

Series Editor
Krzysztof Iniewski
Emerging Technologies CMOS Inc.
Vancouver, British Columbia, Canada

PUBLISHED TITLES:

Atomic Nanoscale Technology in the Nuclear Industry
Taeho Woo

Biological and Medical Sensor Technologies
Krzysztof Iniewski

Building Sensor Networks: From Design to Applications
Ioanis Nikolaidis and Krzysztof Iniewski

**Cell and Material Interface: Advances in Tissue Engineering,
Biosensor, Implant, and Imaging Technologies**
Nihal Engin Vrana

Circuits at the Nanoscale: Communications, Imaging, and Sensing
Krzysztof Iniewski

CMOS: Front-End Electronics for Radiation Sensors
Angelo Rivetti

**CMOS Time-Mode Circuits and Systems: Fundamentals
and Applications**
Fei Yuan

Design of 3D Integrated Circuits and Systems
Rohit Sharma

Electrical Solitons: Theory, Design, and Applications
David Ricketts and Donhee Ham

Electronics for Radiation Detection
Krzysztof Iniewski

Electrostatic Discharge Protection: Advances and Applications
Juin J. Liou

**Embedded and Networking Systems:
Design, Software, and Implementation**
Gul N. Khan and Krzysztof Iniewski

Energy Harvesting with Functional Materials and Microsystems
Madhu Bhaskaran, Sharath Sriram, and Krzysztof Iniewski

Gallium Nitride (GaN): Physics, Devices, and Technology
Farid Medjdoub

PUBLISHED TITLES:

FORTHCOMING TITLES:

Radio Frequency Integrated Circuit Design
Sebastian Magierowski

Semiconductor Devices in Harsh Conditions
Kirsten Weide-Zaage and Malgorzata Chrzanowska-Jeske

Smart eHealth and eCare Technologies Handbook
Sari Merilampi, Lars T. Berger, and Andrew Sirkka

Structural Health Monitoring of Composite Structures Using Fiber Optic Methods
Ginu Rajan and Gangadhara Prusty

Tunable RF Components and Circuits: Applications in Mobile Handsets
Jeffrey L. Hilbert

Wireless Medical Systems and Algorithms: Design and Applications
Pietro Salvo and Miguel Hernandez-Silveira

Cell and Material Interface

Advances in Tissue Engineering, Biosensor, Implant, and Imaging Technologies

EDITED BY **NIHAL ENGIN VRANA**
Protip SAS, Strasbourg, France

KRZYSZTOF INIEWSKI MANAGING EDITOR
CMOS Emerging Technologies Research Inc.
Vancouver, British Columbia, Canada

CRC Press
Taylor & Francis Group
Boca Raton London New York

CRC Press is an imprint of the
Taylor & Francis Group, an **informa** business

Contents

Preface

The biomaterials field has mostly been an application-driven field; most of the early research focused on finding suitable materials for applications to develop medical instruments, implants, etc. Due to myriad problems afflicting numerous targets, the requirements for biomaterials have been expanding and sometimes are contradictory. For this reason, a wide range of materials were tested and some types of materials (ceramics, metals, polymers, etc.) were found to be "biocompatible" and have been successfully utilized for clinical applications.

However, our current ambitions and challenges in the field require more than a tolerance from the host body. Especially in regenerative medicine, we need to direct, induce, and communicate with the host body to achieve healing. This brings us to the "cell–material interface," an interface that is created by our action and that holds one of the keys to control cell behavior in ways that would result in desired outcomes. As the biomaterials used for different parts of the body differ greatly, it has been difficult to develop generalized rules regarding cell–material interface, although many inspiring works have been conducted that elucidated some of the basis of cell–material interactions. The motivation for this book was to establish a starting point for elucidating and exploiting the different aspects of cellular interactions with materials for biomedical engineering. We have tried to cover as many aspects of the biomaterials field, such as biosensors, tissue engineering, and controlled delivery systems. We hope that this book will provide a strong overview of cell–material interactions for professionals, undergraduate and graduate students in the field.

2001: A Space Odyssey is one of my all-time favorite books/movies. Sometimes, our field seems to me like the "Monolith" in that movie. We place a black box into a patient's body (the habitat of the patient's cells) without warning the cells. We mean well, but that does not mean that the body will not protect itself against our intrusion. So it is up to us, biomedical engineers, to develop the necessary language with the host tissue in the form of physical, chemical, and biological properties of our engineered materials. A deeper understanding of cell–material interface is a step forward in this sense.

Acknowledgments

I thank the members of INSERM UMR 1121 Biomaterials and Tissue Engineering Unit for their help during the preparation of this book. I thank my wife, Alix Vrana, for her patience and her help with images during a rather busy period of our lives. I dedicate this book to my daughter, Luna Ender Vrana, for bringing the sense of wonder back in my life and inducing me to look at everyday things differently.

I acknowledge the following grants: EuroTransBio BiMOT Project (ETB-2012-32) and the European Union's Seventh Framework Programme for Research and Technological Development and Demonstration under Grant Agreement no. 602694 (IMMODGEL) and no. 606624 (NanoTi). The research being carried out within these projects mostly provides me with the insights related to cell–material interface.

Editors

Dr. Nihal Engin Vrana is director of fundamental research of Protip Medical. He earned his PhD in 2009 at Dublin City University as a Marie Curie ESR fellow. His major research interests are titanium implants, tissue engineering, cell encapsulation, immunomodulation, real-time monitoring of implants, and cell–biomaterials interactions. His unit includes two postdocs, a research engineer, and a PhD student. He is the scientific coordinator of two European projects (EuroTransBio Bimot and FP7 IMMODGEL). He has published 32 articles in peer-reviewed academic journals (with more than 500 citations, h index: 12) and 4 book chapters and holds 2 European patents (1 more in progress). His awards include the Parlar Foundation Thesis of the Year (2006) award, the ESB Translational Research award (2011), the 2nd Aegean R&D Patent competition 1st place award (2012), and the IFOS Outstanding Paper award (2013). He can be reached at e.vrana@protipmedical.com.

Krzysztof (Kris) Iniewski is managing R&D at Redlen Technologies Inc., a start-up company in Vancouver, Canada. Redlen's revolutionary production process for advanced semiconductor materials enables a new generation of more accurate, all-digital, radiation-based imaging solutions. Kris is also a founder of Emerging Technologies CMOS Inc. (www.etcmos.com), an organization of high-tech events covering communications, microsystems, optoelectronics, and sensors. In his career, Dr. Iniewski held numerous faculty and management positions at the University of Toronto, University of Alberta, SFU, and PMC-Sierra Inc. He has published over 100 research papers in international journals and conferences. He holds 18 international patents granted in the USA, Canada, France, Germany, and Japan. He is a frequent invited speaker and has consulted for multiple organizations internationally. He has written and edited several books for CRC Press, Cambridge University Press, IEEE Press, Wiley, McGraw-Hill, Artech House, and Springer. His personal goal is to contribute to healthy living and sustainability through innovative engineering solutions. In his leisurely time, Kris can be found hiking, sailing, skiing, or biking in beautiful British Columbia. He can be reached at kris.iniewski@gmail.com.

Contributors

Hani A. Alhadrami
Faculty of Applied Medical Sciences
and
Centre of Innovation in Personalised
  Medicine
King Fahd Medical Research Centre
King Abdulaziz University
Jeddah, Saudi Arabia

Jorge Almodovar
Department of Chemical Engineering
University of Puerto Rico, Mayaguez
Mayaguez, Puerto Rico

Heleine V. Ramos Avilez
Department of Chemical Engineering
University of Puerto Rico, Mayaguez
Mayaguez, Puerto Rico

Erkan Turker Baran
BIOMATEN, Center of Excellence
  in Biomaterials and Tissue
  Engineering
Middle East Technical University
Ankara, Turkey

David A. Castilla Casadiego
Department of Chemical Engineering
University of Puerto Rico, Mayaguez
Mayaguez, Puerto Rico

Alexandru Gudima
Medical Faculty Mannheim
Institute of Transfusion Medicine and
  Immunology
University of Heidelberg
Mannheim, Germany

Anwarul Hasan
Faculty of Engineering and
  Architecture
Department of Mechanical Engineering
American University of Beirut
Beirut, Lebanon

and

Department of Medicine
Center for Biomedical Engineering
Brigham and Women's Hospital
Harvard Medical School
and
Harvard-MIT Division of Health
  Sciences and Technology
Massachusetts Institute of Technology
Cambridge, Massachusetts

and

Department of Mechanical and
  Industrial Engineering
College of Engineering
Qatar University
Doha, Qatar

Kenneth Hung
Department of Mechanical Engineering
University of Connecticut
Storrs, Connecticut

Mohammad Ariful Islam
Department of Anesthesiology,
  Perioperative and Pain Medicine
Brigham and Women's Hospital
Harvard Medical School
Boston, Massachusetts

Ayad Jaffa
Faculty of Medicine
Department of Biochemistry and
    Molecular Genetics
American University of Beirut
Beirut, Lebanon

Minkle Jain
School of Materials Science
Japan Advanced Institute of Science
    and Technology
Ishikawa, Japan

Dilek Keskin
BIOMATEN, Center of Excellence in
    Biomaterials and Tissue Engineering
and
Department of Engineering Sciences
Middle East Technical University
Ankara, Turkey

Mahshid Kharaziha
Biomaterials Research Group
Department of Materials Engineering
Isfahan University of Technology
Isfahan, Iran

Beste Kinikoglu
Department of Medical Biology
School of Medicine
Acibadem University
Istanbul, Turkey

Julia Kzhyshkowska
Medical Faculty Mannheim
Institute of Transfusion Medicine and
    Immunology
University of Heidelberg
and
German Red Cross Blood Service
    Baden-Württemberg-Hessen
Mannheim, Germany

and

Laboratory for Translational Cellular
    and Molecular Biomedicine
Tomsk State University
Tomsk, Russia

Caitriona Lally
Centre for Medical Engineering Research
School of Mechanical and
    Manufacturing Engineering
Dublin City University
Dublin, Ireland

Kazuaki Matsumura
School of Materials Science
Japan Advanced Institute of Science
    and Technology
Ishikawa, Japan

Garrett Brian McGuinness
Centre for Medical Engineering Research
School of Mechanical and
    Manufacturing Engineering
Dublin City University
Dublin, Ireland

Adnan Memic
Center of Nanotechnology
King Abdulaziz University
Jeddah, Saudi Arabia

Neerajha Nagarajan
Bioengineering Program
University of Notre Dame
Notre Dame, Indiana

Mehdi Nikkhah
Harrington Department of Biomedical
    Engineering
School of Biological and Health
    Systems Engineering
Arizona State University
Tempe, Arizona

Richard O'Connor
Centre for Medical Engineering Research
School of Mechanical and
    Manufacturing Engineering
Dublin City University
Dublin, Ireland

Irina Pascu
Centre for Medical Engineering
    Research
School of Mechanical and
    Manufacturing Engineering
Dublin City University
Dublin, Ireland

Robin Rajan
School of Materials Science
Japan Advanced Institute of Science
    and Technology
Ishikawa, Japan

Vladimir Riabov
Medical Faculty Mannheim
Institute of Transfusion Medicine and
    Immunology
University of Heidelberg
Mannheim, Germany

and

Laboratory for Translational Cellular
    and Molecular Biomedicine
Tomsk State University
Tomsk, Russia

Harpinder Saini
Harrington Department of Biomedical
    Engineering
School of Biological and Health
    Systems Engineering
Arizona State University
Tempe, Arizona

John Saliba
Faculty of Engineering and
    Architecture
Department of Mechanical
    Engineering
American University of Beirut
Beirut, Lebanon

Feba S. Sam
Harrington Department of Biomedical
    Engineering
School of Biological and Health
    Systems Engineering
Arizona State University
Tempe, Arizona

Amir Sanati-Nezhad
Department of Medicine
Center for Biomedical Engineering
Brigham and Women's Hospital
Harvard Medical School
and
Harvard-MIT Division of Health
    Sciences and Technology
Massachusetts Institute of Technology
Cambridge, Massachusetts

Aysen Tezcaner
BIOMATEN, Center of Excellence in
    Biomaterials and Tissue Engineering
and
Department of Engineering Sciences
Middle East Technical University
Ankara, Turkey

Nihal Engin Vrana
Protip Medical
and
Faculty of Dentistry
University of Strasbourg
Strasbourg, France

Pinar Zorlutuna
Bioengineering Program
and
Department of Aerospace and
    Mechanical Engineering
University of Notre Dame
Notre Dame, Indiana

List of Abbreviations

2D	Two-dimensional
3D	Three-dimensional
4D	Four-dimensional
aa	Amino acid
ABM	Agent-based models
ADSC	Adipose-derived stem cell
AFM	Atomic force microscope
Ag	Silver
APTMS	(3-aminopropyltrimethoxy) silane
ART	Assisted reproductive techniques
ASP	Angelica sinensis polysaccharide
Au	Gold
β-TCP	Beta tricalcium phosphate
BADSCs	Brown adipose derived stem cells
BCECs	Bovine corneal endothelial cell
BMP-2	Bone morphogenetic protein 2
BMP-7	Bone morphogenetic protein 7
BSA	Bovine serum albumin
CA	Cellulose acetate
CaP	Calcium phosphate
CECs	Corneal endothelial cells
CIJ	Continuous inkjet printing
CMs	Cardiomyocytes
CNS	Central nervous system
COOH-PLL	Carboxylated-ε-poly-L-lysine
CPA	Cryoprotectant
CPHF	Chitosan–poly (ethylene glycol) hydrogel films
CS	Chondroitin sulfate
CVD	Cardiovascular disease
DAT	Decellularized adipose tissue
DBCO-Dex	Dibenzylcyclooctyne-substituted dextran
DCP	Dicalcium phosphate
DHT	Dehydrothermal treatment
DIJ	Drop on demand inkjet printing
DLP	Double-layered particles
DMAEMA	2-(dimethylamino)ethyl methacrylate
DN	Double Network
DNA	Deoxyribonucleic acid
DPL	Dip Pen Lithography
DWES	Direct-write electrospinning
ECs	Endothelial cells
ECM	Extracellular matrix

E. coli	*Escherichia coli*
EDC	1-Ethyl-3-[3-dimethylaminopropyl]carbodiimide hydrochloride
EG	Ethylene glycol
EGF	Epidermal growth factor
ELR	Elastin-like recombinamer
ERK	Extracellular signal–regulated kinase
ESCs	Embryonic stem cells
FAK	Focal adhesion kinase
FAT	Focal adhesion targeting
FBCs	Fetal bovine chondrocytes
FBS	Fetal bovine serum
FDM	Fused deposition modeling
FEM	Finite element modeling
FE-SEM	Field emission scanning electron microscopy
FGF-2	Fibroblast growth factor 2
FN	Fibronectin
FS	Force spinning
GAG	Glycosaminoglycans
GG-BAG	Gellan-gums (bioactive-glass-reinforced)
GF	Growth factor
HA	Hyaluronic acid/hydroxy apatite
HaCaT	Human adult low calcium high temperature
HEMA	Poly(2-hydroxylethyl methacrylate)
HES	Hydroxyl ethyl starch
HF	Hollow fiber
hiPSCs	Human-induced pluripotent stem cells
HLF	Human ligament fibroblasts
hMSCs	Human mesenchymal stem cells
hNSCs	Human neural stem cells
hPAMAM	Hyperbranched polyamidoamine
hPSCs	Human pluripotent stem cell
Hz	Hertz
IFN-γ	Interferon-γ
IH	Intimal hyperplasia
IL-1RA	Interleukin 1 receptor antagonist
IL-8	Interleukin 8
LbL	Layer-by-layer
LCST	Lower critical solution temperature
MAA	Methacrylic acid
MAC	Methacrylamide chitosan
MAP2	Microtubule-associated protein 2
MAPK	Mitogen-activated protein kinase
MCS	Methacrylated chondroitin sulfate
μ-CP	Microcontact Printing
μ-CT	Micro-computer tomography
MeGG	Methacrylated Gellan Gum

MGC	Methacrylated glycol chitosan
MI	Myocardial infarction
MIP	Mercury intrusion porosimetry
miRNA	Micro-ribonucleic acid
MMP	Matrix metalloproteinase
MTS	(3-(4,5-dimethylthiazol-2-yl)-5-(3-carboxymethoxyphenyl)-2-(4-sulfophenyl)-2H-tetrazolium)
MTT	Methylthiazol tetrazolium
MVN	Microvascular network
MWCNTs	Multi-walled carbon nanotubes
N	Newton
NFC	Nanofibrillar cellulose
NSPCs	Neural stem/progenitor cells
OTS	Octadecyltricholorosilane
OU	Organoid units
Pa	Pascal
PA	Peptide amphiphiles
PAN	Polyacrylonitrile
PBT	Poly-(butylene terephthalate)
PCL	Poly(ϵ-capro-lactone)
PCL-HEMA	Poly (ϵ-caprolactone)-2-hydroxylethyl methacrylate
PCNU	Polycarbonate urethane
PDMAAm	Poly(N,N'-dimethyl acrylamide)
PDMS	Polydimethylsiloxane
PEEK	Polyether ether ketone
PEG	Poly-(ethylene glycol)
PEGT	Poly-(ethyleneglycol-terephthalate)
PEO	Polyethylene oxide
PES	Polyethersulfone
PEU	Polyester urethane
PGA	Poly glutamic acid
PHB	Poly(3-hydroxybuterate)
PLA	Poly-lactic acid
PLA-PEG	Poly-lactide–poly(ethylene glycol)
PLGA	Poly(lactic acid-co-glycolic acid)
PLL	Poly-L-lysine
PLLA	Poly-L-lactic acid
P(LLA-CL)	Poly(L-lactic-co-ϵ-caprolactone)
PLT	Powder layer thickness
PMMA	Poly(methyl methacrylate)
pNIPAAM	Poly(N-isopropyl acrylamide)
pN$_3$Phe	*Para*-azido-phenylalanine
PNT	Peptide nanotubes
PRP	Platelet-rich plasma
PS	Polystyrene
PU	Polyurethane

PVA	Poly(vinyl acetate)/poly(vinyl alcohol)
PVDF	Poly(vinylidene difluoride)
RAFT	Reversible addition-fragmentation chain transfer
RGD	Arginine-glycine-aspartic acid
RM	Regenerative medicine
RNA	Ribonucleic acid
RNT	Rosette nanotube
ROCK	Rho-kinase
ROS	Reactive oxygen species
rpm	Rotations per minute
RT-PCR	Reverse transcription polymerase chain reaction
RTK	Receptor tyrosine kinases
SA	Self-assembled
SAM	Self-assembled monolayer
SBF	Simulated body fluid
SCS	2-N,6-O-sulfated chitosan
SEM	Scanning electron microscopy
SF	Silk fibroin
SL	Stereolithography
SLS	Selective laser sintering
SMCs	Smooth muscle cells
SWNTs	Single-walled carbon nanotubes
TCPS	Tissue culture polystyrene
TE	Tissue engineering
TEBV	Tissue-engineered blood vessels
TEM	Transmission electron microscopy
TESI	Tissue-engineered small intestine
TGF-β1	Transforming growth factor β1
TP	Tecophilic
TUJ1	Neuron-specific class III beta-tubulin
UV	Ultraviolet
VEGF	Vascular endothelial growth factor

1 Editorial
Introduction to Cell/Material Interface

Nihal Engin Vrana

CONTENTS

Cell/material interface is a seemingly self-explanatory definition that denotes how cells, particularly human cells, interact with a given surface designed for a biomedical application (Anselme et al. 2010). However, when we look at the definition of the word interface, it becomes evident that it is necessary to elaborate on what is really meant by cell/material interface.

Interface is most commonly used for the components of a computing system. A touchscreen is our interface with many devices that we use on a daily basis through which we can send commands. This exchange of information does not necessarily have to be between the user and the device, as different parts of the devices need to have interfaces to each other for functioning properly and in synchrony. The original concept, however, derives from chemistry where it describes the surface between different phases of a given material (e.g., for water, the surface between a body of water and water vapor), different phases of matter in general (such as any liquid with air), or immiscible liquids (such as oil and water). Here, the interface is more of a discontinuity between two systems, a surface that denotes the change from one condition to another. This interface is not formed to exchange information; it is the boundary where the interaction, if there is any, happens. So, where does the interface between the materials and cells stand with respect to these definitions?

The interface between the cells and materials falls somewhere in between these two conditions. The interface is the surface where the material comes into contact with the cells and cells and the materials are mostly two separate entities (with certain exceptions, the material can be engulfed if it is smaller than the size of the cell, such as in the case of nanoparticles). However, unlike in the case of the chemical interfaces, here there is a definitive exchange of information, as the cell processes the signals from the material and decides how to interact with it according to its physical, chemical, and biological properties (Stevens and George 2005). Here, the second line of information is encrypted into the material by design if the material in question has been developed for a biomedical application. Depending on the information we would like to convey to the cells, we need to select the right type of material in the right physical form that would induce the cells to behave in the way we want. This requires an in-depth understanding of the physical and chemical underlying principles of the cell/material interactions and the main governing parameters that decide the final outcome (Harunaga and Yamada 2011; Glinel et al. 2012; Fujie et al. 2014).

This is one of the main focuses of the biomaterials field that designs and develops new materials or modifies existing materials for biomedical applications. The needs are application-specific, which determines the material selection. For example, for hip implants the requirement is to have a biomaterial that would be strong enough to bear the load of the body without failure and thus the obvious selection is the biocompatible metals, such as titanium and its alloys (Ryan et al. 2006). For vascular prosthesis, the conditions are completely different, where a reasonably elastic material that does not induce blood coagulation and deposition from the blood to avoid clogging and that can withstand the circumferential stress applied by the pulsatile blood flow is essential. Here, the choice is generally synthetic polymers that have been shown to have these properties, such as Dacron and PTFE (Kakisis et al. 2005).

The definition of biocompatibility has evolved over the years as the application of materials for medical purposes has diversified significantly (Williams 2008). Originally, the aim was to have bioinert materials that do not induce any adverse immune reactions by the body and can resist the aggressive physiological conditions to be able to achieve their role as a permanent or temporary implant. However, this definition would not cover the structures and systems developed for tissue engineering and regenerative medicine. In regenerative medicine, the aim is the complete integration of the material with the body and its population with the host cells (Sacks et al. 2009). Here, bioinertness would be an obstacle and the real need is biodegradability and remodellability by the host cells. Thus, the main choices of material are natural or synthetic biodegradable polymers in porous forms or as hydrogels (Hollister 2005). All these examples demonstrate the vast array of interfaces formed between the cells and the materials.

Biomaterials are biocompatible in the sense that they can interact with cells in desirable ways. But just because the materials themselves are compatible with the target cells does not mean that the interaction is similar to the interaction of cells with their surroundings under physiological conditions. A human cell rarely comes across titanium, chitosan, or poly(lactic acid) in its lifetime unless these materials are intentionally placed within the host to take over a function. This fact makes the definition of the cell/material interface and the governing laws of this interface harder as this is an artificial, material-type and cell-type-dependent interface. Also, no matter how well the interaction of the material with the cells of the target tissue, for example, bone tissue, is studied, the immune response with several cellular components, including neutrophils, macrophages, etc., should be taken into account for the final application. So, it is important to understand the interactions from a basic chemical and physical point of view by using the cell–extracellular matrix (ECM) interactions as a template (Zamir and Geiger 2001). Just as in the case of cell/ECM interactions, cell interface with the materials is bidirectional, especially for enzymatically degradable materials, that is, cells actively change the properties of the material they are in contact with via synthesis, degradation, and remodeling while the changes they exert will have an effect on their behavior in turn. Even though the bulk properties of the material such as its mechanical properties are of great importance for cell behavior, much of the interactions of the cells with a given material is governed by the interfacial properties of the material. Due to this, a big part of current biomaterials research is dedicated to the definition of cellular microenvironments (Stegemann et al. 2005; Yeung et al. 2005).

Cell migration is a prime demonstration of the intertwined effects of the microenvironment and the bulk properties (Vautier et al. 2003). Cells generally migrate more on the surfaces where they find it hard to attach and they stay in search for the optimal attachment conditions. This is generally formulated in the way that the cell area is inversely proportional to the migration speed of cells, but the way this translates to 3D movement is not straightforward. For tumor cell migration in 3D (e.g., within a hydrogel where cells are in contact with the surrounding gel material in every direction), the movement is governed by the enzymatic activity of the cells and the presence of ligands (local interface). But if the ligand amount is kept constant, the movement is governed by the stiffness of the 3D structure (bulk property). Cell migration also provides a good example of how the properties of the interface are transferred to the cell, in this case via cell cytoskeleton, and how they directly are related to the amount of deformation the cells can exert on the material they are in contact with (which is determined by the bulk mechanical properties of the material) (Zaman et al. 2006). The counterbalance between the traction forces by the cells and the adhesivity of the surface is one of the tools that can be used to control cell movement in 3D for tissue engineering applications. As described by Zaman et al., this can be broken down to three interrelation parameters between the interface of cells with the material: (1) cell adhesiveness to the material and cell-generated force; (2) cell adhesiveness to the material and matrix stiffness; and (3) traction force and steric hindrances created by the material. In order to fully exploit these interactions, not only the physical properties of the material but also its temporal interactions with the cells need to be engineered. But such a design requires a much better and quantitative definition of the cell/material interface. This would also require real-time data. Time-lapse videos are an effective way to analyze the temporal aspect of cell interaction with surfaces as it is possible to monitor the cellular preferences in a continuous fashion (Miller et al. 2011). This provides a better understanding of the process compared to the end-point analysis of cell behavior, which focuses more on the differences between different surfaces rather than the actual interaction of the cell with a given surface.

Currently, such effects are quantified in an empirical and specific material-based fashion. When a new biomaterial is designed, it is first tested for its interaction with cells. This is generally done by checking the spreading, proliferation, migration, or differentiation of the cells. For example, Hjortnaes et al. (2015) demonstrated that utilization of a methacrylated gelatin/hyaluronic acid hydrogel keeps Valvular interstitial cells in a quiescent form whereas gelatin-only gels result in their differentiation to myofibroblast form, similar to their spontaneous differentiation in 2D conditions. However, it is not very easy to deconvolute the effect of hyaluronic acid presence from a biological point of view (the evolutionarily conserved interactions between the interstitial cells and hyaluronic acid), chemical point of view (addition of a sugar-based polymer in an amino acid–based polymer environment, the resulting co-cross-linked network and the noncovalent interactions between the two polymers) and physical properties (mechanical properties, swelling properties of the gel structure, etc.).

One of the biggest challenges in the determination of the effects of interfacial properties is the high level of interdependence of several properties with each other. Chemical composition of the surface directly affects the hydrophilicity and protein adsorption properties. Similarly, surface roughness has a direct effect on the protein adsorption and hydrophilicity (Park et al. 2011; Schmidt et al. 2011; Gomes et al. 2012). There is also constant change of interface properties due to the release of proteolytic or other enzymes (such as hyaluronidase) that can have direct effects on the interface properties. The understanding of these interactions necessitates new high-throughput methods that can show the underlying principles between these interactions. The emerging field of materiomics is trying to answer these questions by simultaneous monitoring of different combinations of material systems and their overall biological effect and the emergent properties of cell/material systems (Cranford et al. 2013).

Together with the analysis of the effects of the biological building blocks at different length scales (from nano to macro) on cells, in the near future we might have a full picture of the cellular interactions with natural and synthetic materials. But, as the materiomics is still in its infancy, in this book, we will mostly focus on the material properties that have been shown to be important parameters in controlling cell behavior. As they are mentioned heavily in the upcoming chapters, some of these properties will be briefly described and exemplified here: hydrophilicity and topography. The upcoming chapters will provide more details on specific properties pertinent to the applications they cover.

Surface hydrophilicity or hydrophobicity refers to the quantitative interaction of a given surface with water. On highly hydrophilic surfaces, water can freely spread over the surface instantaneously and result in the complete coverage of the surface. Whereas on hydrophobic surfaces, water molecules, due to their limited interactions with the surface, cannot move freely and they form a drop that is related to the level of hydrophobicity of the surface (Zhao et al. 2005). A method based on the application of small liquid drops on surfaces followed by the quantification of the angle between the solid and the liquid phases via image analysis (the sessile drop method) has become the standard method for the determination of the surface hydrophilicity of materials. Although the sessile drop method is the most commonly used method, dynamic contact angle measurements generally provide a better understanding of the surface.

Stem cells are of particular interest for tissue engineering and regenerative medicine applications, thus their interaction with material surfaces have been widely studied. Protecting the stemness of the stem cells and finely directing their differentiation to a given cell type is essential for creating artificial tissues. The interactions of the stem cells with their microenvironment (stem cell niche) largely determine the outcomes (Jha et al. 2011; Wu et al. 2011). The stem cell niche itself has more components than the material aspect as the interaction of the cell with the surrounding cells and also soluble bioactive agents such as growth factors and hormones are also important determinants of its behavior. Artificial biomaterials mainly mimic the interaction of stem cells with the ECM surrounding them. Recent years have seen a mounting interest in the effect of interfacial properties such as interfacial energy, surface micro/nano topography, and surface hydrophilicity/hydrophobicity on stem cell behavior. For example, it has been shown that hydrophobicity of the underlying matrix can have a positive effect on stem cell self-renewal and attachment.

To be able to deconvolute the bulk effects from hydrophobicity, Ayala et al. (2011) developed synthetic matrices where the hydrophobicity of the surface was only controlled by the changes of the alkyl chain length, which means that the chemistry the cells are exposed to and the mechanical properties stayed the same. This was achieved by copolymerization of acrylamide with acryoryl amino acid groups with pendant CH_2 groups of varying length (which is denoted as C1, C2 up to C10). As can be seen from Figure 1.1, these chains have a direct effect on cell spreading. With a linear increase in surface hydrophobicity, cell spreading has improved up to a certain level after which hydrophobic surfaces are not amenable to cell attachment either. The surface has also effects on cell attachment strength, movement, and osteogenic differentiation. An optimal contact angle for surfaces that are neither very hydrophilic nor very hydrophobic has been observed in many systems and although there is not a single optimal contact angle, a range of 50°–60° has been considered as a suitable range.

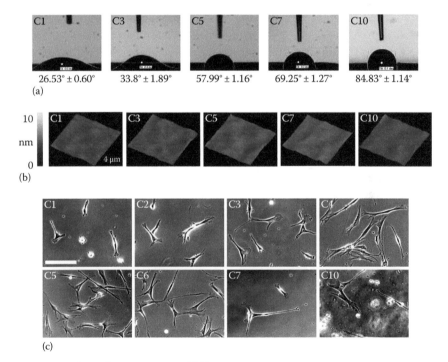

FIGURE 1.1 Effect of surface hydrophilicity/phobicity on cell attachment. The type of the alky chains used determines the extent of cell attachment. (a) The measurement of contact angles on the surfaces with different alkyl chain length. (b) Atomic force microscopy (AFM) images of the surfaces, the alkyl chain length does not significantly effect the surface roughness. (c) Stem cell behavior on the designed surfaces, the highest spreading was achieved in intermediate chain lengths. (Reprinted from Ayala, R. et al., *Biomaterials*, 32(15), 3700, 2011. With permission.)

This study shows that by using this optimal range for the interface, many properties of cells in contact with the surface can be modulated. Hydrophobicity or hydrophilicity of a surface has direct effects on protein adsorption, depending on the hydrophobic or hydrophilic nature of the protein. This property also affects the way in which configuration the protein will be adsorbed (Ker et al. 2011; Assal et al. 2013). For example, it has been observed that fibronectin, an important ECM component having a role in cell adhesion, adsorption on the gels with an intermediate hydrophobicity was significantly higher than that of more hydrophilic or more hydrophobic ones.

In 3D structures, porosity is a very important parameter in order to enable population of thick structures by cells. Porous structures have been used as tissue engineering scaffolds and also porosity has been introduced to the surfaces of several orthopedic implants to improve their anchorage by the surrounding bone tissue (Karageorgiou and Kaplan 2005; Li et al. 2007). Here, another interesting parameter, which is the discontinuity between the interfaces due to the porosity, comes into play, when the effect of the material interface on cellular movement and proliferation is considered. Our group has been working on titanium microbead–based porous structures over 10 years, where complex 3D openly porous structures can be obtained by sintering titanium microbeads of a given size (from 150 to 500 μm). In this case, the porosity of the structure is a function of the bead size (Vrana et al. 2014). Recently, we have shown that the distance between the beads in 3D significantly affects cell movement and ECM secretion both in vitro and in vivo (Figure 1.2). Cells' ability to move between the beads defines the

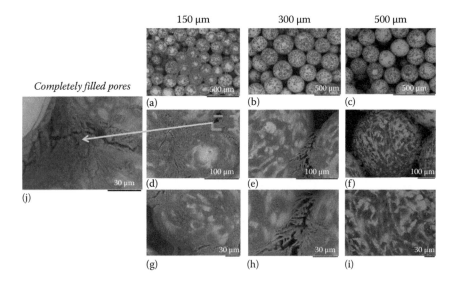

FIGURE 1.2 The effect of bead size on porous titanium implant population. As the bead size decreases, the ability of cells to form contacts between the beads increases; this leads to a faster population of the 3D structure. SEM microgrpahs of fibroblast population of porous titanium implants where the average bead size is 150 μm (a, d, g), 300 μm (b, e, h) and 500 μm (c, f, i) respectively. Close-up of a completely filled pore in implants formed of 150 μm microbeads (j). (Used from Vrana, N.E. et al., *Adv. Healthcare Mater.*, 3(1), 79, 2014. With permission.)

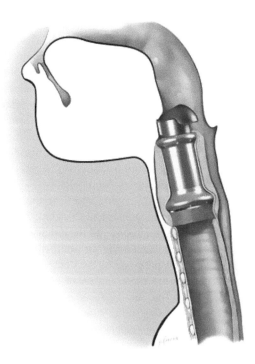

FIGURE 1.3 ENTegral Artificial Larynx, which has already been implanted in humans, uses titanium microbead-based connector to ensure the connection between the implant and the patient's trachea and the surrounding soft tissue. (Courtesy of Protip Medical, Strasbourg, France.)

extent of their overall 3D movement and distribution. We have exploited this feature in our Artificial Larynx (ENTegral™) (Debry et al. 2014) for enabling the faster integration of the implant in humans (Figure 1.3).

Another aspect of many tissues is their multicellular and compartmental nature (Li et al. 2000; Liu et al. 2010). Different parts of the tissue have distinct characteristics important for their function. This complicates the design of scaffolds for these tissues and, generally, necessitates structures formed of several components. For example, cornea, the outermost layer of the eye, is composed of five distinct layers. From the outer limit to inward, first there is an nonkeratinized, stratified, squamous epithelial layer lined with Bowman's layer followed by a layer of thick, highly ordered corneal stroma filled with corneal fibroblasts (keratocytes), a thin Descemet's membrane, and, finally, a corneal endothelial lining. Light transmittance is an important function of cornea that is provided by the physical structure of cornea and the specific activities of the corneal cells. These different structures can only be mimicked by distinctly different biomaterial designs in different parts of the artificial cornea (Nishida et al. 2004). For mimicking thick stroma of cornea (about 400 µm), collagen have been widely used in conjunction with foams containing glycosaminoglycans such as chondroitin sulfate, which is naturally present in cornea. Acun and Hasirci (2014) used a combination of a collagen/chondroitin sulfate foam (stromal replacement) and electrospun collagen fibers (as Bowman's layer replacement) in order to produce a split thickness cornea. In this setting, the 400 μm-thick foam is used as a 3D structural material for corneal keratocytes and the epithelial cells are seeded onto the thin electrospun layer so that the epithelial layer and stromal layer can be kept separate but having contact with each other due to the porous nature of the electrospun fibers. In a related study, Kilic et al. (2014) tried to achieve the same effect by stacks of patterned collagen films, and this study showed that the patterned structures indeed induced improved transparency in a cell-specific manner (improved transparency for corneal keratocytes but no improvement for another fibroblast-type 3T3 mouse fibroblasts). These studies show the importance of discrete interfaces within 3D structures to induce functional responses from cells.

Nanoscale topography is another route of controlling the cellular behavior while concomitantly affecting the bacterial attachment (Yu et al. 2011), thus playing a role in the race to the surface between microbes and the eukaryotic cells in implant scenarios. One such surface is the cicada wing–inspired titanium nanowires. Cicada wings are bactericidal due to the presence of the dense nanopillar arrays on them, which kills bacteria by perforating them. This can be mimicked by development of titanium oxide nanowires on titanium surfaces by alkali treatments; by controlling the treatment conditions, the size of the nanowires can be controlled (Diu et al. 2014). These patterns show selective bacteriocidal activity toward mobile bacteria,

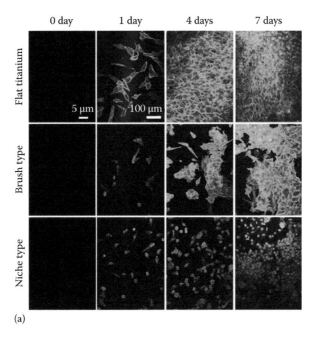

(a)

FIGURE 1.4 Effect of titanium nanowires on the movement and proliferation of human MG63 osteoblast-like cells. The interface formed by the nanowires decreases the movement and proliferation. (a) Flourescent microscope images of MG63 osteoblast like cells on flat titanium surfaces, brush type titanium oxide nanowires and niche type wires over 14 days.

(Continued)

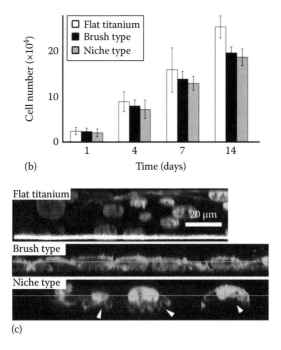

(b)

(c)

FIGURE 1.4 (*Continued*) Effect of titanium nanowires on the movement and proliferation of human MG63 osteoblast-like cells. The interface formed by the nanowires decreases the movement and proliferation. (b) Quantification of average cell numbers on the surfaces, by day 14 there were more cells on flat titanium surfaces. (c) Cross-sections of the MG63/ Titanium surfaces with confocal microscopy, showing the different mode of interactions of cells with flat titanium and nanowire containing surfaces. (Used from Diu, T. et al., *Sci. Rep.*, 4, 7122, 2014. With permission.)

while where restricting the movement of human osteoblasts and affecting their proliferation, without having a cytotoxic effect (Figure 1.4). Such multifunctional surfaces will also be covered in the upcoming chapters.

1.1 BOOK OUTLINE

Many critical health-care problems such as organ failure, chronic diseases that necessitate replacement of tissues, extensive damages to tissues due to accidents, etc., do not have remedies by standard pharmaceutical methods. Organ/tissue replacement or organ function replacement generally necessitates structures like implants or extracorporeal devices, where cells of the body interact with materials that in their natural state they would only interact during an insult (such as an injury). These new interfaces created by biomedical engineering are crucial for both solving the health problems that haunt our societies, but scientifically they are also important and interesting to understand how human cells of different origins interact with them. Most of the time, success or failure of an implant or a medical device can directly be linked to the interface it has with the body, that is, host cells. Aseptic loosening of hip implants and dental implants, neointimal hyperplasia within the stents, and clogging of artificial heart valves are some common examples where the implant surfaces triggers unwanted reactions that lead to additional complications. Thus, a comprehensive understanding of cell/material interface is crucial for future biomedical engineers for designing better implants, in vivo sensors, imaging reagents, and engineered tissues.

This book aims to provide the necessary background for appreciating and exploiting the cell–material interface for biomedical engineering applications. The book has been designed to give the reader the foundational information to appreciate the intricacies of the cell/material interactions while providing robust examples of how these interactions can be exploited for biomedical applications. Thus, each chapter not only provides specialized information about a field of biomedical engineering but also strives to include basic information to provide a better view of the underlying principles.

As we focus on cell/material interface, it is inevitable that ECM, the main structure cells are in contact with, comes up. So, most chapters cover different aspects of ECM and its interaction with the cells.

In *Chapter 2*, McGuinness et al. start with the material side of the interface, providing an overview of the extensive fabrication techniques that are utilized for production of cell interfacing materials. This allows the book to establish techniques such as electrospinning and hydrogel production methods, which are mentioned extensively in the later parts. They reviewed the 3D printing technologies that are becoming more and more relevant for biomedical applications, particularly as cell niches or scaffolds.

In *Chapter 3*, Tezcaner et al. look at the cell/material interface from a size-scale point of view and cover the interfaces at nanoscale. They describe the relevant cell/cell microenvironment (niche) interactions happening at nano level and how these interactions can be controlled for engineering tissues by use of nanotechnology and nanobiomaterials. They describe the ECM components at nanoscale, particularly for the stem cells and their niche. They also provide insights into the use of self-assembling nanostructures such as natural or synthesized peptides and also the use of inorganic nanomaterials (such as carbon nanotubes, bioglasses) for biomedical purposes.

In *Chapter 4*, Anwarul et al. cover most of the ECM interactions with cells and then go on to explain these interactions in the context of cell-based biosensors. This involves a review of the interactions of the cell surface proteins with ECM molecules, development of focal adhesions, and the underlying processes. This is necessary to understand how cell-based sensors can function as predictive tools for drug development and what are the necessities to provide the cells with right signals to ensure an environment that is physiologically relevant. They give specific examples of use of cellular systems as biosensors and also in tissue engineering applications.

In *Chapter 5*, Zorlutuna et al. cover the micropatterning of proteins for creating interfaces that can control the cell behavior at an individual cell level. Overall, surface engineering is an integral part of creating material/cell interfaces and the methods that are used to create such substrates are important tools to achieve cell control. They describe techniques based on photolithography, soft lithography, self-assembled monolayers, and direct write techniques used for patterning of proteins on surfaces with high fidelity and their respective advantages and disadvantages. They describe the processes involved in the absorption of proteins and their relevance to the cell surface interactions and the uses of such substrates for precise control of cell behavior.

In *Chapter 6*, Kinikoglu et al. focus on the interface properties in engineering skin tissue. Skin tissue engineering is one of the fields where another interface, air–liquid interface, is an important component, and this chapter explains how air–liquid interface can be used for keratinization of artificial skin substitutes and the methods to achieve functional tissue engineered skin. They describe the main parameters that influence the outcome of artificial skin tissue development, such as biopolymer chemistry, protein immobilization, and surface hydrophilicity, particularly from the cell/scaffold interface point of view.

In *Chapter 7*, Matsumura et al. define the conditions of cryoprotection and the material considerations for the successful protection of cells. They describe the cells' interface with their surrounding (such as forming ice crystal) below 0°C conditions, which results in a solid and mostly abrasive material. The development of the right conditions for cell preservation, particularly of nontoxic cryoprotectants, is an essential step in the biomaterials research, as they define the initial conditions of the cells used.

In *Chapter 8*, Nikkhah et al. focus on cardiac tissue engineering, and as cardiac tissue is a highly anisotropic tissue, the micropatterning methods that have been used to direct cardiac cells. They describe the structure/function relationship for the cardiac tissue and how the behavior of cardiac cells is governed by their interface with each other and the surrounding extracellular matrix. In this regard, they provide an overview of the current techniques to achieve mimicking the structure of cardiac tissue via 2D and 3D patterning methods.

In *Chapter 9*, Almodovar et al. provide the necessary background of the roles of polysaccharides in biology and their use as biomaterials. As mostly charged molecules, polysaccharides have been widely used in the development of LbL structures. LbL methodology as a way of developing interfaces that are controllable at nanoscale is explained extensively in this chapter according to its ability to control cellular behavior from adhesion to differentiation. From these surface coatings, they move to 3D structures and cell material interfaces in such structures, such as polysaccharide-based hydrogels and also polysaccharide-based nanofibers. They briefly discuss the novel sources of polysaccharides and their use as biomaterials.

Finally, in *Chapter 10*, Kzhyshkowska et al. review the cell/material interface from an immunology perspective. Cells of immune system, particularly macrophages, play an important role in the initial response to the biomaterials by the host and their reactions orchestrate the inflammatory processes. In this chapter, they describe the origin of macrophages, their subpopulations, their activation and roles in foreign body response. Understanding the biological processes leading to adverse reactions to implanted materials can significantly ameliorate the clinical outcomes and this chapter closes the book with the description of possible therapeutic control of macrophages.

ACKNOWLEDGMENTS

This project has received funding from EuroTransBio BiMot (ETB-2012-032), the European Union's Seventh Framework Programme for research, technological development, and demonstration under grant agreement nos. 606624 (NanoTi) and 602694 (IMMODGEL).

REFERENCES

Acun, A. and V. Hasirci. 2014. Construction of a collagen-based, split-thickness cornea substitute. *Journal of Biomaterials Science, Polymer Edition* 25(11):1110–1132.

Anselme, K., P. Davidson, A.M. Popa, M. Giazzon, M. Liley, and L. Ploux. 2010. The interaction of cells and bacteria with surfaces structured at the nanometre scale. *Acta Biomaterialia* 6(10):3824–3846.

Assal, Y., M. Mie, and E. Kobatake. 2013. The promotion of angiogenesis by growth factors integrated with ECM proteins through coiled-coil structures. *Biomaterials* 34(13):3315–3323.

Ayala, R., C. Zhang, D. Yang, Y. Hwang, A. Aung, S.S. Shroff, F.T. Arce, R. Lal, G. Arya, and S. Varghese. 2011. Engineering the cell–material interface for controlling stem cell adhesion, migration, and differentiation. *Biomaterials* 32(15):3700–3711.

Cranford, S.W., J. de Boer, C. van Blitterswijk, and M.J. Buehler. 2013. Materiomics: An-omics approach to biomaterials research. *Advanced Materials* 25(6):802–824.

Debry, C., A. Dupret-Bories, N.E. Vrana, P. Hemar, P. Lavalle, and P. Schultz. 2014. Laryngeal replacement with an artificial larynx after total laryngectomy: The possibility of restoring larynx functionality in the future. *Head & Neck* 36(11):1669–1673.

Diu, T., N. Faruqui, T. Sjöström, B. Lamarre, H.F. Jenkinson, B. Su, and M.G. Ryadnov. 2014. Cicada-inspired cell-instructive nanopatterned arrays. *Scientific Reports* 4:7122.

Fujie, T., Y. Mori, S. Ito, M. Nishizawa, H. Bae, N. Nagai, H. Onami, T. Abe, A. Khademhosseini, and H. Kaji. 2014. Micropatterned polymeric nanosheets for local delivery of an engineered epithelial monolayer. *Advanced Materials* 26(11):1699–1705.

Glinel, K., P. Thebault, V. Humblot, C.M. Pradier, and T. Jouenne. 2012. Antibacterial surfaces developed from bio-inspired approaches. *Acta Biomaterialia* 8(5):1670–1684.

Gomes, S., I.B. Leonor, J.F. Mano, R.L. Reis, and D.L. Kaplan. 2012. Natural and genetically engineered proteins for tissue engineering. *Progress in Polymer Science* 37(1):1–17.

Harunaga, J.S. and K.M. Yamada. 2011. Cell-matrix adhesions in 3D. *Matrix Biology* 30(7–8):363–368.

Hjortnaes, J., G. Camci-Unal, J.D. Hutcheson, S.M. Jung, F.J. Schoen, J. Kluin, E. Aikawa, and A. Khademhosseini. 2015. Directing valvular interstitial cell myofibroblast-like differentiation in a hybrid hydrogel platform. *Advanced Healthcare Materials* 4(1):121–130.

Hollister, S.J. 2005. Porous scaffold design for tissue engineering. *Nature Materials* 4(7):518–524.

Jha, A.K., X. Xu, R.L. Duncan, and X. Jia. 2011. Controlling the adhesion and differentiation of mesenchymal stem cells using hyaluronic acid-based, doubly crosslinked networks. *Biomaterials* 32(10):2466–2478.

Kakisis, J.D., C.D. Liapis, C. Breuer, and B.E. Sumpio. 2005. Artificial blood vessel: The Holy Grail of peripheral vascular surgery. *Journal of Vascular Surgery* 41(2):349–354.

Karageorgiou, V. and D. Kaplan. 2005. Porosity of 3D biomaterial scaffolds and osteogenesis. *Biomaterials* 26(27):5474–5491.

Ker, E.D.P., A.S. Nain, L.E. Weiss, J. Wang, J. Suhan, C.H. Amon, and P.G. Campbell. 2011. Bioprinting of growth factors onto aligned sub-micron fibrous scaffolds for simultaneous control of cell differentiation and alignment. *Biomaterials* 32(32):8097–8107.

Kilic, C., A. Girotti, J.C. Rodriguez-Cabello, and V. Hasirci. 2014. A collagen-based corneal stroma substitute with micro-designed architecture. *Biomaterials Science* 2(3):318–329.

Li, J.P., P. Habibovic, M. van den Doel, C.E. Wilson, J.R. de Wijn, C.A. van Blitterswijk, and K. de Groot. 2007. Bone ingrowth in porous titanium implants produced by 3D fiber deposition. *Biomaterials* 28(18):2810–2820.

Li, Y.W., W.H. Liu, S.W. Hayward, G.R. Cunha, and L.S. Baskin. 2000. Plasticity of the urothelial phenotype: Effects of gastro-intestinal mesenchyme/stroma and implications for urinary tract reconstruction. *Differentiation* 66(2–3):126–135.

Liu, J., Z. Bian, A.M. Kuijpers-Jagtman, and J.W. Von den Hoff. 2010. Skin and oral mucosa equivalents: Construction and performance. *Orthodontics & Craniofacial Research* 13(1):11–20.

Miller, E.D., K. Li, T. Kanade, L.E. Weiss, L.M. Walker, and P.G. Campbell. 2011. Spatially directed guidance of stem cell population migration by immobilized patterns of growth factors. *Biomaterials* 32(11):2775–2785.

Nishida, K., M. Yamato, Y. Hayashida, K. Watanabe, K. Yamamoto, E. Adachi, S. Nagai et al. 2004. Corneal reconstruction with tissue-engineered cell sheets composed of autologous oral mucosal epithelium. *New England Journal of Medicine* 351(12):1187–1196.

Park, H., B.L. Larson, M.D. Guillemette, S.R. Jain, C. Hua, G.C. Engelmayr, and L.E. Freed. 2011. The significance of pore microarchitecture in a multi-layered elastomeric scaffold for contractile cardiac muscle constructs. *Biomaterials* 32(7):1856–1864.

Ryan, G., A. Pandit, and D.P. Apatsidis. 2006. Fabrication methods of porous metals for use in orthopaedic applications. *Biomaterials* 27(13):2651–2670.

Sacks, M.S., F.J. Schoen, and J.E. Mayer. 2009. Bioengineering challenges for heart valve tissue engineering. *Annual Review of Biomedical Engineering* 11:289–313.

Schmidt, J.J., J. Jeong, and H. Kong. 2011. The interplay between cell adhesion cues and curvature of cell adherent alginate microgels in multipotent stem cell culture. *Tissue Engineering Part A* 17(21–22):2687–2694.

Stegemann, J.P., H. Hong, and R.M. Nerem. 2005. Mechanical, biochemical, and extracellular matrix effects on vascular smooth muscle cell phenotype. *Journal of Applied Physiology* 98(6):2321–2327.

Stevens, M.M. and J.H. George. 2005. Exploring and engineering the cell surface interface. *Science* 310(5751):1135–1138.

Vautier, D., J. Hemmerle, C. Vodouhe, G. Koenig, L. Richert, C. Picart, J.C. Voegel, C. Debry, J. Chluba, and J. Ogier. 2003. 3-D surface charges modulate protrusive and contractile contacts of chondrosarcoma cells. *Cell Motility and the Cytoskeleton* 56(3):147–158.

Vrana, N.E., A. Dupret-Bories, P. Schultz, C. Debry, D. Vautier, and P. Lavalle. 2014. Titanium microbead-based porous implants: Bead size controls cell response and host integration. *Advanced Healthcare Materials* 3(1):79–87.

Williams, D.F. 2008. On the mechanisms of biocompatibility. *Biomaterials* 29(20):2941–2953.

Wu, W., R. Allen, J. Gao, and Y.D. Wang. 2011. Artificial niche combining elastomeric substrate and platelets guides vascular differentiation of bone marrow mononuclear cells. *Tissue Engineering Part A* 17(15–16):1979–1992.

Yeung, T., P.C. Georges, L.A. Flanagan, B. Marg, M. Ortiz, M. Funaki, N. Zahir, W. Ming, V. Weaver, and P.A. Janmey. 2005. Effects of substrate stiffness on cell morphology, cytoskeletal structure, and adhesion. *Cell Motility and the Cytoskeleton* 60(1):24–34.

Yu, B., K.M. Leung, Q. Guo, W.M. Lau, and J. Yang. 2011. Synthesis of Ag-TiO$_2$ composite nano thin film for antimicrobial application. *Nanotechnology* 22(11):115603.

Zaman, M.H., L.M. Trapani, A.L. Sieminski, D. MacKellar, H. Gong, R.D. Kamm, A. Wells, D.A. Lauffenburger, and P. Matsudaira. 2006. Migration of tumor cells in 3D matrices is governed by matrix stiffness along with cell-matrix adhesion and proteolysis. *Proceedings of the National Academy of Sciences of the United States of America* 103(29):10889–10894.

Zamir, E. and B. Geiger. 2001. Molecular complexity and dynamics of cell-matrix adhesions. *Journal of Cell Science* 114(20):3583–3590.

Zhao, G., Z. Schwartz, M. Wieland, F. Rupp, J. Geis-Gerstorfer, D.L. Cochran, and B.D. Boyan. 2005. High surface energy enhances cell response to titanium substrate microstructure. *Journal of Biomedical Materials Research Part A* 74A(1):49–58.

2 Scaffold Processing Technologies for Tailored Cell Interactions

Garrett Brian McGuinness, Irina Pascu,
Caitriona Lally, and Richard O'Connor

CONTENTS

2.1   INTRODUCTION

Tissue engineering holds forth the promise of new regenerative therapies to address many forms of diseases and injuries, with the potential to greatly improve outcomes in fields such as wound care, ophthalmology, orthopedics, cardiac and vascular surgery, and organ replacement.[1–9] This chapter focuses on a selection of clinical needs amenable to tissue engineering solutions, the most promising scaffold fabrication technologies currently under investigation, their capabilities and limitations, and engineering insights into the regenerative process.

Tissue engineering scaffolds are intended to provide a structure and environment for attached or encapsulated cells to proliferate, differentiate, and synthesize extracellular matrix for new tissue. Scaffold properties should support and help control the biological processes necessary for tissue generation, ideally mimicking the ECM environment. In order to do this, their properties and features must be tailored over several size scales.[1] Fabrication technologies such as 3D printing, selective laser sintering, electrospinning, hydrogel fabrication protocols, and many others have been intensively investigated with respect to these requirements.

Scaffold properties that are important across various applications include surface topography, surface chemical composition, elastic properties, internal porous microarchitecture, degradation profile, and also protein adsorption and desorption characteristics.[10] These properties are clearly a function of biomaterial selection, in the first place, but are also very directly affected by the processing technologies employed to create the scaffold architecture.

This chapter is ultimately concerned with cell–material interactions in various types of tissue engineering scaffolds and how scaffold creation technologies can be exploited and improved to drive further progress in tissue engineering.

2.2   PROCESSING TECHNOLOGIES

2.2.1   ELECTROSPINNING

2.2.1.1   Fundamental Principles

Martin and Cockshott have reported the first use of electrospinning technique for biomaterial applications in 1977[11] while its application to scaffolds for tissue engineering has emerged more recently.[12] Electrospinning is a simple technique capable of producing nano- and microfibrous biomaterials, and allowing a degree of control over fiber morphology and alignment.[12] These nanofibrous materials have a high surface area to volume ratio, making them attractive for tissue engineering applications.

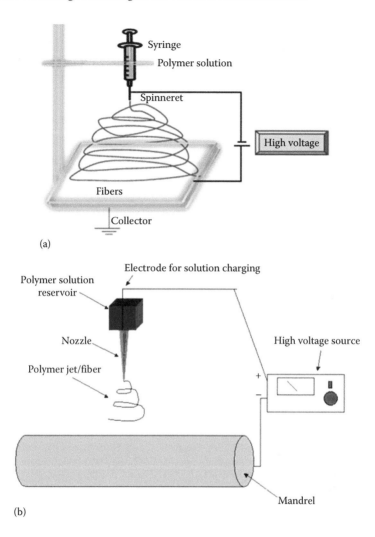

FIGURE 2.1 (a) Electrospinning of 2D membranes using standard electrospinning rig. (From Teo, W.E. and Ramakrishna, S., *Nanotechnology*, 17(14), R89, 2006.) (b) Electrospinning of aligned fiber using rotating drum collector. (Reprinted with permission from Matthews, J.A. et al., *Biomacromolecules*, 3(2), 232. Copyright 2002 American Chemical Society.)

The process typically involves a polymer solution or melt that is subjected to a high voltage (of the order of kilovolts) and brought into the vicinity of a grounded or oppositely charged collector at a controlled rate of flow (see Figure 2.1a).[13] A Taylor cone of the charged polymer forms and, when a threshold voltage is reached, a fiber is ejected.[14,15] As it leaps toward the collector, the fiber may begin to whip. This further reduces the diameter of the fiber, down to the nanoscale in some cases. If a solution rather than a melt has been used as the precursor, the solvent should evaporate before reaching the collector. Otherwise, a flattened fiber morphology will be observed. With a standard flat plate collector, the deposited fibers will be

TABLE 2.1
Key Parameters Affecting Electrospinning Processes

Polymer/Solution	Process Parameters	Environmental
Molecular weight	Voltage	Temperature
Solvent type	Tip-to-collector distance	Humidity
Concentration	Flow rate	
Viscosity	Capillary diameter	
Surface tension	Collection technique	
Electrical conductivity	Electric field	
Additives		
pH		

randomly aligned. The key material and process parameters affecting fiber diameter and morphology are listed in Table 2.1.

One of the most important aspects of tissue engineering is to introduce cells to a biomaterial structure in order to provide a scaffold on which they can anchor, migrate, and proliferate three-dimensionally. Cells binding to scaffolds with microscale architectures attach and spread according to the 3D architectures in question. Scaffolds with nanoscale architectures have bigger surface areas for absorbing proteins and present more binding sites to cell membrane receptors. The adsorbed proteins can further change the conformations, exposing additional binding sites, expected to provide an edge over microscale architectures for tissue generation applications.[16]

Electrospinning generates loosely connected 3D porous mats with high porosity and high surface area, which can mimic extracellular matrix structure and therefore makes itself an excellent candidate for use in tissue engineering. This technique has great potential for many tissues such as vessels,[17,18] bone,[19,20] neural tissue,[21,22] cartilage,[23,24] and tendons/ligaments.[25] It is believed that fibers formed by electrospinning have the ability to mimic, in a limited way, the natural extracellular matrix (ECM) structure in terms of variability of fiber diameter, topology, texture, and mechanical properties.[16] Natural ECM separates different tissues, forms a supportive meshwork around cells, and provides anchorage to the cells. It is made up of proteins and glycosaminoglycans (GAGs), which are carbohydrate polymers.

2.2.1.2 Collector Systems

Many factors can influence the formation of the fibrous structures obtained via electrospinning including solution properties, spinning properties, and the collector system used. Considerable effort has been spent on developing dynamic collector systems in order to increase the control of the fiber architecture within electrospun materials. Fiber alignment, in particular, is a key characteristic of an electrospun scaffold and greatly determines the mechanical properties of the structure.[25,26] Fiber alignment has also been shown to be an important factor for delivering contact guidance cues to cells when used in tissue engineering applications. Cells proliferate along the direction of deposited fibers and aligned fibers have been shown to promote ECM deposition compared to random fibers, increasing the mechanical properties of a tissue scaffold.[25,26]

Increased fiber alignment is often achieved through the use of bespoke collectors, such as metallic wire frames and rotating drums. Rotating collectors such as those shown in Figure 2.1b have been widely used to form highly orientated fibrous sheets and tubular scaffolds. Tubular scaffolds are often created with the view to developing tissue scaffolds for blood vessel applications. Increased fiber alignment is achieved during the deposition process due to the relative movement between the landing fiber and the rotating surface of the drum. This movement applies tensile forces to the landing fibers, pulling them into alignment. Matthews et al.[27] showed that at low rotational speeds of approximately 500 rpm deposited fibers still exhibit a randomly aligned nature upon deposition but at rotational speeds above 4500 rpm fibers aligned with the axis of rotation of the drum.

2.2.1.3 Nanoyarns

Over the past decade, a number of approaches have been developed for the production of highly orientated bundles of electrospun fibers referred to as electrospun yarns or nanoyarns.[26,28–31] Electrospun yarns offer the ability to create diverse textiles through fabric production techniques such as weaving, knitting, and embroidery while retaining the inherent benefits that electrospun nanofibers offer. Nanoyarns are ideal candidates for use in advanced applications as filtration devices, nanocomposite materials, sensors, and biomedical devices.

Considerable focus has centered on the development of medical textiles with the aim of producing cost-effective, mass-producible tissue scaffolds. Yarns are thought to be superior to traditional electrospun membranes for this application due to the increased control in mechanical and morphological properties offered along with the potential to produce complex 3D structures. The highly aligned substructure of yarns benefits cellular proliferation while multiscale porosity and permeability levels have been shown to increase cellular infiltration rates compared to alternative scaffold fabrication techniques.[26]

Ko et al.[32] investigated the production of poly(lactic acid) (PLA) and polyacrylonitrile (PAN) yarns containing single-walled carbon nanotubes (SWNTs). Here, a system comprised of rollers and twisting electrodes shown in Figure 2.2a were used to successfully gather and bundle PAN nanofibers with diameters ranging from 50–100 nm to form a continuous yarn. SWNTs were successfully embedded and orientated within the fibers to provide reinforcement to the yarns. Continuous twisted yarns were produced by Ali et al.[33] using the collection system shown in Figure 2.2b. In this set-up, dual oppositely charged spinnerets were placed in front of a rotating stainless steel funnel. Poly(vinylidene difluoride) (PVDF–HFP) was electrospun onto the edge of the rotating funnel forming a nanofiber web. The web was subsequently drawn off forming a hollow fiber cone that was collected on a rotating mandrel. It was observed that by increasing the rotational speed of the funnel both fiber and yarn diameters could be decreased. Additionally, increasing the rotational speed resulted in higher twist angles within the yarns, which increased the tensile strength and elongation to break of the yarns; however, at too high of a rotational speed a decrease in tensile strength was observed while elongation to break continued to increase.

Other collection techniques have employed liquid reservoirs as collecting targets for electrospun yarns. Smit et al.[34] produced nanoyarns from poly(vinyl acetate) (PVA), PVDF–HFP and PAN by electrospinning the materials onto a still water basin.

By lifting the deposited material off the surface of the water, the electrospun membranes collapsed into a continuous fiber bundle, which was subsequently collected on a rotating mandrel shown in Figure 2.2c. Further developments to this process utilized dynamic liquid collection techniques to produce nanoyarns. Teo et al.[35] employed water vortices within a collector tank to induce self-bundling of deposited electrospun materials. The vortex was formed within the collection tank by allowing liquid to drain from an outlet located in the base of the tank while a recirculating pump maintained a constant fluid height (Figure 2.2d). As the liquid drained, deposited fibers were drawn through the aperture with the flowing water and wound onto a mandrel. Wu et al.[36] showed that mouse fibroblast cultured on poly(L-lactic-co-ε-caprolactone) P(LLA-CL) nanoyarns created using this dynamic liquid technique showed increased proliferation and infiltration rates compared to those cultured on 2D P(LLA-CL) membranes. The cells exhibited elongated morphologies, grew along the nanoyarns, and also bridged gaps between adjacent yarn segments. Further studies by Wu ct al.[37] again showed with pig iliac endothelial cells (PIECs) and MC3T3-E1 pre-osteoblastic cells that nanoyarns accelerated proliferation rates of

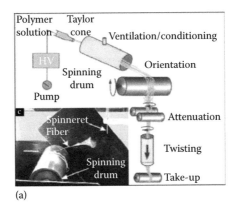

(a)

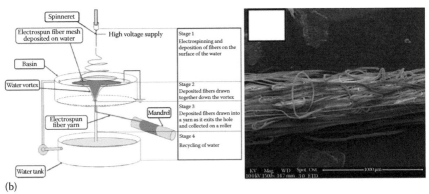

(b)

FIGURE 2.2   (a) Continuous electrospinning of PLA and PAN yarns embedded with SWNTs. (Reprinted from Ko, F. et al., *Adv. Mater.*, 15(14), 1161, 2003. With permission.) (b) Electrospinning of twisted PVDF-HFP yarns using rotating funnel. (Reprinted from Ali, U. et al., *J. Text. Inst.*, 103(1), 80, 2012. With permission.)   *(Continued)*

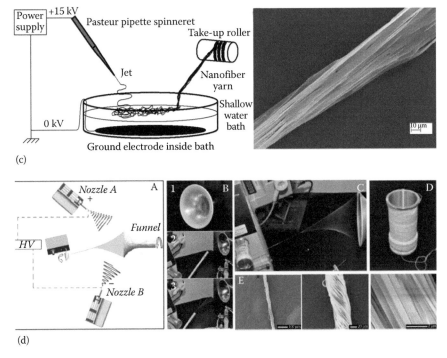

(c)

(d)

FIGURE 2.2 (*Continued*) (c) Electrospinning of yarn collected by water bath. (Reprinted from *Polymer*, 46(8), Smit, E., Büttner, U., and Sanderson, R.D., Continuous yarns from electrospun fibers, 2419–2423, Copyright 2005, with permission from Elsevier.) (d) Electrospinning of yarn using dynamic liquid collection. (Reprinted from *Polymer*, 48(12), Teo, W.-E., Gopal, R., Ramaseshan, R., Fujihara, K., and Ramakrishna, S., A dynamic liquid support system for continuous electrospun yarn fabrication, 3400–3405, Copyright 2007, with permission from Elsevier.)

the cells compared to those cultured on 2D membranes. After 7 days of culturing, both PIECs and MC3T3-E1 penetrated the full depth of the nanoyarn scaffolds while no apparent infiltration was seen on the conventional 2D scaffolds.

Nanoyarns show a promising future within the textile industry and may offer the solution to developing commercially successful tissue engineering scaffolds. Their properties provide the potential to bridge the gap between the nanoscale world and conventional processing methods. Future research is expected to look toward increasing the mechanical properties of these yarns and their manipulation into 3D structures for advanced tissue engineering applications.

2.2.2 SOLID FREEFORM FABRICATION

Technologies originally conceived for rapid prototyping or additive manufacturing offer the possibility of manufacturing predefined 3D shapes from a wide range of precursor materials, including some that are appropriate as tissue engineering scaffolds. The chance to generate complex shapes, based on the geometry of a defect at the site of an injury, or on the need to reconstruct an anatomical shape, is highly promising for tissue engineering. Perhaps more importantly, there may be the potential to reproduce anatomical microarchitecture, such as that of cancellous bone, for example, or any desired porous microstructure with cell-sized features. The main solid freeform fabrication technologies available are selective laser sintering, 3D printing, fused deposition modeling, and stereo lithography.

2.2.2.1 Selective Laser Sintering

Selective laser sintering involves the use of laser energy to sinter geometric patterns in successive layers of polymeric, metallic, or blended powders, creating a solid part with the intended 3D form. The process is illustrated schematically in Figure 2.3.[38]

In tissue engineering, this technology has been mainly directed toward the creation of bone scaffolds using biocompatible polymers, blends, or composites utilizing PMMA, PCL, PVA, PHBV, and PEEK. A common approach is to incorporate a bioceramic such as particles of hydroxyapatite (HA) to form a biocomposite and access the ensuing osteoconductive properties. Lee and Barlow first reported SLS processing of HA that had been spray coated with PMMA before being infiltrated by an inorganic phosphoric acid–based cement and heated to burn off the polymer resulting in porous bioceramic materials.[39] Subsequent work elaborated the mechanical and dimensional capabilities of this approach, including some early *in vivo* canine studies,[40] demonstrating biocompatibility and bone ingrowth. In order to avoid the necessity for solvents, Tan et al. investigated the effect of the three main process parameters on blends of PEEK and HA powders, obtaining composite structures with up to 40 wt% HA, concluding that higher HA content compromised structural integrity.[41] Since these scaffolds are intended to be polymer/ceramic composites, with HA present to enhance bioactivity, no burning off of polymer was carried out. Others have worked with poly(vinyl alcohol) (PVA),[42] which is not biodegradable, as the polymer and many studies have utilized polycaprolactone (PCL),[43] which has a slow degradation profile *in vivo*. PHB (poly(3-hydroxybuterate)), a biodegradable

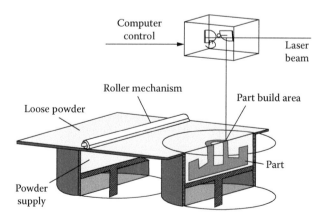

FIGURE 2.3 Schematic of the selective laser sintering process. (Reprinted from Bourell, D.L. et al., *CIRP Ann. Manuf. Technol.*, 60(1), 275, 2011. With permission.)

polymer of natural origin, has also been used to create geometrically controlled scaffolds with square-shaped pores using SLS and characterized with respect to pore area and compressive mechanical strength.[44] Powder layer thickness (PLT) had a greater effect on scaffold properties than the laser scan spacing (SS) and lower values of PLT or SS resulted in increased compressive mechanical properties. Eosoly et al. created composites of poly-ε-caprolactone and hydroxyapatite using SLS and studied the effects of laser fill power, outlined laser power, scan spacing, and part orientation on the accuracy and mechanical properties of the fabricated scaffolds.[45] Dimensional accuracy was strongly dependent on manufacturing direction (orientation of struts relative to the layering of the powder) and scan spacing. Further studies on the effects of cell culture (MC 3T3 osteoblast-like cells) on the scaffold's physical and mechanical properties were then considered, with evidence of an initial loss of compressive strength being subsequently reversed, and substantial morphological changes to the scaffolds.[46] Elastic moduli were unaffected, and μ-CT analysis showed a smoothening of the scaffolds surfaces under cell culture conditions.[46]

2.2.2.2 3D Printing

Three-dimensional printing typically involves printing a binder or a glue onto layers of polymeric, ceramic, or metallic powder. Following the printing of one pattern, a new powder layer of defined thickness is spread over the working area, and the printing head traces out the cross-sectional pattern for the next layer (Figure 2.4).

This technology has been extensively explored with respect to bone tissue scaffold applications, frequently employing calcium phosphate for its osteoconductive properties and moderate strength.[47,48] High temperatures can be used to bind the printed structure post hoc through sintering, or binders can be used to facilitate a low-temperature process (enabling the incorporation of bioactive molecules of natural origin).

Seitz et al. used 3D printing as part of a process chain to create sintered hydroxyapatite scaffolds with computer-controlled microarchitecture,[49] achieving inner-channel dimensions down to 450 μm and wall structure thicknesses of 330 μm.

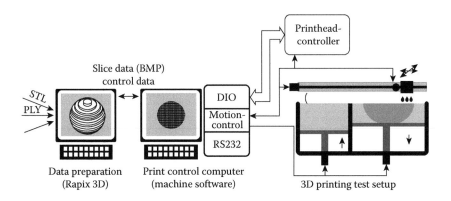

FIGURE 2.4   3D Printing process chain used by Seitz et al. (Reprinted from Seitz, H. et al., *J. Biomed. Mater. Res. B: Appl. Biomater.*, 74(2), 782, 2005. With permission.)

Zhou et al. have considered the effects of a process and material parameters on 3D printing of calcium sulfate with a water-based binder, including powder bed packing, powder particle size, CaP:CaSO$_4$ ratio, and the type of CaP (HA or β-TCP).[50] The scaffolds containing HA showed, generally, better compressive strength than their counterparts containing β-TCP.

Inzana et al. have recently developed a low-temperature process with a phosphoric acid–based binder solution for printing scaffolds with calcium phosphate and collagen.[51] Binder concentration was adjusted to limit cytotoxicity and produced scaffolds with microporosity in the 20–50 μm range. The presence of collagen in the binder solution significantly improved the scaffold's flexural strength. *In vivo* implantation in murine femoral defects showed evidence of new bone formation at 9 weeks, but no host–host bridging. Future incorporation of growth factors in this process may enable full healing of critically sized defects.

2.2.2.3 Fused Deposition Modeling

Fused deposition modeling (FDM) is distinguished from the previous two technologies by the fact that the material is directly deposited by a nozzle under 3D control. The feedstock polymer is usually in the form of a thin filament, which is melted and deposited. This is a relatively simple process and very popular for general rapid prototyping purposes. The process does not involve any solvent, chemical binders, or curing reactions. It is a low-cost process but can be slow, potentially requiring days to complete a large or complex structure.

FDM has been explored with respect to bone tissue engineering,[52,53] utilizing polymers such as polycaprolactone,[52] polypropylene,[53] polymethylmethacrylate,[54] and incorporating additives such as hydroxyapatite,[55] tricalcium phosphate,[53] or bioactive glasses.[56]

2.2.3 HYDROGELS

2.2.3.1 Fundamental Principles

Hydrogels are soft, hydrated materials consisting of cross-linked networks of polymeric chains. These polymer networks swell to many times their dry weight in water and exhibit rubber-like mechanical properties. The polymer networks may be formed either through covalent bonds between chains, physical cross-linking assisted by weak bonds such as hydrogen bonds, or by the formation of crystallites linking polymer chains. Their structure makes them potentially suitable for use as soft tissue scaffolds, but a limitation to their use in load-bearing applications has been their poor mechanical strength relative to the native tissue. Hydrogels based on either synthetic polymer chains, natural polymers, or composites have been extensively studied in relation both as direct articular cartilage substitutes and as support systems for cell-based therapies. Some of the most common material platforms as well as promising recent research advances are summarized in the following sections.

Many hydrogels share some of the physical characteristics of soft biological tissues, which are also highly hydrated, such as high diffusivity and nonlinear mechanical response characteristics under large strains. The key determinants of hydrogel properties include the molecular structure of the repeat chains, the nature and arrangement of the cross-linking, the water or fluid content, and any patterning or featuring at higher size scales. At the molecular level, hydrogel behavior is influenced by interactions between functional groups on its chains and the surrounding water molecules, affecting, for example, the conformations of the chain between cross-links. Chain lengths between network junctions also affect behavior, particularly mechanically, as would particulate reinforcement or the existence of interpenetrating networks of different polymers. Depending on the processing methods, gels may incorporate features such as pores, channels, surface texturing, or even bespoke 3D structuring. The final functionality of a hydrogel scaffold will depend, therefore, not only on the chemistry of the constituent polymers and additives but also on the precise route by which the gel is fabricated.

2.2.3.2 Cryogelation Processes

Cryogelation, meaning the formation of gels via physical cross-links formed during freezing and thawing cycles, is another promising technique for hydrogel scaffold preparation.[57,58] This technique is particularly associated with poly(vinyl alcohol), usually referred to as PVA or PVOH, producing cryogels with many interesting properties for scaffold applications.[59–65] In particular, the mechanical properties of PVA gels mimic the classic soft tissue tensile response characteristic and can be tailored through choice of the freezing and thawing rates and the number of cycles.[66–70] Relatively high strengths and failure strains can be achieved, which is important for load-bearing tissue applications, such as for cartilage or blood vessel scaffolds. The mechanical properties of the gels arise from the microstructure, specifically crystallites formed during the temperature cycles that act as physical cross-linking sites, assisted by other physical cross-linking effects such as entanglement.[71–76] Further cycles applied after the first freeze–thaw cycle mainly induce secondary crystallites in the amorphous region between the original cross-links, rather than introduce new cross-link junctions.[71] Crystals of the order of 28 Å have been measured after the first freeze–thaw cycle, increasing to 34 Å after the third freeze–thaw cycles.[77] Inter-crystallite distances of the order of 200 Å have been detected.[77]

Since PVA cryogels are hydrophilic and do not therefore present ideal surfaces for cell attachment, it is usually necessary to incorporate additives. Biomacromolecules such as gelatin, chitosan, and bacterial cellulose have been incorporated in PVA cryogels, leading to improved cell attachment without unduly compromising tissue-like mechanical properties.[61,62,78]

For cartilage applications, secure attachment of the scaffold or biomaterial to bone is highly desirable. The PVA cryogelation process can be modified to produce gels with advantageous properties in this respect. For example, nanohydroxyapatite (nHAp), a bioactive substance found in bone, has also been incorporated in PVA cryogels either directly or by *in situ* nHAp synthesis.[79–82] The process can also be modified to produce a functionally graded gel with multiple layers of different compositions and with different freeze–thaw histories.[81,82] Compressive mechanical properties and frictional properties, both key cartilage substitute requirements, can be fine-tuned using these parameters.

The cryogelation process has also been combined with cell encapsulation to provide cell-laden scaffolds and gels for therapeutic purposes.[64,65,83] The main difficulty with such encapsulation processes is the narrow cell survival window combined with the appropriate temperature cycle rates for effective cryogelation. Qi et al. used a freeze–thaw cryogelation protocol with PVA and pancreatic islets to produce macroencapsulated islets for diabetes treatment.[83] More recently, as part of a vascular tissue engineering study, carboxylated ε-poly-L-lysine (COOH–PLL) in combination with fetal bovine serum has been used with some success to retain vascular smooth muscle cell viability in a PVA gelatin cryogelation process.[64,65]

2.2.3.3 Photopolymerization

Hydrogel cross-links can be chemical or physical in nature, where chemical cross-linking involves the formation of covalent bonds to join chains. Photopolymerization is suitable for cross-linking some polymers and involves the use of a visible or ultraviolet light source, with or without the addition of a photoinitiator, to form the covalent bonds. Forming gel networks from polymeric solutions using a light source has many advantages. In particular, there is the opportunity to inject the biomaterial in the form of a liquid precursor that can conform to adjacent cavities or structures, followed by *in vivo* gelation after exposure to an endoscopic light source. In general, photopolymerization may also open up the possibility of encapsulating cells suspended in a solution, and so can be a route to delivering cell-laden scaffolds *in vivo*.

2.2.3.4 Double Network Hydrogels

One persistent issue with using many hydrogels in scaffold applications is lack of mechanical strength.[84] However, when prepared in a specific way, it is possible to produce hydrogels with two coexisting networks that exhibit strength properties significantly in excess of those associated with single network gels of either of the constituent polymers.[85] The original discovery of this effect showed strengths of the order of 20 MPa for hydrogels of PAMPS, poly(2-acrylamido-2-methylpropanesulfonic acid), and polyacrylamide containing 60%–90% water, together with high wear resistance due to their low coefficient of friction.

This strengthening effect is invoked when the molar ratio of the first to the second network is of the order of a few to several tens, and if the first network is loosely cross-linked but the second one highly cross-linked. Naturally, this phenomenon has attracted the attention of researchers interested in the development of artificial cartilage and other load-bearing tissues. The two structural parameters that are crucial in obtaining these high-strength gels are the molar ratio of the first to the second network and the cross-linking densities.[85] A dramatic improvement in the mechanical strength of the gel is observed only when the molar ratio of the second network to the first network is in the range of several tens. Another substantial increase in strength is observed when the first network is highly cross-linked and the second is loosely crosslinked.[85,86] This allows collagen or agarose to be used as the first network and synthetic polymers such as poly(2-hydroxylethyl methacrylate) (HEMA) and poly(N,N′-dimethyl acrylamide) (PDMAAm), which are used in contact lenses, as the second network. DN gels are produced simply by synthesizing the second network in the presence of the first network. The first network is immersed into the synthesizing medium containing the second network for 24 h until equilibrium is reached.

Several follow-up studies have attempted to explain the underlying mechanism for the impressive strength of these hydrogels, but a clear understanding of the source of this strength has yet to emerge.[86] Nevertheless, the possibility of such high-performance hydrogels, achievable through very specific combinations of solution and polymerization parameters, is likely to lead to new opportunities for application of hydrogels in advanced biomaterials applications.[87,88]

2.2.3.5 Vascularization of Scaffolds

Organ transplantation is often the only suitable treatment option for those with advanced disease, but donor organ shortages persist. The development of tissue-engineered artificial organs, such as livers, kidneys, hearts, and lungs, is therefore widely recognized to be an important long-term clinical need. Organs typically have considerable structural complexity at all size scales and require intricate vascularization networks to facilitate delivery of oxygen and nutrients. They are therefore inherently much more challenging to generate than simpler tissue structures. The need to create complex architectures with biochemical and mechanical gradients, compartmentalized cell encapsulation, and an interweaving vascular system has led to the pursuit of hydrogel photopolymerization strategies evolved from 3D rapid prototyping technologies.

2.2.3.6 3D Patterning and Prototyping of Hydrogels

A recent and comprehensive review of 3D hydrogel biofabrication strategies for tissue engineering and regenerative medicine is provided by Bajaj et al.[9] Of particular interest are techniques available to selectively target regions of polymeric solution with incident light, either through masking or controlled movement of a light source (e.g., laser). These methods enable the creation of sophisticated hydrogel structures featuring multicompartment cell-laden scaffolds. Figure 2.5 shows the methodology of an early study to build a 3D PEG-based hydrogel/cell network, with features as small as 50 μm containing cells.[89,90] In this way, it is possible to begin to attempt to reproduce the complex architectures of tissues or organs.

Other exciting applications of photopolymerization are connected with digital fabrication technologies such as stereolithography (SL) or two photon laser scanning photolithography.

SL uses a directed laser to selectively traverse designated regions of powder or liquid photopolymer in a layer-wise fashion and is capable of building 3D solid- or gel-like structures. As discussed earlier, it has been widely used as a rapid prototyping technology and has been extensively investigated with respect to both solid- and gel-based tissue-engineered scaffold preparation. A drawback of this approach, however, is the minimum feature size, which is dictated by the beam width of the laser.

Multiphoton laser scanning photolithography can deliver scaffolds with lateral features in the ~1 μm range, 5–6 μm axial range, but with depth of patterning limited to approximately 1 mm. The 3D patterning capabilities of this approach have allowed the Shoichet group to pattern controlled gradients of VEGF (vascular endothelial growth factor) into pre-soaked agarose-based hydrogels that are then washed to flush the unbound growth factor.[91–93] Guided migration of endothelial cells and formation of tubules have been achieved, illustrating the high potential of this approach for vascularization of thick scaffolds in the future.

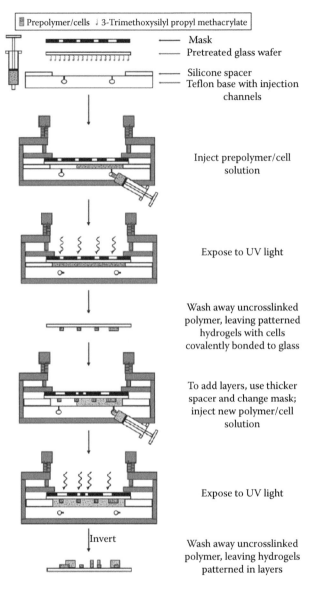

FIGURE 2.5 Process for formation of hydrogels microstructures containing living cells. The apparatus is assembled, including a pretreated glass wafer with reactive methacrylate groups on its surface, and a Teflon base with an inlet and outlet. Once the cells and prepolymer solution are injected, the inlet and outlet are closed, and the unit is exposed to UV light. The resulting patterned hydrogels containing cells are covalently bound to the glass wafer. At this time, a thicker space can be used in conjunction with a new mask to add another layer of cells. This process can be repeated several times. (With kind permission from Springer Science+Business Media: *Biomed. Microdev.*, Three-dimensional photopatterning of hydrogels containing living cells, 4, 2002, 257, Liu, V.A. and Bhatia, S.N., Figure 2.1, Copyright 2002.)

2.3 CELL SCAFFOLD INTERACTIONS AND TISSUE REGENERATION

2.3.1 CARDIOVASCULAR DISEASE

2.3.1.1 Small Diameter Blood Vessels

Cardiovascular disease (CVD) is a major health and economic burden throughout the world, especially in most developed countries where it is the leading cause of death. While great strides have been made in the development of devices and procedures for minimally invasive interventional cardiology, highly or totally occluded arteries can be inaccessible for balloons and stents and require bypass graft surgery. While synthetic grafts are available, and quite suitable for bypassing larger arteries, autologous grafts (veins, arteries) are more successful for small diameter cases but less readily available. Two key issues for small diameter vascular grafts are intimal hyperplasia (IH) and thrombogenesis. Mechanically, it is crucial that bypass vessels have compliance characteristics close to those of the native vessel. A pressing clinical need therefore exists for implantable small diameter blood vessels (synthetic, biological, or hybrid), with suitable compliance and anti-thrombogenicity properties. Tissue engineering offers a possible route to creating such vessels, through controlled generation of *de novo* vascular tissue in blood vessel form. In vascular tissue engineering, bare or cell-seeded tissue-engineered constructs initially act as temporary scaffolds, which either maintain cells *in situ* or attract cells and induce neovascularization. Over time the existing scaffold is remodeled such that it is altered or replaced by the products of cells such as collagen, laminin, elastin, GAGs, or fibronectin. The constructs undergo continuous remodeling and the ultimate goal is for the scaffold and cells to reorganize into near-normal tissue. Encouraging cell growth to an optimum level that minimizes the risk of thrombosis, while preventing excessive IH, is the main goal in the development of any bypass graft or replacement tissue-engineered vascular graft.

Vascular tissue engineering therefore poses a number of challenges, with key issues including scaffold biocompatibility, compliance, thrombogenicity, and the mechanical and biochemical cues that control proliferation of cells and deposition of new extracellular matrix. From the extensive research to date, it is clearly evident that cell behavior is not only influenced by the chemistry and/or mechanical properties of the scaffold materials but also the topological features of the biomaterial at the micrometer and nanometer scale. Mechanical stimuli, such as the strain or shear stress experienced by individual cells in porous scaffolds, are also critical and are affected by structural features at these micro- and nano-size scales. Macroscopic properties such as compliance and strength also have their origins in the structural organization of the scaffold materials, and their microporous architectures.

2.3.1.2 Tissue Engineering of Blood Vessels

From the days of research on developing vascular grafts using materials that produce minimal interaction with the inflowing blood and adjacent tissues, researchers have come a long way to develop constructs at the nanoscale that interact with cells and cause blood vessel formation. Conventional electrospinning produces randomly oriented nanofibers; however, Mo and Weber[94] developed an aligned biodegradable PLLA-CL

(75:25) nanofibrous scaffold using a rotating collector disk for collection of aligned electrospun nanofibers. These aligned nanofibers were explored to fabricate tubular scaffolds that could be used for engineering blood vessels. Their results demonstrated that the nano-sized fibers mimic the dimensions of natural ECM, provide mechanical properties comparable to human coronary artery, and form a well-defined architecture for smooth muscle cell adhesion and proliferation.[94-96] Aligned fibers not only offer structural integrity but also maintain vasoactivity as they provide the necessary mechanical strength that is needed to sustain high pressure of the human circulatory system. Xu et al.[96] studied the response of endothelial cells along with smooth muscle cells (SMCs) on the aligned nanofibers of PLLA-CL, and their results demonstrated that both the cell types showed enhanced adhesion and proliferation rates on the nanofibrous scaffold. In addition, it was observed that the SMCs cytoskeleton organization was along the direction of the nanofibers. These results suggested that aligned nanofibers might provide for a good scaffolding system for vascular tissue engineering.

It is now established that there is a significant effect of nanoscale-textured surface roughness on cell response in terms of cell adhesion and proliferation.[97] It is also known that cells attach and organize very well around fibers with diameters smaller than them.[98] Therefore, Ma et al.[99] processed a conventional polymer, PET, into a nonwoven nanofibrous mat by electrospinning and modified its surface by grafting gelatin. Their study demonstrated enhanced spreading and proliferation of endothelial cells on the modified PET nanofiber mats, while preserving their phenotype. Based on this study, gelatin-modified PET nanofibers could be potential candidates for the engineering of vascular grafts. Boland et al.[100] developed electrospun micro- and nanofibrous scaffolds from natural polymers such as collagen and elastin with the goal of developing constructs for vascular tissue engineering. Their results demonstrated that electrospun collagen and elastin nanofibers were able to mimic the complex architecture required of vascular constructs and were able to provide good mechanical properties that are desired in the environment of the blood stream. Their study indicated that micro and nanofibrous scaffolds synthesized from natural polymers such as collagen and elastin could be useful in the engineering of artificial blood vessels.

Vatankhah et al.[101] fabricated a tubular composite scaffold using electrospinning with biomechanical properties closely simulating those of native blood vessels. They blended a hydrophilic and compliant polyurethane, namely tecophilic (TP) with gelatin (gel) at a weight ratio of 70:30. Furthermore, the hydrophilic properties of the composite scaffold induced non-thrombogenicity while the incorporation of gelatin molecules within the scaffold greatly improved the capacity of the scaffold to serve as an adhesive substrate for vascular SMCs, in comparison to pure TP. The results demonstrated gel's potential feasibility toward functioning as a vascular graft.

A novel approach in treating vasospasm, a common postoperative complication after vascular anastomosis, was presented by Zhu et al.[102] They developed highly flexible and rapidly degradable papaverine-loaded electrospun fibrous membranes to be wrapped around vascular suturing to prevent vasospasm. Poly-L-lactic acid/polyethylene glycol (PLLA/PEG) electrospun fibers containing papaverine maintained a high degree of flexibility and could withstand any folding, and are therefore suitable for wrapping vascular suturing. A rapid release of papaverine, between 2 and 7 days, was achieved by adjusting the proportions of PEG and PLLA. PLLA electrospun fibers containing 40% PEG (PLLA-40%) could control drug release and polymer degradation most effectively during the first 2 weeks post operation. Testing using an *in vivo* rabbit model showed that PLLA-40% fibrous membranes produced significant antispasmodic effect without observable inflammation or hyperplasia, and the fibrous membranes were ideally biodegradable, with no impact on regional blood flow, pressure, vessel diameter, or surrounding tissue hyperplasia. The results confirmed that the fibrous membranes show the potential to greatly reduce postoperative vasospasm and maintain regular vascular morphology during antispasmodic therapy.

2.3.2 MUSCULOSKELETAL INJURY AND DISEASE

2.3.2.1 Bone

Loss or failure of bone as a result of injury or disease remains a common and serious health issue despite much improved medical technology and practice. At present, fracture nonunions and other bone defects are treated with bone grafting procedures, such as allografting or autografting of cancellous bone or vascularized grafts of the fibula and iliac crest. Autologous graft procedures require an additional surgical step to harvest the graft, and there are obvious limitations to the supply of self-donated tissue. Allografts, from donors, have intrinsic potential for disease transmission and the possibility of rejection. Issues with bone volume maintenance can also arise with donor grafts.

Natural bone is composed of a mineral phase and an organic phase. The mineral phase is composed of hydroxyapatite, while the organic phase includes collagen and proteoglycans. It is, in fact, a composite material. Major steps toward the development of tissue-engineered bone grafts have been made over several years, yet many issues remain to be overcome. A wide range of biomaterials have been considered for bone tissue engineering scaffold development, from metals to high-performance synthetic polymers. Composites of ceramics (such as hydroxyapatite) and natural or synthetic polymers have been increasingly targeted. It is clear that bone scaffold biomaterials need to be conducive to cell attachment and be safe for long-term implantation in the body. A challenge has been to produce scaffolds combining the open porous architecture of cancellous bone with the strength requirements associated with musculoskeletal loading. It is known that the behavior of cells (and specifically their differentiation from progenitor cells to osteoblasts or chondrocytes) is dependent on the mechanical stimuli that they experience. These stimuli are, in turn, dependent on scaffold properties as the microarchitecture of the bone tissue engineering scaffold determines how musculoskeletal loads are translated into the deformations and forces applied to attached cells. Precise control over the microporous architecture and the micromechanics of scaffolds is, therefore, necessary to optimize the performance of scaffolds.

2.3.2.2 Electrospun Scaffolds for Bone Tissue Engineering

The design of scaffolds for bone tissue engineering is based on the physical properties of bone tissue such as mechanical strength, pore size, porosity, hardness, and overall 3D architecture. For bone tissue engineering, scaffolds with a pore size in the range of 100–350 μm and porosity greater than 90% are preferred for better cell/tissue in-growth and hence enhanced bone regeneration.[55,103]

Yoshimoto et al.[104] developed nonwoven PCL scaffolds by electrospinning for bone tissue engineering. To understand the influence of mesenchymal stem cells (MSCs) on nanofibers, MSCs derived from bone marrow of neonatal rats were seeded on the nanofibrous scaffold. The results indicated that the MSCs migrated inside the scaffold and produced abundant extracellular matrix in the scaffold. In continuation to this study, Shin et al. tested the PCL nanofibers along with MSCs *in vivo* in a rat model. Their results demonstrated ECM formation throughout the scaffold along with mineralization and type I collagen synthesis.[105] These studies demonstrated that PCL-based nanofibrous scaffolds are potential candidates for bone tissue engineering.

PCL was further studied by Nedjari et al.[106] They elaborated honeycomb nanofibrous scaffolds by electrospinning onto micropatterned collectors either with poly(ε-caprolactone) (PCL) or poly(D,L-lactic acid) (PLA). The unimodal distribution of fiber diameters, observed for PLA, led to relatively flat scaffolds; on the other hand, the bimodal distribution of PCL fiber diameters significantly increased the relief of the scaffolds' patterns due to the preferential deposition of the thick fiber portions on the walls of the collector's patterns via preferential electrostatic interaction. Finally, a biological evaluation demonstrated the effect of the scaffolds' relief on the spatial organization of MG63 osteoblast-like cells. Mimicking hemi-osteons, cell gathering was observed inside PCL honeycomb nests with a size ranging from 80 to 360 μm.

In another study, Ramay and Zhang used HA with β-tricalcium phosphate (β-TCP) to develop biodegradable nanocomposite porous scaffolds.[107] β-TCP/HA scaffolds built from HA nanofibers with β-TCP as a matrix were used to fabricate porous scaffolds by a technique that integrated the gel casting technique with the polymer sponge method.[108] The *in vitro* results demonstrated that incorporation of HA nanofibers as a second component in β-TCP significantly increased the mechanical strength of the porous composite scaffolds. This study introduced nanocomposites with HA nanofibers as a promising scaffolding system for load bearing applications such as bone tissue engineering.

On the other hand, natural polymers showed potential for bone regeneration. One recent study conducted by Lai et al.[109] studied the growth and osteogenic differentiation of human bone marrow mesenchymal stem cells (hMSCs) on nanofibrous membrane scaffolds of chitosan (CS), silk fibroin (SF), and CS/SF blend prepared by electrospinning. The morphology and physicochemical properties of all membrane scaffolds were compared. The influence of CS and SF on cell proliferation was assessed by the MTS assay, whereas osteogenic differentiation was determined from the Alizarin Red staining, alkaline phosphatase activity, and expression of osteogenic marker genes. CS and SF nanofibers enhanced the osteogenic differentiation and proliferation of hMSCs, respectively. Blending CS with SF retained the osteogenesis capacity of CS without negatively influencing the cell proliferative effect of SF. By taking advantage of the differentiation/proliferation cues from individual components, the electrospun CS/SF composite nanofibrous membrane scaffold is suitable for bone tissue engineering.

2.3.2.3 Cartilage

Articular cartilage tissue has a limited capacity for repair due to the reduced availability of chondrocytes and complete absence of progenitor cells near the wound to mediate the repair process. The chondrocytes available for repair are embedded in the dense ECM of the articular surface, which restricts their mobility and hence limits their contribution to the wound healing process.[110] In addition, articular cartilage is an avascular tissue, which further limits its capacity to self-regenerate. To provide a solution to this problem, multiple surgical techniques have been developed, but with limited success.[111] Therefore, tissue engineering as a potential approach to regenerate cartilage tissue holds good promise. One of the methods of engineering cartilage tissue is by 3D scaffolds combined with chondrocytes or progenitor cells.[112]

2.3.2.4 Electrospun Scaffolds for Cartilage Tissue Engineering

Li et al.[113] developed PCL-based nanofibrous scaffolds by electrospinning. These scaffolds were then seeded with fetal bovine chondrocytes (FBC) and studied for their ability to maintain chondrocytes in a mature functional state. The results demonstrated that FBCs seeded on the PCL nanofibers were able to maintain their functional chondrocyte phenotype by expressing cartilage-specific extracellular matrix genes, like aggrecan, collagen type II and IX, and cartilage oligomeric matrix protein. Further, FBCs exhibited a spindle or round shape on the nanofibrous scaffold in contrast to a flat, well-spread morphology as seen when cultured on tissue culture polystyrene. Another interesting finding from this study was that cells in serum-free medium produced more sulfated proteoglycan-rich cartilaginous matrix when compared with those cultured in monolayer on tissue culture polystyrene. These results demonstrated that the bioactivity of FBCs depends on the architecture of the scaffold and the composition of the culture medium. Hence, the PCL nanofibers show potential to be further explored as scaffolds for cartilage tissue engineering.

Nanofibrous PCL and adult bone marrow–derived MSCs were used to test whether the PCL fibers will support *in vitro* MSCs chondrogenesis. The results indicated that PCL nanofibers in the presence of a member of the transforming growth factor-β family caused the differentiation of MSCs to chondrocytes that was comparable to that caused by cell aggregates or pellets. However, since the PCL nanofibrous scaffolds possess better mechanical properties than cell pellets, they show potential to be developed as a scaffolding system for MSCs delivery and hence cartilage tissue engineering.

Another interesting approach for cartilage tissue engineering was presented in the study conducted by Kisiday et al.[114] They developed a self-assembling peptide hydrogel scaffold using the peptide KDK-12 that had a sequence of (AcN-KLDLKLDLKLDL-CNH2) (where K is lysine, D is aspartic acid, and L is leucine). This peptide was seeded with bovine chondrocytes and then allowed to self-assemble into a hydrogel. The chondrocyte-seeded hydrogels were then studied for their ability to support chondrocyte proliferation, ECM production, and phenotype maintenance. Their results demonstrated that the chondrocytes were able to produce cartilage-like ECM, which was rich in proteoglycan and type II collagen (phenotypic markers of chondrocytes). Further, the authors observed that the mechanical properties continuously increased with time, which was indicative of the continuous deposition of glycosaminoglycan-rich matrix by the chondrocytes. In addition, the ability to design the peptide may offer advantages in controlling scaffold degradation, cell attachment, and growth factor delivery. Therefore, the self-assembling peptide hydrogel scaffold may be a suitable candidate for cartilage tissue engineering.

Man et al.[115] studied scaffolds that can both specifically enrich BMSCs and release rhTGF-β1 to promote chondrogenic differentiation of the incorporated BMSCs.

They first fabricated coaxial electrospun fibers using a polyvinyl pyrrolidone/ bovine serum albumin/rhTGF-β1 composite solution as the core fluid and poly(ε-caprolactone) solution as the sheath fluid. Structural analysis revealed that scaffold fibers were relatively uniform with a diameter of 674.4 ± 159.6 nm; the core-shell structure of coaxial fibers was homogeneous and proteins were evenly distributed in the core. Subsequently, the BMSC-specific affinity peptide E7 was conjugated to the coaxial electrospun fibers to develop a co-delivery system of rhTGF-β1 and E7. The results of ^{1}H nuclear magnetic resonance indicate that the conjugation between the E7 and scaffolds was covalent. The rhTGF-β1 incorporated in E7-modified scaffolds could maintain sustained release and bioactivity. Cell adhesion, spreading, and DNA content analyses indicate that the E7 promoted BMSC initial adhesion, and that the scaffolds containing both E7 and rhTGF-β1 (CBrhTE) were the most favorable for BMSC survival. Meanwhile, CBrhTE scaffolds could promote the chondrogenic differentiation ability of BMSCs. Overall, the CBrhTE scaffold could synchronously improve all three of the basic components required for cartilage tissue engineering *in vitro*, which paves the road for designing and building more efficient tissue scaffolds for cartilage repair.

2.3.2.5 Ligaments

Ligaments are bands of dense connective tissue responsible for joint movement and stability. Ligament ruptures result in abnormal joint kinematics and often irreversible damage of the surrounding tissue, leading to tissue degenerative diseases, which do not heal naturally and cannot be completely repaired by conventional clinical methods.[116,117] New tissue engineering methods involving nanofibers have been successfully employed to meet this challenge. In particular, aligned nanofibers enhanced cell response and, hence, were explored as scaffolds for ligament tissue engineering.

2.3.2.6 Electrospun Scaffolds for Ligament Tissue Engineering

Lee et al. studied the effects of PU nanofiber alignment and direction of mechanical stimuli on the ECM generation of human ligament (anterior cruciate) fibroblasts (HLF).[25]

Conventional electrospinning produces randomly oriented nanofibers; however, in this study, aligned electrospun fibers were created using a rotating target. The fibers were then seeded with HLFs to study the influence of alignment on HLF behavior. The results demonstrated that HLFs were spindle shaped, oriented in the direction of nanofibers, and showed enhancement in the synthesis of ECM proteins (collagen) on aligned nanofibers when compared with randomly oriented nanofibers. In addition, the authors also studied the effect of direction of mechanical stimuli on the ECM produced by HLFs. HLFs were seeded on parallel aligned, vertically aligned to the strain direction, and randomly oriented PU nanofibers. The results demonstrated that HLFs were more sensitive to strain in the longitudinal direction. Therefore, this study concluded that aligned nanofibrous scaffolds showed promise for use in ligament tissue engineering.[25]

2.3.2.7 Skeletal Muscle Tissue Engineering

Skeletal muscles are responsible for voluntary movement of the body and once damaged (by disease or trauma) are difficult to regenerate in adults.[118] Moreover, even if skeletal muscle tissue engineering is a challenging study area, it is an exciting alternative to surgical techniques for skeletal muscle regeneration. One such example is the study conducted by Riboldi et al.[119] The authors have explored the use of electrospun microfibers made from degradable polyester urethane (PEU) as scaffolds for skeletal muscle tissue engineering. Based on their preliminary studies using primary human satellite cells (biopsy from a 38-year-old female), C2C12 (murine myoblast cell line), and L6 (rat myoblast cell line), their results indicated that the electrospun microfibers of PEU showed satisfactory mechanical properties and encouraging cellular response in terms of adhesion and differentiation. Based on these studies, the electrospun microfibers of PEU show potential to be further explored as a scaffolding system for skeletal muscle tissue engineering.

2.3.2.8 Electrospinning with Cells

Recent years have seen interest in approaches for directly generating fibers and scaffolds following a rising trend for their exploration in the health sciences. Compared with traditional *in vitro* cell culture materials, 3D nanofibrous scaffolds provide a superior environment for promoting cell functions. Since nanofibrous scaffolds have nanometer pore sizes, cells are unable to penetrate on their own, incorporating them into the scaffold fabrication would ensure better cell distribution.[120] In this direction, Seil and Webster produced biodegradable and cytocompatible poly(DL-lactide-co-glycolide) (PLGA) nanofibers using an electrospinning process. As a model cell line, fibroblasts were periodically sprayed from a pump-action spray bottle onto the developing scaffold. The viability of cells before and after spraying, and after incorporation into the scaffold, was compared. Results indicated that cell spraying and the scaffold fabrication process did not significantly reduce cell viability. These findings, thus, contribute to the understanding of how to produce more physiological relevant cell-seeded nanofibrous scaffolds, an important element for the future of nanotechnology and tissue engineering.[120] Bettahalli et al. evidenced another approach.[121] The study presents the development of a multilayer tissue construct by rolling pre-seeded electrospun sheets (prepared from poly(L-lactic acid) [PLLA] seeded with C2C12 pre-myoblast cells) around a porous multibore hollow fiber (HF) membrane and its testing using a bioreactor. Important elements of the study were (1) the medium permeating through the porous walls of multibore HF acted as an additional source of nutrients and oxygen to the cells, which exerted low shear stress (controllable by transmembrane pressure); (2) application of dynamic perfusion through the HF lumen and around the 3D construct to achieve high cell proliferation and homogenous cell distribution across the layers; and (3) cell migration occurred within the multilayer construct (shown using pre-labeled C2C12 cells), illustrating the potential of using this concept for developing thick and more complex tissues.

This technology's unique ability is to immobilize multiple cell types with a wide range of molecules simultaneously within a fiber during the scaffold generation process. The technology has been shown to generate many cell-laden complex architectures from true 3D sheets to those multicore vessels.[122] A novel study led by Ehler and Jayasinghe showed for the first time the ability to immobilize primary cardiac myocytes within these fibers, looking to develop the electrospinning technology for creating 3D cardiac patches that could be used for repairing, replacing, and rejuvenating damaged, diseased, and/or aging cardiac tissues.[122]

2.4 COMPUTATIONAL MODELS FOR MECHANICAL CHARACTERIZATION OF TISSUE ENGINEERING SCAFFOLDS

Computational modeling can play a significant role in the development of tissue-engineered replacements. It offers a low-cost and ethically sound alternative to animal trials for testing the mechanical compatibility of load-bearing tissue-engineered tissues such as bone, skeletal muscle, and blood vessels with that of the natural host tissue. Computational models also offer the possibility of using stochastic approaches to incorporate the inherent variability in host tissues and of developing patient-specific scaffolds to suit various anatomies and tissue properties.

2.4.1 MECHANOBIOLOGY: BONE TISSUE ENGINEERING

The idea that mechanical stimuli influence tissue growth and differentiation began with Pauwels[123] in 1960. Pauwels identified that physical factors cause stress and deformation of the mesenchymal stem cells, and that different combinations of these stimuli could determine different cell differentiation pathways in bone and cartilaginous tissues. Since 1960, this concept has been adapted and further developed using computational approaches to analyze patterns of bone and cartilage tissue differentiation during fracture healing[124,125] and more recently to predict the optimum scaffold properties to stimulate osteochondral defect repair[126] and induce bone regeneration[127] (see Figure 2.6).

This concept of mechanoregulation of tissue types has lent itself well to mechanobiological modeling whereby the use of finite element models, which characterize the load induced in the tissue, can be used to direct cell growth and differentiation and to investigate the required properties of a scaffold for successful bone tissue engineering[128] (see Figure 2.7). For a more thorough review on the use of computational modeling for bone and cartilage tissue engineered, see articles by Checa et al.[129] and Olivares and Lacroix.[130]

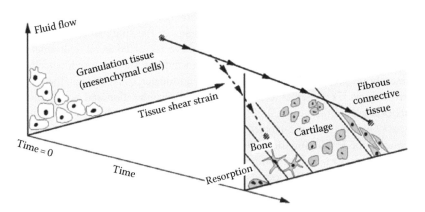

FIGURE 2.6 The mechanoregulation of tissue differentiation concept. (Reprinted from *J. Biomech.*, 38(7), Kelly, D.J. and Prendergast, P.J., Mechano-regulation of stem cell differentiation and tissue regeneration in osteochondral defects, 1413–1422, Copyright 2005, with permission from Elsevier.)

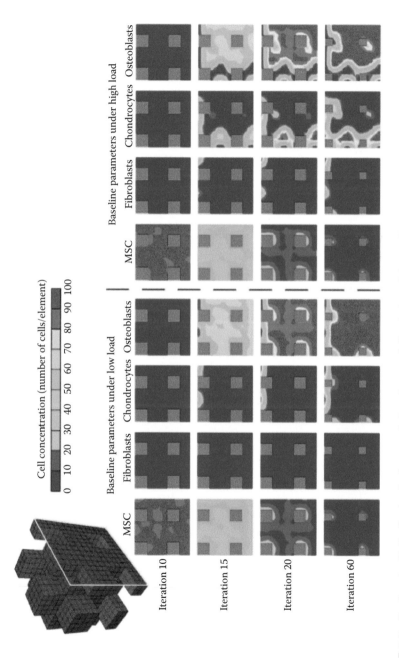

FIGURE 2.7 Predicted cell distribution in the granulation tissue and dissolution of the scaffold biomaterial. (Reprinted from *Biomaterials*, 28(36), Byrne, D.P., Lacroix, D., Planell, J.A., Kelly, D.J., and Prendergast, P.J., Simulation of tissue differentiation in a scaffold as a function of porosity, Young's modulus and dissolution rate: Application of mechanobiological models in tissue engineering, 5544–5554, Copyright 2007, with permission from Elsevier.)

While considerable advances have been made in bone and cartilage tissue engineering using computational modeling approaches, highly deformable soft tissues such as arteries and vascular scaffolds have been less well explored.

2.4.2 MECHANOBIOLOGY: VASCULAR TISSUE ENGINEERING

In vascular tissue engineering, the mechanical properties of synthetic bypass grafts have been strongly linked to the development of intimal hyperplasia, whereby the compliance mismatch between the graft and the artery limits the long-term success of the bypass procedure (see Figure 2.8).[131]

While surface modifications and cell seeding can tackle thrombosis within vascular grafts, to reduce the risk of intimal hyperplasia, the mechanical response of small diameter tissue-engineered blood vessels (TEBVs) needs to be close to that of native tissue to provide the optimum environment for vascular cells to grow without inducing intimal hyperplasia.[132]

Numerical modeling tools, such as finite element modeling (FEM), enable engineers to virtually design optimized scaffolds with the required microstructure and stiffness to closely match the properties of a host vessel and support optimum cell infiltration and growth. However, the fact that arteries are not merely mechanical structures, but living tissue, introduces complexity into such models that cannot be addressed by FEM alone. To successfully develop a tissue-engineered artery requires a strong fundamental knowledge of cellular interactions with the local environment and how these interactions, which span multiple length scales, contribute to overall organ function.[133]

The complex interplay between the ECM and supported cells (SMCs/MSCs) lends itself well to multiscale mechanobiological models at the cell and tissue level. Similar to granulation tissue in bone, SMC phenotype and growth has been demonstrated to be controlled by cyclic strain amplitude such that low strain amplitude promotes SMC proliferation (Figure 2.9).[134]

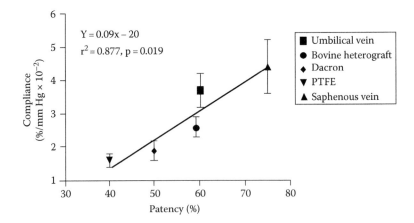

FIGURE 2.8 Patency data reported based on compliance of various biological and prosthetic grafts. (From Sarkar, S. et al., *Eur. J. Vasc. Endovasc. Surg.*, 31(6), 627, 2006.)

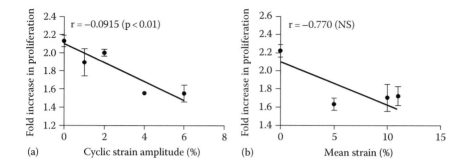

FIGURE 2.9 Correlation analyses of proliferation data for (a) cyclic strain amplitude and (b) mean cyclic strain. (With kind permission from Springer Science+Business Media: *Biomech. Model. Mechanobiol.*, Cyclic strain amplitude dictates the growth response of vascular smooth muscle cells in vitro: Role in in-stent restenosis and inhibition with a sirolimus drug-eluting stent, 12(4), 2013, 671, Colombo, A., Guha, S., Mackle, J.N., Cahill, P.A., and Lally, C., Figure 5, Copyright 2013.)

SMCs also have the ability to synthesize and remodel collagen in response to stretch.[135] This therefore creates a dynamic environment within vascular conduits whereby mechanical stimuli control cell activity and collagen synthesis, altering scaffold architecture and compliance, and thus influencing cell growth. Using low cyclic strain to encourage SMC growth and collagen fiber production and remodeling in a decellularized construct could provide a means of generating a mechanically and biologically compatible TEBV. While the mechanical properties of the vascular conduit are critical for success of such a remodeling approach to a TEBV, quantifying cell activity in response to the mechanical environment is also critical. Multiscale modeling approaches can incorporate the cell-level activity and use insights into cell growth and even differentiation to alter the macroscopic mechanical properties of a scaffold. In addition, changes to the mechanical environment are critical for controlling cell activity. In the following section, the use of numerical modeling techniques, such as finite element modeling and multiscale computational approaches in vascular tissue engineering, will be discussed in terms of their applicability for designing and optimizing tissue engineering blood vessels, which are both mechanically and biologically compatible with a host vessel.

2.4.2.1 Multiscale Vascular Mechanobiological Models

Since the mechanical properties of TEBVs are key to cell growth within the scaffold and the performance of the scaffold itself as a viable bypass graft, scaffolds that have shown significant potential include materials that are biologically and biomechanically similar to the native arterial tissue such as decellularized arteries and veins. While the majority of decellularization protocols take from several days[136,137] to weeks,[138] this has implications for the mechanical properties of the arteries that can degrade and alter over time.[139] A short-term decellularization protocol for small caliber porcine coronary arteries has now been developed and implemented to create load-bearing decellularized scaffolds with mechanical properties close to that of fresh arteries. While these scaffolds may offer significant potential as a viable scaffold that is mechanically compatible with a diseased vessel, their success needs to be determined following cell seeding and implantation into the body.

Finite element modeling can be used to assess the mechanical environment within the scaffold post implantation. Previous work by Zahedmanesh et al.[140] outlined a method to assess the mechanical suitability of bacterial cellulose as a TEBV scaffold, whereby the scaffold was tested in uniaxial extension tests, and data were used to develop a constitutive model for the tissue that could be implemented into a finite element model. While Zahedmanesh et al.[140] demonstrated the close match between the compliance of this scaffold to that of natural scaffolds, the constitutive model data could be used to generate a virtual anastomosis to a host vessel with a range of different mechanical properties to test the degree of compliance mismatch between the vessel and its potential influence on cell behavior. This influence on cell behavior could be ascertained by looking at *in vitro* test results on the response of vascular cells to mechanical stimuli[134,141] or by using numerical mechanobiological models that can predict complex collective cell behavior from inputs on the growth of individual cell groups.

Using a multiscale mechanobiological framework for vascular tissue engineering, such as that presented in Zahedmanesh and Lally,[142] can provide key insights into the optimum scaffold-host vessel combination to encourage scaffold repopulation without the risk of excessive intimal hyperplasia. In these models, stresses or stains in the scaffold, calculated within the finite element models, are used as the controlling input to cell behavior in agent-based models (ABM) or lattice models. These coupled FE-ABM models used a large strain hyperelastic material model to represent the vascular scaffold in the simulations, and the strains within the scaffold dictated the cell and ECM production based on data from *in vitro* cell experiments.[134,143–145] Realistic patterns of cell infiltration and cell and ECM growth could be predicted using these state-of-the-art models. The strain environment, dictated by the scaffold stiffness and pressure conditions, was found to be critical in terms of the degree of intimal hyperplasia within the scaffold and provided critical insights into the optimum scaffold properties for bypassing arteries (see Figure 2.10).[142]

Mechanobiological models, using these cell- and tissue-based rules, can subsequently be used to provide vital insights into the optimum mechanical environment for VSMCs, ECs, and even MSCs to encourage differentiation and repopulation without excessive cell growth in a vascular conduit and also provide critical insights into ECM production.[146]

Models of this nature facilitate parameter variation studies, to investigate how alterations in the initial properties of arterial scaffolds (e.g., scaffold architecture, stiffness, porosity, cell seeding density, etc.), influence outcomes.[128] Directed by the *in silico* models, cell seeded tubular scaffolds can be created with the optimum initial architecture and stiffness, and subjected to the optimum loading regime, to create biologically and mechanically compatible TEBVs.

While considerable insights have already been gained from the use of mechanobiological models for tissue engineering, these models could be further advanced. The models could explicitly model the underlying collagen fiber architecture within scaffolds and alterations within this architecture could be controlled by an empirically informed remodeling algorithm, similar to those used in soft tissue growth

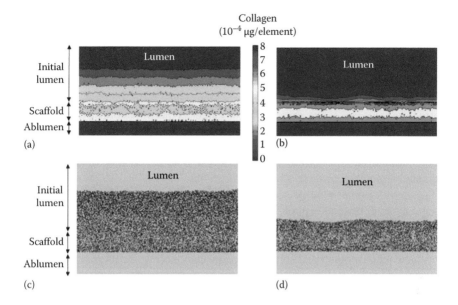

FIGURE 2.10 ECM synthesis and viable SMCs (a, c) TEBV under static pressure (100 mmHg) and (b, d) TEBV under pulsatile pressure. (With kind permission from Springer Science+Business Media: *Biomech. Model. Mechanobiol.*, A multiscale mechanobiological modeling framework using agent-based models and finite element analysis: Application to vascular tissue engineering, 11(3–4), 2012, 363, Zahedmanesh, H. and Lally, C., Figure 12, © 2012.)

and remodeling.[147,148] This would enable cells to alter the fiber network based on experimentally established rules, thereby changing the mechanical environment and further influencing cell growth and ECM production or degradation. In addition, as computational capabilities improve, the possibility of virtually designing patient-specific tissue-engineered materials will become real and the preclinical testing of the long-term viability of these tissues could be fully explored before implantation. For a more thorough exploration of computational modeling approaches in vascular disease and tissue engineering, see Reference 151.

Undoubtedly, however, multiscale mechanobiological models, informed by well-designed *in vitro* experiments, can enable significant advancements in the field of regenerative medicine by providing a framework with which to determine the ideal mechanical properties of an engineered tissue prior to *in vivo* implantation.

2.5 OUTLOOK AND FUTURE TRENDS

The development of enhanced tissue engineering scaffolds will require ongoing development of the major fabrication processes outlined in this chapter. The need is for scaffolds that provide the appropriate physical structure and biochemical cues for tissue regeneration. This requires the fabrication of scaffolds with precisely controlled and reproducible physical architecture, as well as associated mechanical stimuli under loading, to regulate cell and tissue response. Future scaffolds should also aim to incorporate the functionality associated with moieties from the macromolecular structures of native tissue components, again in a reproducible fashion. The potential of nanoparticle delivery modes for growth factors and other biochemical cues will also require continued exploration.

Alongside the advantages offered by electrospinning for certain, simpler, tissue types, the potential for digitally controlled techniques for both solid and hydrogel scaffold forms is clear. The ability to pattern internal networks of cell adhesion peptides within a hydrogel block offers exciting possibilities for vascularization of such scaffold with capillaries formed by endothelial cells. Such techniques, utilizing two photon lithography, also have a significant potential for the creation of more complex cell-laden organ structures.

Successful *engineering* of scaffolds also requires the parallel development of modeling capabilities to explain, predict, and optimize the functionality of the systems under development. The techniques based on continuum mechanics and the finite element method, combined with agent-based modeling, are being used to investigate mechanobiological effects in tissue engineering scaffolds across multiple scales. Such models will be invaluable in the optimization of future scaffold designs.

REFERENCES

1. Lanza, R., R. Langer, and J.P. Vacanti, eds. *Principles of Tissue Engineering*. Academic Press, San Diego, CA, 2011.
2. Abrigo, M., S.L. McArthur, and P. Kingshott. Electrospun nanofibers as dressings for chronic wound care: Advances, challenges, and future prospects. *Macromolecular Bioscience* 14(6) (2014): 772–792.
3. Liu, X.-Y., J. Chen, Q. Zhou, J. Wu, X.-L. Zhang, L. Wang, and X.-Y. Qin. In vitro tissue engineering of lamellar cornea using human amniotic epithelial cells and rabbit cornea stroma. *International Journal of Ophthalmology* 6(4) (2013): 425.
4. Johnstone, B., M. Alini, M. Cucchiarini et al. Tissue engineering for articular cartilage repair—The state of the art. *European Cells and Materials* 25 (2013): 248–267.
5. Broderick, J.M., D.J. Kelly, and K.J. Mulhall. Optimizing stem cell engineering for orthopaedic applications. *Journal of the American Academy of Orthopaedic Surgeons* 22(1) (2014): 63–65.
6. Seifu, D.G., A. Purnama, K. Mequanint, and D. Mantovani. Small-diameter vascular tissue engineering. *Nature Reviews Cardiology* 10(7) (2013): 410–421.
7. Sauer, I.M., N. Raschzok, and P. Neuhaus. The artificial liver, in vivo tissue engineering, and organ printing. *Textbook of Organ Transplantation* (2014): 545–553.
8. Billiet, T., M. Vandenhaute, J. Schelfhout, S. Van Vlierberghe, and P. Dubruel. A review of trends and limitations in hydrogel-rapid prototyping for tissue engineering. *Biomaterials* 33(26) (2012): 6020–6041.
9. Bajaj, P., R.M. Schweller, A. Khademhosseini, J.L. West, and R. Bashir. 3D biofabrication strategies for tissue engineering and regenerative medicine. *Annual Review of Biomedical Engineering* 16(1) (2014): 247–276.
10. Williams, D.F. On the mechanisms of biocompatibility. *Biomaterials* 29(20) (2008): 2941–2953.
11. Cockshott, I.D. and G.E. Martin. Fibrillar product of electrostatically spun organic material. U.S. Patent 4,043,331, issued August 23, 1977.
12. Rim, N.G., C.S. Shin, and H. Shin. Current approaches to electrospun nanofibers for tissue engineering. *Biomedical Materials* 8(1) (2013): 014102.

13. Bhardwaj, N. and S.C. Kundu. Electrospinning: A fascinating fiber fabrication technique. *Biotechnology Advances* 28(3) (2010): 325–347.
14. Li, D. and Y. Xia. Electrospinning of nanofibers: Reinventing the wheel? *Advanced Materials* 16(14) (2004): 1151–1170.
15. Pham, Q.P., U. Sharma, and A.G. Mikos. Electrospinning of polymeric nanofibers for tissue engineering applications: A review. *Tissue Engineering* 12(5) (2006): 1197–1211.
16. Zanatta, G., D. Steffens, D.I. Braghirolli, R.A. Fernandes, C.A. Netto, and P. Pranke. Viability of mesenchymal stem cells during electrospinning. *Brazilian Journal of Medical and Biological Research* 45(2) (2012): 125–130.
17. Teebken, O.E., C. Puschmann, B. Rohde, K. Burgwitz, M. Winkler, A.M. Pichlmaier, J. Weidemann, and A. Haverich. Human iliac vein replacement with a tissue-engineered graft. *Vasa* 38(1) (2009): 60–65.
18. Stitzel, J., J. Liu, S.J. Lee et al. Controlled fabrication of a biological vascular substitute. *Biomaterials* 27(7) (2006): 1088–1094.
19. Thomas, V., M.V. Jose, S. Chowdhury, J.F. Sullivan, D.R. Dean, and Y.K. Vohra. Mechano-morphological studies of aligned nanofibrous scaffolds of polycaprolactone fabricated by electrospinning. *Journal of Biomaterials Science, Polymer Edition* 17(9) (2006): 969–984.
20. Sui, G., X. Yang, F. Mei, X. Hu, G. Chen, X. Deng, and S. Ryu. Poly-L-lactic acid/hydroxyapatite hybrid membrane for bone tissue regeneration. *Journal of Biomedical Materials Research Part A* 82(2) (2007): 445–454.
21. Schnell, E., K. Klinkhammer, S. Balzer et al. Guidance of glial cell migration and axonal growth on electrospun nanofibers of poly-ε-caprolactone and a collagen/poly-ε-caprolactone blend. *Biomaterials* 28(19) (2007): 3012–3025.
22. Yang, F., R. Murugan, S. Wang, and S. Ramakrishna. Electrospinning of nano/micro scale poly(L-lactic acid) aligned fibers and their potential in neural tissue engineering. *Biomaterials* 26(15) (2005): 2603–2610.
23. Zhao, G., S. Yin, G. Liu et al. *In vitro* engineering of fibrocartilage using CDMP1 induced dermal fibroblasts and polyglycolide. *Biomaterials* 30(19) (2009): 3241–3250.
24. Sahoo, S., H. Ouyang, J.C.-H. Goh, T.E. Tay, and S.L. Toh. Characterization of a novel polymeric scaffold for potential application in tendon/ligament tissue engineering. *Tissue Engineering* 12(1) (2006): 91–99.
25. Lee, C.H., H.J. Shin, I.H. Cho, Y.-M. Kang, I. Kim, K.-D. Park, and J.-W. Shin. Nanofiber alignment and direction of mechanical strain affect the ECM production of human ACL fibroblast. *Biomaterials* 26(11) (2005): 1261–1270.
26. Xu, C.Y., R. Inai, M. Kotaki, and S. Ramakrishna. Aligned biodegradable nanofibrous structure: A potential scaffold for blood vessel engineering. *Biomaterials* 25(5) (2004): 877–886.
27. Matthews, J.A., G.E. Wnek, D.G. Simpson, and G.L. Bowlin. Electrospinning of collagen nanofibers. *Biomacromolecules* 3(2) (2002): 232–238.
28. Teo, W.E. and S. Ramakrishna. A review on electrospinning design and nanofibre assemblies. *Nanotechnology* 17(14) (2006): R89.
29. Dalton, P.D., D. Klee, and M. Möller. Electrospinning with dual collection rings. *Polymer* 46(3) (2005): 611–614.
30. Wang, X., K. Zhang, M. Zhu, H. Yu, Z. Zhou, Y. Chen, and B.S. Hsiao. Continuous polymer nanofiber yarns prepared by self-bundling electrospinning method. *Polymer* 49(11) (2008): 2755–2761.
31. Bazbouz, M.B. and G.K. Stylios. Novel mechanism for spinning continuous twisted composite nanofiber yarns. *European Polymer Journal* 44(1) (2008): 1–12.
32. Ko, F., Y. Gogotsi, A. Ali, N. Naguib, H. Ye, G.L. Yang, C. Li, and P. Willis. Electrospinning of continuous carbon nanotube-filled nanofiber yarns. *Advanced Materials* 15(14) (2003): 1161–1165.

33. Ali, U., Y. Zhou, X. Wang, and T. Lin. Direct electrospinning of highly twisted, continuous nanofiber yarns. *Journal of the Textile Institute* 103(1) (2012): 80–88.
34. Smit, E., U. Büttner, and R.D. Sanderson. Continuous yarns from electrospun fibers. *Polymer* 46(8) (2005): 2419–2423.
35. Teo, W.-E., R. Gopal, R. Ramaseshan, K. Fujihara, and S. Ramakrishna. A dynamic liquid support system for continuous electrospun yarn fabrication. *Polymer* 48(12) (2007): 3400–3405.
36. Wu, J., S. Liu, L. He, H. Wang, C. He, C. Fan, and X. Mo. Electrospun nanoyarn scaffold and its application in tissue engineering. *Materials Letters* 89 (2012): 146–149.
37. Wu, J., C. Huang, W. Liu et al. Cell infiltration and vascularization in porous nanoyarn scaffolds prepared by dynamic liquid electrospinning. *Journal of Biomedical Nanotechnology* 10(4) (2014): 603–614.
38. Bourell, D.L., M.C. Leu, K. Chakravarthy, N. Guo, and K. Alayavalli. Graphite-based indirect laser sintered fuel cell bipolar plates containing carbon fiber additions. *CIRP Annals—Manufacturing Technology* 60(1) (2011): 275–278.
39. Lee, G. and J.W. Barlow. Selective laser sintering of bioceramic materials for implants. In *Proceedings of the Solid Freeform Fabrication Symposium*, Austin, TX, pp. 376–380, 1993.
40. Vail, N.K., L.D. Swain, W.C. Fox, T.B. Aufdlemorte, G. Lee, and J.W. Barlow. Materials for biomedical applications. *Materials & Design* 20(2) (1999): 123–132.
41. Tan, K.H., C.K. Chua, K.F. Leong, C.M. Cheah, P. Cheang, M.S. Abu Bakar, and S.W. Cha. Scaffold development using selective laser sintering of polyetheretherketone–hydroxyapatite biocomposite blends. *Biomaterials* 24(18) (2003): 3115–3123.
42. Wiria, F.E., C.K. Chua, K.F. Leong, Z.Y. Quah, M. Chandrasekaran, and M.W. Lee. Improved biocomposite development of poly(vinyl alcohol) and hydroxyapatite for tissue engineering scaffold fabrication using selective laser sintering. *Journal of Materials Science: Materials in Medicine* 19(3) (2008): 989–996.
43. Williams, J.M., A. Adewunmi, R.M. Schek, C.L. Flanagan, P.H. Krebsbach, S.E. Feinberg, S.J. Hollister, and S. Das. Bone tissue engineering using polycaprolactone scaffolds fabricated via selective laser sintering. *Biomaterials* 26(23) (2005): 4817–4827.
44. Pereira, T.F., M.A.C. Silva, M.F. Oliveira et al. Effect of process parameters on the properties of selective laser sintered poly(3-hydroxybutyrate) scaffolds for bone tissue engineering. *Virtual and Physical Prototyping* 7(4) (2012): 275–285.
45. Eosoly, S., D. Brabazon, S. Lohfeld, and L. Looney. Selective laser sintering of hydroxyapatite/poly-ε-caprolactone scaffolds. *Acta Biomaterialia* 6(7) (2010): 2511–2517.
46. Eosoly, S., N.E. Vrana, S. Lohfeld, M. Hindie, and L. Looney. Interaction of cell culture with composition effects on the mechanical properties of polycaprolactone-hydroxyapatite scaffolds fabricated via selective laser sintering (SLS). *Materials Science and Engineering C* 32(8) (2012): 2250–2257.
47. Curodeau, A., E. Sachs, and S. Caldarise. Design and fabrication of cast orthopedic implants with freeform surface textures from 3-D printed ceramic shell. *Journal of Biomedical Materials Research* 53(5) (2000): 525–535.
48. Warnke, P.H., H. Seitz, F. Warnke et al. Ceramic scaffolds produced by computer-assisted 3D printing and sintering: Characterization and biocompatibility investigations. *Journal of Biomedical Materials Research Part B: Applied Biomaterials* 93(1) (2010): 212–217.
49. Seitz, H., W. Rieder, S. Irsen, B. Leukers, and C. Tille. Three-dimensional printing of porous ceramic scaffolds for bone tissue engineering. *Journal of Biomedical Materials Research Part B: Applied Biomaterials* 74(2) (2005): 782–788.
50. Zhou, Z., F. Buchanan, C. Mitchell, and N. Dunne. Printability of calcium phosphate: Calcium sulfate powders for the application of tissue engineered bone scaffolds using the 3D printing technique. *Materials Science and Engineering C* 38 (2014): 1–10.

51. Inzana, J.A., D. Olvera, S.M. Fuller et al. 3D printing of composite calcium phosphate and collagen scaffolds for bone regeneration. *Biomaterials* 35(13) (2014): 4026–4034.

52. Hutmacher, D.W., T. Schantz, I. Zein, K.W. Ng, S.H. Teoh, and K.C. Tan. Mechanical properties and cell cultural response of polycaprolactone scaffolds designed and fabricated via fused deposition modeling. *Journal of Biomedical Materials Research* 55(2) (2001): 203–216.

53. Kalita, S.J., S. Bose, H.L. Hosick, and A. Bandyopadhyay. Development of controlled porosity polymer-ceramic composite scaffolds via fused deposition modeling. *Materials Science and Engineering C* 23(5) (2003): 611–620.

54. Espalin, D., K. Arcaute, D. Rodriguez, F. Medina, M. Posner, and R. Wicker. Fused deposition modeling of patient-specific polymethylmethacrylate implants. *Rapid Prototyping Journal* 16(3) (2010): 164–173.

55. Hutmacher, D.W. Scaffolds in tissue engineering bone and cartilage. *Biomaterials* 21(24) (2000): 2529–2543.

56. Korpela, J., A. Kokkari, H. Korhonen, M. Malin, T. Närhi, and J. Seppälä. Biodegradable and bioactive porous scaffold structures prepared using fused deposition modeling. *Journal of Biomedical Materials Research Part B: Applied Biomaterials* 101(4) (2013): 610–619.

57. Lozinsky, V.I., I.Y. Galaev, F.M. Plieva, I.N. Savina, H. Jungvid, and B. Mattiasson. Polymeric cryogels as promising materials of biotechnological interest. *Trends in Biotechnology* 21(10) (2003): 445–451.

58. Alves, M.-H., B.E.B. Jensen, A.A.A. Smith, and A.N. Zelikin. Poly(vinyl alcohol) physical hydrogels: New vista on a long serving biomaterial. *Macromolecular Bioscience* 11(10) (2011): 1293–1313.

59. Baker, M.I., S.P. Walsh, Z. Schwartz, and B.D. Boyan. A review of polyvinyl alcohol and its uses in cartilage and orthopedic applications. *Journal of Biomedical Materials Research Part B: Applied Biomaterials* 100(5) (2012): 1451–1457.

60. Lozinsky, V.I. and F.M. Plieva. Poly(vinyl alcohol) cryogels employed as matrices for cell immobilization. 3. Overview of recent research and developments. *Enzyme and Microbial Technology* 23(3) (1998): 227–242.

61. Mathews, D.T., Y.A. Birney, P.A. Cahill, and G.B. McGuinness. Vascular cell viability on polyvinyl alcohol hydrogels modified with water-soluble and-insoluble chitosan. *Journal of Biomedical Materials Research Part B: Applied Biomaterials* 84(2) (2008): 531–540.

62. Liu, Y., N.E. Vrana, P.A. Cahill, and G.B. McGuinness. Physically crosslinked composite hydrogels of PVA with natural macromolecules: Structure, mechanical properties, and endothelial cell compatibility. *Journal of Biomedical Materials Research Part B: Applied Biomaterials* 90(2) (2009): 492–502.

63. Vrana, N.E., P.A. Cahill, and G.B. McGuinness. Endothelialization of PVA/gelatin cryogels for vascular tissue engineering: Effect of disturbed shear stress conditions. *Journal of Biomedical Materials Research Part A* 94(4) (2010): 1080–1090.

64. Vrana, N.E., A. O'Grady, E. Kay, P.A. Cahill, and G.B. McGuinness. Cell encapsulation within PVA-based hydrogels via freeze-thawing: A one-step scaffold formation and cell storage technique. *Journal of Tissue Engineering and Regenerative Medicine* 3(7) (2009): 567–572.

65. Vrana, N.E., K. Matsumura, S.-H. Hyon, L.M. Geever, J.E. Kennedy, J.G. Lyons, C.L. Higginbotham, P.A. Cahill, and G.B. McGuinness. Cell encapsulation and cryostorage in PVA–gelatin cryogels: Incorporation of carboxylated ε-poly-L-lysine as cryoprotectant. *Journal of Tissue Engineering and Regenerative Medicine* 6(4) (2012): 280–290.

66. Mathews, D.T., Y.A. Birney, P.A. Cahill, and G.B. McGuinness. Mechanical and morphological characteristics of poly(vinyl alcohol)/chitosan hydrogels. *Journal of Applied Polymer Science* 109(2) (2008): 1129–1137.

67. Wan, W.K., G. Campbell, Z.F. Zhang, A.J. Hui, and D.R. Boughner. Optimizing the tensile properties of polyvinyl alcohol hydrogel for the construction of a bioprosthetic heart valve stent. *Journal of Biomedical Materials Research* 63(6) (2002): 854–861.

68. Stammen, J.A., S. Williams, D.N. Ku, and R.E. Guldberg. Mechanical properties of a novel PVA hydrogel in shear and unconfined compression. *Biomaterials* 22(8) (2001): 799–806.

69. Chu, K.C. and B.K. Rutt. Polyvinyl alcohol cryogel: An ideal phantom material for MR studies of arterial flow and elasticity. *Magnetic Resonance in Medicine* 37(2) (1997): 314–319.

70. Pazos, V., R. Mongrain, and J.C. Tardif. Polyvinyl alcohol cryogel: Optimizing the parameters of cryogenic treatment using hyperelastic models. *Journal of the Mechanical Behavior of Biomedical Materials* 2(5) (2009): 542–549.

71. Willcox, P.J., D.W. Howie, K. Schmidt-Rohr, D.A. Hoagland, S.P. Gido, S. Pudjijanto, L.W. Kleiner, and S. Venkatraman. Microstructure of poly(vinyl alcohol) hydrogels produced by freeze/thaw cycling. *Journal of Polymer Science Part B: Polymer Physics* 37(24) (1999): 3438–3454.

72. Ricciardi, R., C. Gaillet, G. Ducouret, F. Lafuma, and F. Lauprêtre. Investigation of the relationships between the chain organization and rheological properties of atactic poly(vinyl alcohol) hydrogels. *Polymer* 44(11) (2003): 3375–3380.

73. Ricciardi, R., F. Auriemma, C. Gaillet, C. De Rosa, and F. Lauprêtre. Investigation of the crystallinity of freeze/thaw poly(vinyl alcohol) hydrogels by different techniques. *Macromolecules* 37(25) (2004): 9510–9516.

74. Valentín, J.L., D. López, R. Hernández, C. Mijangos, and K. Saalwachter. Structure of poly(vinyl alcohol) cryo-hydrogels as studied by proton low-field NMR spectroscopy. *Macromolecules* 42(1) (2008): 263–272.

75. Lozinsky, V.I., L.G. Damshkaln, B.L. Shaskol'skii, T.A. Babushkina, I.N. Kurochkin, and I.I. Kurochkin. Study of cryostructuring of polymer systems: 27. Physicochemical properties of poly(vinyl alcohol) cryogels and specific features of their macroporous morphology. *Colloid Journal* 69(6) (2007): 747–764.

76. Lozinsky, V.I., L.G. Damshkaln, I.N. Kurochkin, and I.I. Kurochkin. Study of cryo-structuring of polymer systems: 28. Physicochemical properties and morphology of poly(vinyl alcohol) cryogels formed by multiple freezing-thawing. *Colloid Journal* 70(2) (2008): 189–198.

77. Ricciardi, R., G. Mangiapia, F. Lo Celso, L. Paduano, R. Triolo, F. Auriemma, C. De Rosa, and F. Lauprêtre. Structural organization of poly(vinyl alcohol) hydrogels obtained by freezing and thawing techniques: A SANS study. *Chemistry of Materials* 17(5) (2005): 1183–1189.

78. Millon, L.E. and W.K. Wan. The polyvinyl alcohol–bacterial cellulose system as a new nanocomposite for biomedical applications. *Journal of Biomedical Materials Research Part B: Applied Biomaterials* 79(2) (2006): 245–253.

79. Fenglan, X., L. Yubao, W. Xuejiang, W. Jie, and Y. Aiping. Preparation and character-ization of nano-hydroxyapatite/poly(vinyl alcohol) hydrogel biocomposite. *Journal of Materials Science* 39(18) (2004): 5669–5672.

80. Pan, Y. and D. Xiong. Friction properties of nano-hydroxyapatite reinforced poly(vinyl alcohol) gel composites as an articular cartilage. *Wear* 266(7) (2009): 699–703.

81. Yusong, P., S. Qianqian, P. Chengling, and W. Jing. Prediction of mechanical properties of multilayer gradient hydroxyapatite reinforced poly(vinyl alcohol) gel biomaterial. *Journal of Biomedical Materials Research Part B: Applied Biomaterials* 101(5) (2013): 729–735.

82. Yusong, P., W. Jing, and P. Chengling. Preparation and characterisation of functional gradient nanohydroxyapatite reinforced polyvinyl alcohol gel biocomposites for articu-lar cartilage. *Micro & Nano Letters* 7(9) (2012): 880–884.

83. Qi, Z., C. Yamamoto, N. Imori et al. Immunoisolation effect of polyvinyl alcohol (PVA) macroencapsulated islets in type 1 diabetes therapy. *Cell Transplantation* 21(2–3) (2012): 525–534.
84. Calvert, P. Hydrogels for soft machines. *Advanced Materials* 21(7) (2009): 743–756.
85. Gong, J.P., Y. Katsuyama, T. Kurokawa, and Y. Osada. Double-network hydrogels with extremely high mechanical strength. *Advanced Materials* 15(14) (2003): 1155–1158.
86. Gong, J.P. Why are double network hydrogels so tough? *Soft Matter* 6(12) (2010): 2583–2590.
87. Haque, Md.A., T. Kurokawa, and J.P. Gong. Super tough double network hydrogels and their application as biomaterials. *Polymer* 53(9) (2012): 1805–1822.
88. Kwon, H.J., K. Yasuda, J.P. Gong, and Y. Ohmiya. Polyelectrolyte hydrogels for replacement and regeneration of biological tissues. *Macromolecular Research* 22(3) (2014): 227–235.
89. Liu Tsang, V. and S.N. Bhatia. 3D tissue fabrication. *Advanced Drug Delivery Reviews* 56(11) (2004): 1635– 1647.
90. Liu, V.A. and S.N. Bhatia. Three-dimensional photopatterning of hydrogels containing living cells. *Biomedical Microdevices* 4(4) (2002): 257–266.
91. Aizawa, Y., R. Wylie, and M. Shoichet. Endothelial cell guidance in 3D patterned scaffolds. *Advanced Materials* 22(43) (2010): 4831–4835.
92. Wylie, R.G. and M.S. Shoichet. Three-dimensional spatial patterning of proteins in hydrogels. *Biomacromolecules* 12(10) (2011): 3789–3796.
93. Wylie, R.G., S. Ahsan, Y. Aizawa, K.L. Maxwell, C.M. Morshead, and M.S. Shoichet. Spatially controlled simultaneous patterning of multiple growth factors in three-dimensional hydrogels. *Nature Materials* 10(10) (2011): 799–806.
94. Mo, X. and H.-J. Weber. Electrospinning P (LLA-CL) nanofiber: A tubular scaffold fabrication with circumferential alignment. In *Macromolecular Symposia*, Vol. 217(1), Wiley-VCH Verlag, Weinheim, Germany, pp. 413–416, 2004.
95. Mo, X.M., C.Y. Xu, M.E. Al Kotaki, and S. Ramakrishna. Electrospun P (LLA-CL) nanofiber: A biomimetic extracellular matrix for smooth muscle cell and endothelial cell proliferation. *Biomaterials* 25(10) (2004): 1883–1890.
96. Xu, C., R. Inai, M. Kotaki, and S. Ramakrishna. Electrospun nanofiber fabrication as synthetic extracellular matrix and its potential for vascular tissue engineering. *Tissue Engineering* 10(7–8) (2004): 1160–1168.
97. Webster, T.J., L.S. Schadler, R.W. Siegel, and R. Bizios. Mechanisms of enhanced osteoblast adhesion on nanophase alumina involve vitronectin. *Tissue Engineering* 7(3) (2001): 291–301.
98. Xu, C., F. Yang, S. Wang, and S. Ramakrishna. In vitro study of human vascular endothelial cell function on materials with various surface roughness. *Journal of Biomedical Materials Research Part A* 71(1) (2004): 154–161.
99. Ma, Z., M. Kotaki, T. Yong, W. He, and S. Ramakrishna. Surface engineering of electrospun polyethylene terephthalate (PET) nanofibers towards development of a new material for blood vessel engineering. *Biomaterials* 26(15) (2005): 2527–2536.
100. Boland, E.D., J.A. Matthews, K.J. Pawlowski, D.G. Simpson, G.E. Wnek, and G.L. Bowlin. Electrospinning collagen and elastin: Preliminary vascular tissue engineering. *Frontiers in Bioscience: A Journal and Virtual Library* 9 (2004): 1422–1432.
101. Vatankhah, E., M.P. Prabhakaran, D. Semnani, S. Razavi, M. Morshed, and S. Ramakrishna. Electrospun tecophilic/gelatin nanofibers with potential for small diameter blood vessel tissue engineering. *Biopolymers* 101 (2014): 1165–1180.
102. Zhu, W., S. Liu, J. Zhao, S. Liu, S. Jiang, B. Li, H. Yang, C. Fan, and W. Cui. Highly flexible and rapidly degradable papaverine-loaded electrospun fibrous membranes for preventing vasospasm and repairing vascular tissue. *Acta Biomaterialia* 10(7) (2014): 3018–3028.

103. Bruder, S.P. and A.I. Caplan. Bone regeneration through cellular engineering. In *Principles of Tissue Engineering*, Lanza, R., Langer, R., and Vacanti, J.P. (eds.), Academic Press, San Diego, CA, 2000, pp. 683–696.

104. Yoshimoto, H., Y.M. Shin, H. Terai, and J.P. Vacanti. A biodegradable nanofiber scaffold by electrospinning and its potential for bone tissue engineering. *Biomaterials* 24(12) (2003): 2077–2082.

105. Shin, M., H. Yoshimoto, and J.P. Vacanti. In vivo bone tissue engineering using mesenchymal stem cells on a novel electrospun nanofibrous scaffold. *Tissue Engineering* 10(1–2) (2004): 33–41.

106. Nedjari, S., S. Eap, A. Hébraud, C.R. Wittmer, N. Benkirane-Jessel, and G. Schlatter. Electrospun honeycomb as nests for controlled osteoblast spatial organization. *Macromolecular Bioscience* 14 (2014): 1580–1589.

107. Ramay, H.R. and M. Zhang. Biphasic calcium phosphate nanocomposite porous scaffolds for load-bearing bone tissue engineering. *Biomaterials* 25(21) (2004): 5171–5180.

108. Ramay, H.R. and M. Zhang. Preparation of porous hydroxyapatite scaffolds by combination of the gel-casting and polymer sponge methods. *Biomaterials* 24(19) (2003): 3293–3302.

109. Lai, G.-J., K.T. Shalumon, S.-H. Chen, and J.-P. Chen. Composite chitosan/silk fibroin nanofibers for modulation of osteogenic differentiation and proliferation of human mesenchymal stem cells. *Carbohydrate Polymers* 111 (2014): 288–297.

110. McPherson, J.M. and R. Tubo. Articular cartilage injury. In *Principles of Tissue Engineering*, Lanza, R., Langer, R., and Vacanti, J.P. (eds.), Academic Press, San Diego, CA, 2000, pp. 697–709.

111. Colwell, C.W., D.D. D'Lima, and M. Lotz. Articular cartilage repair. In *Clinical Orthopaedics and Related Research* (ed. Brighton, C.T.), Lippincott Williams and Wilkins, Pennsylvania, PA, 2001, p. 391S.

112. Tuli, R., W.-J. Li, and R.S. Tuan. Current state of cartilage tissue engineering. *Arthritis Research and Therapy* 5(5) (2003): 235–238.

113. Li, W.-J., R. Tuli, C. Okafor, A. Derfoul, K.G. Danielson, D.J. Hall, and R.S. Tuan. A three-dimensional nanofibrous scaffold for cartilage tissue engineering using human mesenchymal stem cells. *Biomaterials* 26(6) (2005): 599–609.

114. Kisiday, J., M. Jin, B. Kurz, H. Hung, C. Semino, S. Zhang, and A.J. Grodzinsky. Self-assembling peptide hydrogel fosters chondrocyte extracellular matrix production and cell division: Implications for cartilage tissue repair. *Proceedings of the National Academy of Sciences* 99(15) (2002): 9996–10001.

115. Man, Z., L. Yin, Z. Shao et al. The effects of co-delivery of BMSC-affinity peptide and rhTGF-β1 from coaxial electrospun scaffolds on chondrogenic differentiation. *Biomaterials* 35(19) (2014): 5250–5260.

116. Lin, V.S., M.C. Lee, S. O'Neal, J. McKean, and K.L. Paul Sung. Ligament tissue engineering using synthetic biodegradable fiber scaffolds. *Tissue Engineering* 5(5) (1999): 443–451.

117. Goulet, F., D. Rancourt, R. Cloutier et al. Tendons and ligaments. In *Principles of Tissue Engineering*, Lanza, R., Langer, R., and Vacanti, J.P. (eds.), Academic Press, San Diego, CA, 2000, pp. 711–722.

118. DiEdwardo, C.A., P. Petrosko, T.O. Acarturk, P.A. DiMilla, W.A. LaFramboise, and P.C. Johnson. Muscle tissue engineering. *Clinics in Plastic Surgery* 26(4) (1999): 647–56.

119. Riboldi, S.A., M. Sampaolesi, P. Neuenschwander, G. Cossu, and S. Mantero. Electrospun degradable polyesterurethane membranes: Potential scaffolds for skeletal muscle tissue engineering. *Biomaterials* 26(22) (2005): 4606–4615.

120. Seil, J.T. and T.J. Webster. Spray deposition of live cells throughout the electrospinning process produces nanofibrous three-dimensional tissue scaffolds. *International Journal of Nanomedicine* 6 (2011): 1095–1099.

121. Bettahalli, N.M.S., N. Groen, H. Steg, H. Unadkat, J. Boer, C.A. Blitterswijk, M. Wessling, and D. Stamatialis. Development of multilayer constructs for tissue engineering. *Journal of Tissue Engineering and Regenerative Medicine* 8(2) (2014): 106–119.

122. Ehler, E. and S.N. Jayasinghe. Cell electrospinning cardiac patches for tissue engineering the heart. *Analyst* 139(18) (2014): 4449–4452.

123. Pauwels, F. Eine neue Theorie über den Einfluß mechanischer Reize auf die Differenzierung der Stützgewebe. *Zeitschrift für Anatomie und Entwicklungsgeschichte* 121(6) (1960): 478–515. (Translated by Maquet, P. and Furlong, R. A new theory concerning the influence of mechanical stimuli in the differentiation of supporting tissues. In *Biomechanics of the Locomotor Apparatus*, pp. 375–407, 1980.)

124. Carter, D.R., P.R. Blenman, and G.S. Beaupre. Correlations between mechanical stress history and tissue differentiation in initial fracture healing. *Journal of Orthopaedic Research* 6(5) (1988): 736–748.

125. Lacroix, D. and P.J. Prendergast. A mechano-regulation model for tissue differentiation during fracture healing: Analysis of gap size and loading. *Journal of Biomechanics* 35(9) (2002): 1163–1171.

126. Kelly, D.J. and P.J. Prendergast. Mechano-regulation of stem cell differentiation and tissue regeneration in osteochondral defects. *Journal of Biomechanics* 38(7) (2005): 1413–1422.

127. Boccaccio, A., A. Ballini, C. Pappalettere, D. Tullo, S. Cantore, and A. Desiate. Finite element method (FEM), mechanobiology and biomimetic scaffolds in bone tissue engineering. *International Journal of Biological Sciences* 7(1) (2011): 112.

128. Byrne, D.P., D. Lacroix, J.A. Planell, D.J. Kelly, and P.J. Prendergast. Simulation of tissue differentiation in a scaffold as a function of porosity, Young's modulus and dissolution rate: Application of mechanobiological models in tissue engineering. *Biomaterials* 28(36) (2007): 5544–5554.

129. Checa, S., C. Sandino, D.P. Byrne, D.J. Kelly, D. Lacroix, and P.J. Prendergast. Computational techniques for selection of biomaterial scaffolds for tissue engineering. In *Advances on Modeling in Tissue Engineering*, Fernandes, P.R. and Bártolo, P.J. (eds.) Springer, the Netherlands, pp. 55–69, 2011.

130. Olivares, A.L. and D. Lacroix. Computational methods in the modeling of scaffolds for tissue engineering. In *Computational Modeling in Tissue Engineering*, Geris, L. (ed.), Springer, Berlin, Germany, pp. 107–126, 2013.

131. Sarkar, S., H.J. Salacinski, G. Hamilton, and A.M. Seifalian. The mechanical properties of infrainguinal vascular bypass grafts: Their role in influencing patency. *European Journal of Vascular and Endovascular Surgery* 31(6) (2006): 627–636.

132. Zaucha, M.T., R. Gauvin, F.A. Auger, L. Germain, and R.L. Gleason. Biaxial biomechanical properties of self-assembly tissue-engineered blood vessels. *Journal of the Royal Society Interface* 8(55) (2011): 244–256.

133. Ingber, D.E. Mechanobiology and diseases of mechanotransduction. *Annals of Medicine* 35(8) (2003): 564–577.

134. Colombo, A., S. Guha, J.N. Mackle, P.A. Cahill, and C. Lally. Cyclic strain amplitude dictates the growth response of vascular smooth muscle cells in vitro: Role in in-stent restenosis and inhibition with a sirolimus drug-eluting stent. *Biomechanics and Modeling in Mechanobiology* 12(4) (2013): 671–683.

135. Stegemann, J.P., H. Hong, and R.M. Nerem. Mechanical, biochemical, and extracellular matrix effects on vascular smooth muscle cell phenotype. *Journal of Applied Physiology* 98(6) (2005): 2321–2327.

136. Sheridan, W.S., G.P. Duffy, and B.P. Murphy. Mechanical characterization of a customized decellularized scaffold for vascular tissue engineering. *Journal of the Mechanical Behavior of Biomedical Materials* 8 (2012): 58–70.

137. Zhao, Y., S. Zhang, J. Zhou, J. Wang, M. Zhen, Y. Liu, J. Chen, and Z. Qi. The development of a tissue-engineered artery using decellularized scaffold and autologous ovine mesenchymal stem cells. *Biomaterials* 31(2) (2010): 296–307.
138. Williams, C., J. Liao, E.M. Joyce, B. Wang, J.B. Leach, M.S. Sacks, and J.Y. Wong. Altered structural and mechanical properties in decellularized rabbit carotid arteries. *Acta Biomaterialia* 5(4) (2009): 993–1005.
139. Chow, M.-J. and Y. Zhang. Changes in the mechanical and biochemical properties of aortic tissue due to cold storage. *Journal of Surgical Research* 171(2) (2011): 434–442.
140. Zahedmanesh, H., J.N. Mackle, A. Sellborn, K. Drotz, A. Bodin, P. Gatenholm, and C. Lally. Bacterial cellulose as a potential vascular graft: Mechanical characterization and constitutive model development. *Journal of Biomedical Materials Research Part B: Applied Biomaterials* 97(1) (2011): 105–113.
141. Morrow, D., C. Sweeney, Y.A. Birney, P.M. Cummins, D.Walls, E.M. Redmond, and P.A. Cahill. Cyclic strain inhibits Notch receptor signaling in vascular smooth muscle cells in vitro. *Circulation Research* 96(5) (2005): 567–575.
142. Zahedmanesh, H. and C. Lally. A multiscale mechanobiological modelling framework using agent-based models and finite element analysis: Application to vascular tissue engineering. *Biomechanics and Modeling in Mechanobiology* 11(3–4) (2012): 363–377.
143. DiMilla, P.A., J.A. Stone, J.A. Quinn, S.M. Albelda, and D.A. Lauffenburger. Maximal migration of human smooth muscle cells on fibronectin and type IV collagen occurs at an intermediate attachment strength. *Journal of Cell Biology* 122(3) (1993): 729–737.
144. Hahn, M.S., M.K. McHale, E. Wang, R.H. Schmedlen, and J.L. West. Physiologic pulsatile flow bioreactor conditioning of poly(ethylene glycol)-based tissue engineered vascular grafts. *Annals of Biomedical Engineering* 35(2) (2007): 190–200.
145. Peirce, S.M., E.J. Van Gieson, and T.C. Skalak. Multicellular simulation predicts microvascular patterning and in silico tissue assembly. *The FASEB Journal* 18(6) (2004): 731–733.
146. Zahedmanesh, H., H. Van Oosterwyck, and C. Lally. A multi-scale mechanobiological model of in-stent restenosis: Deciphering the role of matrix metalloproteinase and extracellular matrix changes. *Computer Methods in Biomechanics and Biomedical Engineering* 17(8) (2014): 813–828.
147. Nagel, T. and D.J. Kelly. Remodelling of collagen fibre transition stretch and angular distribution in soft biological tissues and cell-seeded hydrogels. *Biomechanics and Modeling in Mechanobiology* 11(3–4) (2012): 325–339.
148. Creane, A., E. Maher, S. Sultan, N. Hynes, D.J. Kelly, and C. Lally. Prediction of fibre architecture and adaptation in diseased carotid bifurcations. *Biomechanics and Modeling in Mechanobiology* 10(6) (2011): 831–843.
149. Zahedmanesh, H. and C. Lally. Multiscale modeling in vascular disease and tissue engineering. In *Multiscale Computer Modeling in Biomechanics and Biomedical Engineering*, Gefen, A. (ed.), Springer, Berlin, Germany, pp. 241–258, 2013.

3 Nanomaterial Cell Interactions in Tissue Engineering

Erkan Turker Baran, Dilek Keskin, and Aysen Tezcaner

CONTENTS

3.1 NANOMATERIAL CELL INTERACTIONS IN TISSUE ENGINEERING

The emergence of nanotechnology brought new ideas toward solutions of existing problems in biomedical area. One of the biggest challenges was development of scaffolds that can provide a better extracellular matrix (ECM) for cells in tissue engineering. Although cells in our tissues generally vary in size between 5 and 20 µm, it is known that they reside in their niche and their interactions with this microenvironment are at submicron level. Besides, the perfect intracellular and extracellular organizations at micron level are created by building blocks having nanometer sizes. So, the nanoscience and nanotechnology have received more attention toward finding these molecules and their special roles on cell attachment.

3.1.1 Cell–Nanomaterial Interactions

ECM provides the immediate microenvironment of cells *in vivo* composed of a three-dimensional complex and dynamic network of macromolecules (proteins, glycoproteins, proteoglycans, and other macromolecules), soluble and ECM-bound factors. ECM has a hierarchical organization from nano to macro scale. Bone, for example, is a composite structure composed of a matrix consisting of organic and inorganic phases. Collagen type I constitutes main part of organic phase, and it is composed of a macromolecule of 300 nm in length and 1.5 nm in diameter. The inorganic phase of bone involves stacked nanohydroxyapatite crystals in a staggered manner, and it is embedded in the collagenous matrix. The fundamental subunit of bone, called osteon, consists of concentric layers, or lamellae of collagen fibers that surround a central canal, named Haversian canal. The orientation of collagen fibers in each lamella is parallel to each other in which collagen molecules are stacked in a staggered manner.

Mimicking functional *in vivo* structures is still one of the key challenges faced by scientists in engineering tissues. It is now widely accepted that these nanostructures have an important role in functioning of tissues. The fibrous structural elements like collagen have a dimension in nanometer range; thereby it is not surprising that nanomaterials, which mimic native ECM closely, cause positive cellular responses.

Considerable progress has been made from the days where ECM was considered to be a static structural framework until recently. Although some of the mechanisms responsible for the role of microenvironment on the behavior of cells have been elucidated, there are still unknown mechanisms. To understand cell–nanomaterial interactions and the role of nanomaterials in engineering tissues studying the extracellular matrix is important. ECM is discussed in detail in the next section.

3.1.2 Extracellular Matrix

Apart from serving as a structural framework for cell adhesion, the important role of ECM on differentiation of cells and engineering functional tissues has been well documented by many studies (Jeon et al., 2013; Prodanov et al., 2013; Levett et al., 2014; Pittrof et al., 2012). ECM supports the cells *in vivo*, and there is a continuous bi-directional cross-talk between this structural framework and cells. Cells interact with ECM mainly through integrins and also proteoglycan receptors. There are different types of cell–ECM interactions, namely, those (1) that are mediated by integrin and proteoglycan receptors important for adhesion/de-adhesion processes taking place during cell migration; (2) that are growth factor or cytokine receptor mediated which affect cell proliferation, survival, induction, and maintenance of differentiation; and (3) that are mediated by receptors responsible for the processes related with apoptosis and epithelial-to-mesenchymal transitions.

Integrins are transmembrane cell surface receptors composed of α and β subunits. They can bind to different types of ECM components and undergo bi-directional signaling. These ECM–cell interactions play a key role in the regulation of cellular processes taking place during development and wound healing (Petreaca and Martins-Green, 2007).

ECM components have biochemical, structural, and functional diversity. Such diversity renders tissue specific physical, biochemical, and biomechanical properties to ECM (Gattazzo et al., 2014). Matrix stiffness, porosity, and topography are among the physical properties that affect anchorage related functions (i.e., cell division, cell polarity, and cell migration) (Hynes, 2009; Brown and Badylak, 2014). With changes in composition and structure, the components in native ECM are arranged into supramolecular structures like fibers and meshes creating different topographies. The effect of topography on stem cell behavior has been also reported in several studies (Park et al., 2007; Biggs et al., 2009; Jose et al., 2010; Schwartz and Chen, 2013). Biochemically, ECM interacts with cells either directly through receptors or indirectly by noncanonical growth factor presentation to cells. Mechanical microenvironment of cells is also influential on cell behavior.

Understanding the cellular response to mechanical stimuli is critical for the success of biomedical devices and tissue-engineered constructs. There are a variety of physical signals that can make up the mechanical environment in the ECM. Among these are tissue deformation-based strain or fluid flow, streaming potentials with ion movement along the membranes, pressure changes, and the piezoelectric field effect (You and Jacobs, 2005).

Biomechanical properties of ECM are also influential on cell behavior. Stiffness is one of the important properties of ECM by which cells can sense their external environment. The degree that ECM resists deformation is called matrix stiffness. Homeostasis in stiffness in tissues should be maintained for proper functioning. It is known that matrix stiffness plays an important role during tissue morphogenesis and throughout life (Daley and Yamada, 2013). Additionally, *in vitro* studies revealed that stiffness is one of the important factors determining the differentiation of stem cells (Engler et al., 2006; Tse and Engler, 2011; Wang et al., 2012; Gershlak et al., 2013). When cells detect an external mechanical signal, they convert this stimulus into a chemical stimulus, which leads to a cellular response through activation of a number of pathways. This process is called mechanotransduction. There are mainly two mechanisms responsible for the detection of mechanical cues in the extracellular matrix by the cells (You and Jacobs, 2005). The first mechanism involves the detection of mechanical signals through channels and receptors on the cell membranes, which leads to transduction of mechanical forces into cells with the activation of one or more signaling pathways. These events lead to further changes in cells like cytoskeleton reorganization and focal adhesion complex formation (Figure 3.1). The focal adhesion complexes (integrins, adaptors, signaling molecules) are physically connected to cytoskeletal elements, nuclear matrix, and nuclear envelope, thereby influencing the response of cells. Focal adhesion kinase (FAK) plays a key role in integrin-mediated signaling, and its activation triggers extracellular signal regulated kinase (ERK)/mitogen-activated protein kinase (MAPK) signaling pathway (Salasznyk et al., 2007a,b), which upregulates gene transcription for cell cycling and replication. The second mechanism involves the deformation of the cell membrane by mechanical forces, which results with the activation of mechanosensitive ion channels and/or receptors. Multiple cross-talks between different signaling pathways take place during mechanical stimuli detection.

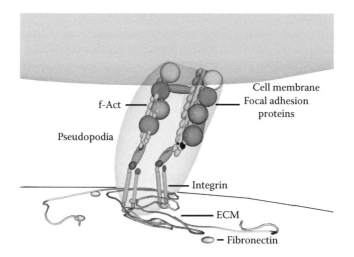

FIGURE 3.1 Mechanotransduction events start at focal adhesion complex. The fibronectin molecule presenting cell attachment sequences is recognized by integrin receptor, and it starts the cascade events of signal transduction from cell membrane to nucleus mediated by cytoskeleton proteins and associated messenger proteins as well as other intracellular messenger proteins.

Different cells experience different mechanical environments. Mechanical microenvironment of bone cells involves direct strain, pressure, shear stress, and streaming potentials, whereas chondrocyte cells are subject to hydrostatic pressure and compression. The ability of cells to respond to mechanical cues is very important for their function. *In vitro* studies revealed that the cellular response to mechanical forces differs among cell types. Cyclic pressurization of osteoblasts at 1 Hz resulted with increased expression of early osteoblastic differentiation markers (Nagatomi et al., 2001) and collagen synthesis for new bone formation (Nagatomi et al., 2003), whereas this same mechanical stress did not have any effect on the endothelial cells (Nagatomi et al., 2001). Endothelial cells are exposed to fluid flow shear stresses that regulate the functions of these cells. Some prominent examples of cell types that respond to fluid-induced shear stresses are chondrocytes and osteoblasts. In bone, load-induced fluid flow of extracellular fluid takes place between mineralized matrix spaces connected through long channels where osteocytes reside (You and Jacobs, 2005). In a recent study, Matsugaki et al. (2013) were able to establish anisotropic bone matrix architecture in which osteoblasts were aligned in the direction of continuous mechanical cyclic stretch applied during a long cultivation period. The findings of this study also pointed the fact that mechanical environment decides the organization of the bone matrix.

Well-orchestrated dynamic interplays between these properties ensure maintenance of homeostasis, function of cells, and stem cell pool in the tissues. There are several recent studies that have focused on the investigation of nanostructural cues together with fluid flow on the osteoblastic behavior (Salvi et al., 2010; Prodanov et al., 2013). Salvi et al. investigated cellular response to fluid flow on nano-topographies and showed the positive synergistic effect of these two parameters on increasing mechanosensitivity of stem cells. Prodanov et al. have studied the effect of combining parallel nano-grooved polystyrene surfaces with pulsed fluid flow on the cell behavior (Prodanov et al., 2013). They showed that interstitial fluid forces and structural cues contributed to the expression of bone-specific markers by MC3T3-E1 osteoblast-like cells. These studies show that controlling more properties of ECM-like structures will bring us closer to functional engineered tissues. Decellularized tissues and organs have been also used for reconstruction of complex tissues and organs purposes in both preclinical and clinical trials (el-Kassaby et al., 2008; Remlinger et al., 2010; Cebotari et al., 2011; Quinn et al., 2011; Brown and Badylak, 2014). The rationale behind the use of decellularized matrices is that they maintain 3D architecture and their composition has functional relevance to the native tissue, thereby these inductive native scaffolds promote regeneration, functioning, and integration of the engineered tissues. The decellularization of tissues and their applications have been discussed in several reviews (Crapo et al., 2011; Song and Ott, 2011; Cheng et al., 2014).

Another point that should be considered for the role of ECM on the fate of cells is that ECM is degraded and remodeled, which is vital for normal tissue homeostasis. ECM components when degraded and/or modified by the cells create new recognition sites. These cryptic peptides called matricryptins or matrikines (i.e., endostatin, angiostatin, hyaluronic acid fragments, etc.) have been shown to affect several cellular processes like angiogenesis, antiangiogenesis, chemotaxis, and adhesion with mechanisms relying mainly on integrin, toll-like receptor, and scavenger receptor signaling pathways (Davis et al., 2000; Davis, 2010). The activity and parent molecule of some cryptic peptides were discussed in a review of Brown and Badylak (2014). One of the best-known matricryptic peptide that finds wide applications in tissue engineering is the Arg-Gly-Asp peptide that is present mainly in fibronectin and also in collagen, vitronectin, and osteopontin.

It is a well-accepted fact that mimicking complex ECM closely is crucial for obtaining functional engineered tissues and organs. However, creating functional tissues still remains a big challenge because there is a need for creation of the microenvironment with the structural, biochemical, and temporal cues in the right configuration, distribution, and concentration for cells.

3.1.3 UNDERSTANDING AND ENGINEERING STEM CELL NICHES

Stem cells are key cell sources for regenerative medicine and tissue engineering applications. Today, it is known that all tissues have their own stem cell reservoir, which decreases with age and in disease states. Stem cells are undifferentiated cells capable of self-renewal and giving rise to daughter cells committed to a given lineage (Weissman, 2000). Stem cells reside in a specific microenvironment called "stem cell niche," and this tissue-specific niche has the signal that allows stem cell survival and maintenance of stem cell identity and stem cell pool. Supportive cells, blood vessels, and nerves are present together with stem cells that are either active or quiescent in these specialized microenvironments (Morrison and Spradling, 2008). Apart from this, a balance between quiescence, self-renewal, and differentiation is dynamically regulated by these specialized niches. However, it should be noted that not only ECM components have an influence on the fate of stem cells but cells also produce and remodel their microenvironment. Cell–ECM interaction is reciprocal (Petreaca and Martins-Green, 2007; Kurtz and Oh, 2012). As discussed earlier, the properties of stem cell niche are important determinants of stem cell behavior. Properties of stem cell niches and several tissues' niches have been reviewed in detail by several researchers (Lander et al., 2012; Gattazzo et al., 2014; Han et al., 2014). Considerable efforts have been put to engineer physical microenvironments of stem cells. The development of scaffolds mimicking natural ECM of cells combined with the presentation of chemical and physical cues in a spatiotemporal manner has been a research and clinical goal in regenerative medicine and tissue engineering. Studies on the effects of molecular gradients, mechanical force gradients (Guvendiren and Burdick, 2013), and stiffness gradients have shown the importance of creating a dynamically controllable ECM for successful engineering of tissues and organs (Tse and Engler, 2011; Wang et al., 2012; Gershlak et al., 2013). Nanoscale engineering approaches have been mainly under focus in the recent years both for understanding the mechanisms of ECM–cell interactions at healthy and diseased states (Pittrof et al., 2012; Prodanov et al., 2013) and also mimicking ECM for regenerative medicine and tissue engineering applications (Chen et al., 2011; Orlando et al., 2011; Levett et al., 2014). Nanomaterials, which closely resemble *in vivo* ECM, are preferred in designing scaffolds for tissue engineering applications. Many studies showed that the interaction of cells with nanomaterials starts a cross-talk through mechanotransduction, which is an important regulator in the stem cell differentiation (Park et al., 2007; Namgung et al., 2011; Wang et al., 2011). As pointed out by Han et al. (2014), most of the research conducted for studying the effect of physical and biochemical microenvironment on stem cell behavior is conducted with two-dimensional *in vitro* models, and there are also recent studies reporting that mesenchymal stem cells behave differently to matrix stiffness gradients when cultured in 2D (Tse and Engler, 2011) and 3D *in vitro* physical niches (Jeon et al., 2013). However, cells *in vivo* have a 3D microenvironment, so there is a need for research of stem cell behavior in 3D niches to get further insight on cell–ECM interaction mechanisms. State of the art of different approaches for engineering ECM like structures using nanotechnology is discussed in the following sections.

3.2 NANOBIOMATERIALS

Nanomaterials are defined as materials that have at least one dimension within the nanometer range (<100 nm) and whose nanostructured features provide characteristics important to the bulk property of the material. Among different forms of nanomaterials used are nanotubes, nanofibers, nanocomposites, and nanophased and nanostructured materials (Wan and Ying, 2010). With the recognition that the unique physical and chemical characteristics of nanomaterials (e.g., polymer/ceramic or carbon-based nanocomposites, nanophase ceramics, nanofibers, surfaces with topographical cues, etc.) exert a positive effect on cell response, research on their use for regeneration of different tissues has increased enormously in the last decades.

3.2.1 NANOFIBER BIOMATERIALS

It has been recognized that there are insoluble proteins, mainly collagen in the extracellular matrices of cells in tissues, promoting not only cell attachment but also modulate their characteristic morphology and proliferation. Collagen as the major structural protein in ECM and connective tissues has more than 16 types; but most of the collagen in the body is type I, II, and III (Lodish et al., 2000). The 3D structure and composition of collagen fibers is tissue type dependent. But the basic form of these structural proteins found in most of the native cellular environments is "the fibrous structure" that forms the protein "backbone" of ECM making its characteristic 3D form with desired properties like porosity, hydrophilicity, mechanical strength, etc., together with soluble molecules like multiadhesive matrix proteins (i.e., fibronectin, laminin) and proteoglycans or mineral components like hydroxyapatite. These components take special roles in cell–matrix interactions. So, the use of nanofibers in scaffolding made possible the development of ECM-like scaffolds for tissue engineering applications.

Nanofibers provide many advantages over solid scaffold structures: (1) the high surface to volume ratio of the nanofibers that offers more specific contact surfaces while enabling a well-organized pore network for both material transport and even cell movement inside; (2) the diameter range of nanofibers can be changed for different tissues to match the degradation rate of polymer with expected lifetime of the scaffold; and (3) the nanofibers can be aligned to obtain a directional attachment of the cells, which is especially important in functionality of several cell types such as muscle or nerve cells.

Many researchers have developed and investigated nanofibers from natural or synthetic polymers toward this aim. Nanofibrous scaffolds can be produced by a few methods: electrospinning, phase separation, and self-assembly (SA). While the first method was more easily applicable to synthetic polymers, the last one was more suitable for natural ones like proteins. Andric et al. (2011) prepared electrospun PLLA and gelatin/PLLA nanofibers that are wrapped around PGA microfiber core to mimic the unit structure of cortical bone tissue; osteon (Figure 3.2). After mineralizing these scaffolds in 10X simulated body fluid (SBF), they observed successful results on cell attachment (mouse pre-osteoblastic cells, MC3T3). The fact that proliferation on scaffolds during the 4-week period was not different for the two types of nanofibers despite the presence of gelatin was explained with the loss of non-crosslinked gelatin in aqueous environment. Although mineralized scaffolds showed less proliferation than their unmineralized counterparts, the alkaline phosphatase (ALP) activity of the cells for all scaffolds showed their functionality in scaffolds. So, by mimicking the structural organization of bone tissue at nanoscale via electrospinning method they achieved promising results.

Lee et al. (2013) proposed a new method based on (1) direct-write electrospinning (DWES) for generation of lattice-patterned nanofibrous mats of PCL and (2) stacking the mats into a 3D scaffold with organized pores and desired thickness. In order to show the importance of interconnected and organized pore structure of nanofibrous scaffolds, this new design was compared with scaffolds prepared by conventional electrospinning and salt leaching methods using the same polymer. In accordance with their hypothesis, there were more spreading and attachment of NIH3T3 cells on the new nanofibrous

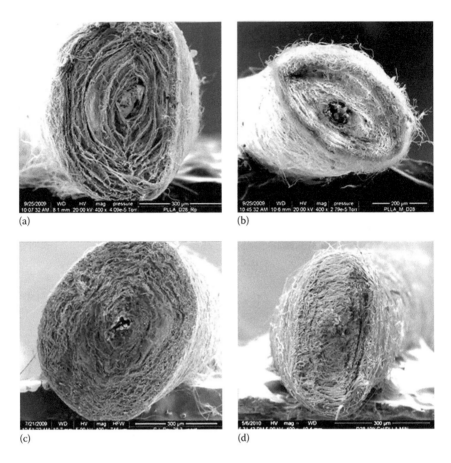

(a)   (b)

(c)   (d)

FIGURE 3.2   ESEM images of scaffold cross sections on day 28 of the degradation study: (a) PLLA, (b) PLL-mineralized, (c) 10% gel/PLLA, and (d) 10% gel/PLLA mineralized. (Reprinted from *Mater. Sci. Eng. C*, 31, Andric, T., Sampson, A.C., Freeman, J.W., Fabrication and characterization of electrospun osteon mimicking scaffolds for bone tissue engineering, 2–8, Copyright (2011), with permission from Elsevier.)

scaffolds while they were mostly aggregated on the scaffold prepared by salt leaching method. Cells on the proposed nanofibrous scaffold were also observed to secrete more ECM than those on scaffolds prepared by the last two conventional methods. Besides, many cells migrated toward inside of these DWES scaffolds, whereas only a few cells or none did so in conventional electrospun and salt-leached scaffolds, respectively.

It was reported that endothelial cells had superior spreading morphology on 100–300 nm fibers than on 1200 nm fibers (Whited and Rylander, 2014). However, the behavior of cells on different diameter fibers might vary for different types of cells. Elastin or elastin-like polypeptides in scaffolds or their modifications has been used in many scaffold researches (Daamena et al., 2007; Nivison-Smith et al., 2010). Use of elastin-like peptides (ELPs) that contain amino acid repeat sequence—VGVAPG—and cross-linking domains was suggested as useful alternative for elastin-derived proteins. Blit et al. (2012) developed aligned nanofibrous polyurethane (PCNU—polycarbonate urethane) scaffolds by electrospinning and cross-linked ELP4 on the surface. They have also prepared polyurethane flat films and random fiber electrospun scaffolds of the same polymer to evaluate the effect of scaffold nanoarchitecture on cell–material interactions with smooth muscle cells (SMC). Electrospun materials showed alignment and spreading of cells according to the alignment of the fibers and highest coverage of the scaffold surfaces were obtained by ELP4 cross-linked surfaces. Similarly, higher number of cells attached to films with ELP4 besides cell alignment being promoted while no cell attachment was seen in the only PCNU flat films. Cells on ELP4 cross-linked materials also revealed a spindle-like morphology, actin filament organization, and expression of markers suggesting their contractile phenotype. They have suggested that introduction of topographical cues associated with the polypeptides or ELP4-mediated cell signaling mechanism might have determined this result. Similarly, Guex et al. (2012) carried out a research on controlling substrate architecture and surface composition to promote muscle tissue development using PCL electrospun fibers. They have investigated the effect of fiber diameter using micro -and nanofibrous substrates that were either aligned or randomly oriented. The effect of surface composition was also changed by plasma coating with ultrathin (about 12 nm) oxygen functionalized hydrocarbon. They reported the similar effect of different sized, aligned fibers on myoblast orientation along parallel fibers, suggesting that this factor is the main trigger for spatial cell orientation. However, when random fibers were compared for fiber size, nanofibrous substrates were mentioned to promote confined cellular organization locally similar to confluent cell cultures on tissue culture plates. Ultrathin coat was successful in terms of maintaining the fiber architecture and soft mechanical properties while improving cell adhesion and growth.

3.2.2 SELF-ASSEMBLY MACROMOLECULES

Self-assembled (SA) macromolecules receive increasing attention in tissue engineering research as they offer an ECM-like nanofibrous environment. Macromolecules that can assemble into ordered nanostructures by noncovalent-type forces in an energy-driven or nondriven state are classified as self-assembling molecules (Mendes et al., 2013a). These types of macromolecules have been increasingly used in building nanofibrous TE substrates and in modulating cell differentiation by bioactive peptide sequences or synthetic, mostly amphiphilic, molecules. In SA approach, the peptide sequences that can self-assemble into nanofibers constitute the majority of that class of SA macromolecules. In a modular approach, the amino acid sequences that create physical assembly and the sequences that modulate cell adhesion and differentiation are often used together by solid phase peptide synthesis (Cai and Heilshorn, 2014). In nature, the SA process is often used to build nanoscale structures, which may provide strong structural materials as seen by assemblies of fibroin, collagen, and silk fibroin (Table 3.1). These materials have already been in large scale of use in tissue engineering applications as they are highly biocompatible materials and their SA mechanism has been investigated and imitated in construction of nanoscale features that can provide both physical and biological cues needed for cell differentiation (Bai et al., 2013; Brown and Barker, 2014).

TABLE 3.1
Potential Nano-Sized Building Blocks That Can Be Used in Tissue Engineering

Nano Building Blocks

Biological   Silk proteins: Fibroin

Glycine   Alanine Glycine Alanine

(a)

(b)

Silk fibroin molecule repeating unit. Cocoons (a) and reconstituted fibroin nanomesh produced by electrospinning on metal grid (b).

Extracellular macromolecules: Collagen

Collagen assembly forming tendon (Courtesy of E.T. Baran.)

Synthetic   Carbon nanotubes

Carbon nanotubes (Courtesy of E.T. Baran.)

Amphiphilic peptides and polymers

Micelle
Hydrophilic

Hydrophobic

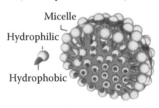

Micelle structure (Courtesy of E.T. Baran)

Inorganic   Hydroxyapatite

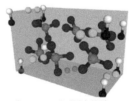
Hydroxyapatite $Ca_5(PO_4)_3(OH)$

Nanohydroxyapatite crystals

One of the most important advantages of using SA peptide molecules as substrate is that they can be decorated with bioactive sequences, such as cell adhesion and stimulating factors. Besides having nanofibrillar structure, the presence of active sequences stimulates cells in a similar way that ECM matrix does. The cytocompatibility of SA peptide–polymer conjugate (DGRFFF–PEG3000) containing the RGD sequence as the cell attachment factor was evaluated by Castelletto et al. (2013). They prepared β-sheet SA, core–shell fibril formation with a fibril width of 73 ± 1 nm and DGRFFF–PEG3000 films at low concentration. They observed that these structures enhanced corneal fibroblast proliferation. Two potent cell adhesion sequences, laminin and fibronectin sequences, were used in peptide-based SA systems for corneal tissue engineering (Uzunalli et al., 2014). The nanofiber mesh structure formed with laminin sequences supported corneal keratocyte cells compared to the fibronectin mimetic nanofibers. The incorporation of neuronal apoptosis inhibitory protein sequence SKPPGTSS into SA peptide system (ac-(RADA)4) is another rational approach in creating 3D cell culture matrix for nerve tissue engineering (Koutsopoulos et al., 2013). This peptide matrix that forms nanofiber gel performed better cell survival rates compared with collagen I and Matrigel®, although initially this was lower in a shorter period. Likewise, the immobilization of substance P (an 11-amino acid neuropeptide and engaged in innervation and direct cellular contacts) with self-assembling RADA16-II was studied for neural regeneration in ischemic hind limb models (Kim et al., 2013). These bioactive peptides were observed to self-assemble into 10 nm nanofibers within ischemic regions and attract mesenchymal cells, which eventually resulted with angiogenesis.

Without requiring complex bioactive sequences in physical gels, the sole structural SA process can be sufficient to sustain certain cell types, such as chondrocytes. In one such approach, an octapeptide sequence (FEFEFKFK)–based gel system was used to culture bovine chondrocytes (Mujeep et al., 2013). It was seen that the viability of bovine chondrocytes was sustained in 3D culture and the chondrogenic marker (collagen type II) expression was increased. Similarly, gel–cell constructs prepared with a structural SA peptide HLT2 (VLTKVKTKV^{DPL}PTKVEVKVLV-NH$_2$) stimulated glycosaminoglycan (GAG) and type II collagen production by chodrocytes under static culture conditions (Sinthuvanich et al., 2012). Three-dimensional encapsulation into structural SA peptide gel can also be suitable environment for mechanical stimulation of cells for enhanced tissue properties. This strategy was used for stimulation of skeletal muscle cells in SPG-178 ([CH$_3$CONH]-RLDLRLALRLDLR-[CONH$_2$]; R: arginine, L: leucine, D: aspartic acid, and A: alanine) peptide (Nagai et al., 2012). The SA formed scaffold with peptide nanofibers, exhibiting a diameter of 10 nm and a length of 500 nm was statically stretched during cell culture. A rapid cytoplasmic phosphorylation of ERK proteins during stretching proved stimulation of cells effectively by mechanical stimuli in nanofibrillar 3D environment.

SA tissue-engineered constructs were investigated as scaffold materials with various mechanical properties to facilitate cell differentiation and tissue growth. In one of such studies, SA macromolecules with variable stiffnesses were compared for augmenting hippocampus neurons (Sur et al., 2013). It has become certain that the soft peptide SA gels facilitated neuron polarity as cell retraction become favorable in soft environment. Stevenson et al. (2013) also similarly investigated the effect of SA gel stiffness and the cell adhesion property on cell behavior by blending of two SA peptide macromolecules. Especially, the matrices containing the RGD binding site (acetyl-**GRGD**SP-GG-FKFEFKFF-CONH$_2$) (R, arginine; G, glycine; D, aspartic acid; F, phenylalanine; E, glutamic acid; K, lysine) facilitated microvascular network (MVN) development while the SA matrix that present higher stiffness inhibited MVN formation.

In addition, SA peptides can mimic mineralized tissue by simulating organic part of the bone, which is constituted by highly organized collagen molecules that nucleate hydroxyapatite (HA) nanocrystals hierarchically. In this way, Vines et al. (2012) mixed hydroxyapatite nanoparticles (HANP, 100 nm) with a solution of peptide amphiphiles incorporating a cell attachment factor (PA-RGDS) for constructing bone mimetic tissue. Their transmission electron microscope examination proved the formation of nanofibers at 8–10 nm in diameter and several microns in length around HANP aggregates, forming a nanomatrix composite at a size of 100–200 nm. With this matrix, it was observed that the osteogenic differentiation of human mesenchymal stem cells was improved greatly by using the highest amount of HANP (66%). In a more advanced study, Sargeant et al. designed peptide amphiphiles (PA, [GS(P) EELLLAAA-C$_{16}$]) by introducing sequences of phosphoserine to nucleate HA formation in the culture medium (Sargeant et al., 2012). The PA nanofibers network intensely aggregated nanocrystalline carbonated HA (approximately 100 nm in diameter) after 7 days in calcium supplemented culture medium. Recently, self-assemblies of peptide nanotubes (PNTs) have emerged with potential applications in biomedical field (Seabraa and Durán, 2013). An RGDSK modified rosette nanotube (RNT) hydrogel was tested for cytocompatibility with bone cells (Zhang et al., 2009). Surface modified surfaces with PNTs significantly increased osteoblast adhesion compared to poly-L-lysine and collagen coating. This study is an indication of the stimulation of cells by nanoscale features of RNTs together with cell adhesion properties.

3.2.3 Inorganic and Organic Nanoparticles and Nanotubes

Carbon nanotubes, nano-sized calcium phosphates (CaP) and bioglasses are among the inorganic nanomaterials, which find wide applications in the biomedical field (Table 3.2).

Carbon nanotubes have unique molecular structure with carbon hexagons in tube shape. Owing to their electrical, thermal, and mechanical properties, their use in biomedical applications has increased in recent years (Saito, 2009). CNTs can be single-walled (SWCNTs) or multi-walled (MWCNTs) with corresponding diameters between 0.8–2 and 2–100 nm, respectively (De Volder et al., 2013). This diameter range of CNTs, together with its flexible length scale, makes them suitable as a biomimetic material of collagen fibers in the ECM of most of the tissues (Newman et al., 2013). Properties like number of walls, length, diameter, chirality, and existence of functional groups can be considered among the factors that may affect interactions between cells/tissues and CNTs (Mwenifumbo et al., 2007). *In vitro* studies conducted with SWNTs in composites with polymers showed their cytocompatibility (Shi et al., 2008). Sitharaman et al. (2008) also investigated SWNT/polymer nanocomposites for bone tissue engineering *in vivo*. They have observed greater bone

TABLE 3.2
Engineering Methods Used for Production of Nano-Sized Scaffolds

Techniques for Nano-Size Scaffold Production

	Principle	Nanostructure Morphology
Electrospinning	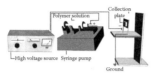 Electrospinning of polymers through high voltage	Cefuroxime loaded PCL nanofibers
Lithography	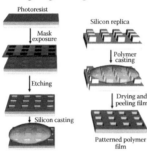 Lithography process used for patterning polymers	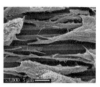 hPSC cells stained for phalloidin (green) and DAPI (blue) on 2 μm grating. Below: hPSC colonies on the 2 μm gratings (Chan et al., 2013).
Self-assembly	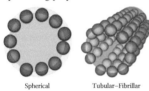	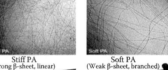 CryoTEM images of both PAs in anaqueous environment show long cylindrical nanofibers formed by both PAs. 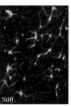 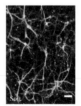 A schematic depicting the differences in molecular arrangement within the stiff and soft PA nanofibers (Sur et al., 2013).

tissue in growth in defects containing nanocomposites at 12 weeks compared to polymer scaffolds. They have suggested osteoconductivity and bioactivity of these nanocomposites owing to the presence of carbon nanotubes.

CNTs were shown to absorb extracellular matrix and serum proteins. This potential of having more interactions with biological milieu was suggested to render them bioactive and biocompatible. Therefore, various functionalization approaches were studied for CNTs. Functional groups like carboxyl or alcohol can be added to the walls or ends of the nanotubes (Hopley et al., 2014). Mu et al. (2009) showed that carboxyl groups inhibited cell proliferation. It has been shown that this result was due to cessation of the cell cycle by suppression of proliferation and differentiation-associated Smad-dependent human bone morphogenetic proteins. Besides, lower proliferative performance of MWCNTs was also correlated with low concentration of fetal bovine serum (FBS) in culture medium (Akasaka et al., 2010). However, Hopley et al. (2014) suggested that covalent bonding during carboxyl functionalization lowers the cell death and toxicity of CNTs.

Pryzhkova et al. (2014) investigated the human pluripotent stem cell (hPSC) behavior on unmodified and on UV/ozone-treated CNT arrays. Treatment rendered CNTs hydrophilic and made them functionalized. They observed that hPSCs (upon seeding as clumps on these CNT arrays) were unable to adhere to these hydrophilic surfaces, and they expressed apoptotic markers. However, when the surfaces were coated with Geltrex they supported survival and growth of the cells. Besides, hPSCs were shown to be able to differentiate to all three embryonic germ layers on these modified CNT arrays. They also showed spontaneous differentiation toward meso-dermal lineage owing to physical characteristics of the CNTs.

CaPs have long been in use for biomedical applications and carbon nanotubes now receive attention due to their properties. However, there is some toxicity concern related with their use. CaPs receive much attention in many fields like nanomedicine, orthopedics, and dentistry due to their compositional similarity to mineral phase of bone, their tunable degradability, and bioactivity. Biocompatibility, osteoconductivity, and osteoinductivity are the properties that make these materials ideal for orthopedic and dental applications. These properties of CaPs have been well-documented in the literature with *in vitro* and *in vivo* findings (Bernstein et al., 2013; Chen et al., 2013; Fricain et al., 2013; Abdal-hay et al., 2014; Kang et al., 2014). CaPs are also used as delivery systems for growth factors and drugs for bone and dental tissue engineering applications and related studies have been recently reviewed by Bose and Tarafder (2012). There are different types of CaPs (monocalcium phosphate, monohydrate [MCPM], dicalcium phosphates [DCP], tricalcium phosphates [TCP], hydroxyapatites [HA]), which differ in their bioactivity and degradation rate. These properties mainly depend on Ca/P ratio, crystallinity, and purity (Bose and Traffer, 2012). The osteoinductivity and osteoconductivity of CaPs differ, and this difference is due to the different physical and chemical properties they possess. Hydroxyapatite and β-TCP are the most widely used CaPs for tissue engineering applications. For bone tissue engineering applications, CaPs are usually used together with polymers for mimicking composite structure of bone.

There are many studies on the use of hydroxyapatite, especially nano-sized ones, which have the closest stoichiometric Ca/P ratio (1.67) to bone. Recent findings have put forth the fact that nano-sized HA powders have improved sinterability and enhanced densification that improves their mechanical properties (Han et al., 2004; Chaudhry et al., 2011; Aminzure et al., 2013), as well as biological responses (Li et al., 2011; Abdal-hay et al., 2014). CaPs and hydroxyapatites due to their brittleness and composite structure of bone are used together with polymers to prepare nanocomposite scaffolds for bone tissue engineering applications. One critical point is the homogeneous distribution of these nanoparticles for obtaining good mechanical properties. In a recent study, the researchers achieved vertically aligned hydroxyapatite plates on electrospun nylon fibers using hydrothermal conditions. With this orientation of HA, the mechanical properties and cellular responses in terms of attachment, proliferation, and osteoblastic differentiation were improved.

However, there are many reports stating that amorphous CaPs (Combes and Rey, 2010; Chai et al., 2012; Hild et al., 2012) as well as CaP-rich mineralized scaffolds have higher osteoinductivity effect as compared to hydroxyapatite-incorporated matrices (Osathanon et al., 2008; Kang et al., 2014). Under physiological conditions where there is a dynamic environment, amorphous materials like CaPs have higher solubility, thereby higher activity compared to crystalline materials.

For tissue engineering applications, cells are cultivated in osteogenic medium after seeding onto scaffolds for obtaining bone-like tissues. A very recent study by Kang et al. (2014) reported the osteogenic differentiation of human-induced pluripotent stem cells (hiPSCs) in cell culture media devoid of osteoinductive soluble factors using inherent material-based cues on mineralized gelatin methacrylate-based scaffolds. This study showed that scaffolds do not provide only structural framework for adhesion, proliferation but could also provide tissue-specific functions such as directing stem cell differentiation.

Organic nanoparticles constitute a promising group of nanobiomaterials for use in tissue engineering as they offer biodegradability and availability of functional groups that can be conjugated by bioactive molecules for further targeting and drug loading (Table 3.1). Organic nanoparticulate systems, however, are mostly aimed on the treatment of specific diseases by controlled release of drugs. Nevertheless, nanoparticles also find application in tissue engineering as they can be used as carriers for various growth factors intended for augmentation of tissue regeneration *in situ*. The controlled release of these bioactive factors over prolonged time from such systems provides continuous stimulation of cells.

Especially, the submicron size particles with encapsulated growth factors have enabled significant advantages in skin regeneration because of their ready absorption by surrounding cells after topical administration. For example, recombinant human endothelial growth factor-loaded solid lipid nanoparticle (mean size 332 nm) preparation, which was indented for restoration of the inflammatory process and re-epithelization of full-thickness wound in db/db mice animal model, showed a high cellular uptake (Gainza et al., 2014). This nanoparticle system significantly improved healing by restoring inflammation, providing re-epithelization and closing the wound. In another study, negatively charged low-molecular-weight heparin (fragmin) and electropositive protamine were complexed into water-insoluble micro/nanoparticles (0.1–3 μm in diameter) in which platelet-rich plasma (PRP) containing high levels of various GFs was encapsulated for repair in a split-thickness skin graft (Takabayashi et al., 2015). Nanoparticle-PRP system was reported to stimulate epithelialization and angiogenesis of wound effectively compared to particle-free formulation and control group. In a similar study, electropositive and RGD peptide functionalized chitosan and electronegative chondroitin sulfate were complexed into nanoparticles with a net positive charge (150–200 nm range, 20 mV) for use in wound healing by promoting adhesion and migration of skin cells (Hansson et al., 2012). Human dermal fibroblasts were adhered on the surface effectively by their cell area that was increased three times due to the enhanced cell–nanoparticle interaction and stimulation of cells by RGD sequence. In addition, growth factor encapsulated nanoparticles also showed their potency in tissue regeneration for other types of tissues. 2-N,6-O-sulfated chitosan (SCS) nanoparticles loaded with bone morphogenic protein-2 (BMP-2) have been recently applied into gelatin scaffolds for bone regeneration and angiogenesis in critically sized rabbit radius defects (Cao et al., 2014). The BMP-2 loaded nanoparticle carrier, which provided 3 weeks of BMP-2 release profile, stimulated peripheral and new vessel formation and accelerated healing of defects in 8 weeks.

Organic nanoparticles and stem cells can be applied to defect site simultaneously to provide synergic healing of the defect and restoration of the tissue. In such a study, the effect of the administration of both IL-1Ra loaded nanoparticles and mesenchymal stem cell was tested for the treatment of acute liver failure (Xiao et al., 2013). For this purpose, IL-1Ra loaded chitosan nanoparticles were conjugated with lactose for targeting liver cells before injection via portal vein of swine animal models; while stem cells were transplanted into liver tissue. They reported that the animal group that received nanoparticle-IL-1Ra showed significant improvement in inflammation, liver function, and transplanted hepatocytes proliferation was significantly higher compared with control groups. For cardiac repair, Zhu et al. proposed hypoxia-regulated vascular endothelial growth factor (VEGF) gene delivery by hyperbranched polyamidoamine (h-PAMAM) dendrimer (150 nm mean size) and skeletal myoblast (SkM) transplantation (Zhu et al., 2013). The gene carrier system was shown to be relatively nontoxic *in vitro*. It resulted in high transfection efficiency and modulated SkMs to express hVEGF165 for 18 days under hypoxia. Consequently, the group reported that intramyocardial transplantation of the transfected SkMs decreased the infarct size and increased blood vessel formation in C57/BL6 mice model. Stem cell commitment by gene transfection strategy was applied for bone tissue engineering as well by utilizing polyethylenimine cationized nanoparticle based on the polysaccharide from Angelica sinensis (ASP) (Deng et al., 2013). The complexation of cationic polymer with plasmid encoding transforming growth factor-beta 1 (TGF-β1) resulted in spherical nanoscaled particles (ranging from 20 to 50 nm), and this was used to transfect bone marrow and human umbilical cord mesenchymal stem cells. Similarly, the nanoparticles formed by complexion of low-molecular-weight protamine and miRNA encoding anti-osteogenic factors (30–50 nm in diameter) was used to transfect human mesenchymal stem cells (hMSCs) to promote osteoblastic differentiation (Suh et al., 2013). The elevated level of alkaline phosphatase activity and increased Alizarin red S staining of synthesized mineral tissue suggested an effective osteogenic differentiation of hMSCs by interfering with negative regulators of this pathway upon transfection with nanoparticle-miRNA system.

3.3 ENGINEERING METHODS

For scaffold development, both top-down (electrospinning, freeze drying) and bottom-up (self-assembly, nanoimprinting, etc.) approaches are being used. Traditional tissue engineering applications mainly used top-down approach. For this approach, cells are seeded on top of a porous scaffold. The idea is that cells will populate the scaffold with time. A bottom-up approach is based on the assembly or directed assembly of the scaffolds from smaller components and cells can be incorporated into the scaffold during this process. Most of the scaffold processing methods used have been adapted from the methods that are used in many engineering applications (i.e., fiber-based techniques, which is used in applications like filtration, composite fabrication, microfluidics, etc.). In recent years, for scaffolding ECM-like structures are under focus. In this section, we will discuss main methods for fabrication of scaffolds with nanoscale features for different tissue engineering applications.

3.3.1 ELECTROSPINNING

Electrospinning (ES) is a method where nanofibers could be produced with homogenous diameters in the desired size range and free of beads (unextended polymer regions) by adjusting the parameters related with polymer solution, electrospinning system, and ambient environmental conditions. Related with polymer solution, concentration and molecular weight of the polymer, type of the polymer and solvents used, could be counted as the main determinants of electrospinnability and properties of final form (Sun et al., 2014). The difficulty in electrospinning of the biopolymers (like collagen, fibroin, chitosan, hyaluronan, etc.) compared to synthetic ones are well documented by many researchers. Maeda et al. (2014) mentioned the difficulty of electrospinning biopolymers alone and, therefore, used PEO (polyethylene oxide) together with chitosan to obtain nanofibers with this method. They also reported that as the molecular weight of PEO increases, the minimum concentration required for electrospinnability decreases. Based on the same problem, they also did not incorporate hyaluronic acid (an important component of ECM) into electrospinning solution but combined coating and electrospinning methods to develop chitosan/hyaluronic acid polyelectrolyte complex nanofibers. Electrospun chitosan/PEO (polyethylene oxide) nanofiber mats were coated with hyaluronic acid after removal of PEO. However, fiber diameters were observed to increase during removal of PEO from the nanofibers.

Recently, co-electrospinning has gained interest to obtain composite fiber structures in different forms like core–shell structures. In this method, two solutions can be fed through different coaxial capillary channels to obtain structures like polymer core shell fibers, hollow polymer core shell fibers, or hollow fibers containing ceramics. Besides, this method might also be applied to load special components like drugs, growth factors, or cells into electrospun fibers (Braghirolli et al., 2014).

Another recent approach for safe incorporation of cells and bioactive agents into electrospinning media is defined as "green electrospinning" that uses water as diluents in suspension or emulsion electrospinning. This method, thus, reduces the use of organic solvents (Pal et al., 2014). In the study by Pal and coauthors, uneven surfaced meshes of PCL were produced by this method using PVA in the emulsions. The efficiency of uneven surfaced PCL meshes produced by this method was compared with uniform surfaced PCL meshes prepared by solution electrospinning for cell attachment and proliferation. They concluded that the formation of uneven surfaces on meshes significantly improved cell proliferation as compared to PCL alone as well as in comparison with unwashed form of the emulsion electrospun fiber meshes.

3.3.2 SELF-ASSEMBLY

The SA of macromolecules is a unique way of forming biocompatible nanostructures by using chemical and physical triggers. Moreover, it would be advantegeous if SA macromolecules can be triggered into insoluble nanostructures by physiological ionic strength and pH if living cells are intended to be encapsulated or *in situ* application is desired with injectable systems (Mendes et al., 2013b). A variety of external chemical and physical triggers may be used to stimulate intermolecular assembly by weak forces such as electrostatic, hydrophobic, hydrogen bonds, and van der Waals interactions. Especially, amino acids (aas), which can form all the aforementioned forces due to the different pendant groups of 20 different types of aas, are ideal monomers in building macromolecular SA components. Fine control over the strength and directionality of these forces are the utmost important features of the SA systems, which defines the geometry of nanostructures (Aida et al., 2012).

α-Helical secondary structure is one of the common units that is used by SA macromolecules to conform into fibrous structures, which is essential for creating ECM-like substrates. Helix-forming heptad repeat sequences adopted from well-known α-helix structures (such as leucine zipper and tropoelastin domains) are usually modified to increase stability of the constructs formed (Zimenkov et al., 2004). For example, leucine zipper peptide containing the six natural heptad sequences was used in SA construction and an amino acid is modified by the introduction of a cysteine residue by site-directed mutagenesis to enable covalent S–S bonds for stable scaffold formation for tissue engineering applications (Huang et al., 2014). β-Sheet formation is another common secondary structure unit formed between peptide strands that are stabilized through rich hydrogen bonds. For example, the synthesized octapeptide, FEFEFKFK, was shown to self-assemble in solution and form β-sheet-rich nanofibers (Mujeep et al., 2013). β-Sheet-based SA formation, however, can give rise to various structural forms depending on pH and the presence of ionizable groups in the secondary structure of macromolecule. Elgersma et al. showed that amyloidogenic aggregate mimetic Ab(16–22) peptides (Lys-Leu-Val-Phe-Phe-Ala-Glu-OH) can show various shapes such as lamellar sheet, helical tape, PNT, or their mix depending on particular pH value (Elgersma et al., 2014).

SA molecules are designed so that they can be triggered by environmental signals, such as pH, ionic strength, and chelating ions to prevent premature gelation before their application. Ionic SA peptides are the ideal group that can form hydrogels upon changes in pH and ionic strength. Peptide P_{11}-4 (H_3CO-QQRFEWEFEQQ-NH_2) was constructed to form β-sheet at low pH, which turns into liquid state at high pH by the deprotonation of the glutamic acid side chains (Carrick et al., 2007). The peptide amphiphile (PA) molecules, which contain both hydrophobic alkyl tail and a hydrophilic peptide molecule on their structure, is another SA peptide groups that can be triggered with ionizable groups. The SA of PA molecules can be triggered by pH change or oppositely charged molecules like DNA, heparin or soluble ions, and β-sheet formation is achieved when these factors neutralize charged residues in PA molecule (Toksoz et al., 2011).

3.3.3 NANOIMPRINTING

Over the last decade, there have been comprehensive developments in studies focused on cell behavior and stem cell differentiation on micro- and nanoscale geometrical patterns, thanks to lithographical methods, which were initially developed for microelectronic industry. By those fabrications techniques, it has become possible to process biomaterials' surfaces with predefined micro/nanoscale features by the replication of physical patterns on silicon wafer into silicone elastomers, which is known as soft lithography. It has been becoming much clearer that the biochemical cues are changing cell morphology by modulating their cytoskeleton with the use of micro/nanopatterned surface platforms that have micron and nanoscale topographies, which, in turn, leads to transduction of mechanical signals into cell nucleus by so-called mechanotransduction events (Figure 3.1). Besides, it has been well characterized that constraining cell morphologies into certain dimensions by using these cues may become an important strategy to direct stem cell lineage into a target phenotype (Seo et al., 2011).

The use of stem cells has provided significant promises and huge potentials in regenerative medicine as they can differentiate into target cell lineages by use of topographical cues. One of the first observations of spread area–dependent stem cell differentiation was by Chen et al. They reported the mesenchymal human stem cells differentiated when these cells were constrained on small and large cell adhesion islands provided by microcontact printing of fibronectin (McBeath et al., 2004). It was shown clearly that cells tend to flatten and spread on large islands and differentiate into osteogenic lineage while constraining them into a cell size by relatively smaller islands forced these cells into adipogenic lineage. By a recent study, it was also shown that 350 nm surface patterns induced human dental pulp stem cells into adipogenic lineage rather than osteogenic one without requiring specific culture media (Kim et al., 2014). The effect of cell shape on lineage development of stem cells was further shown by another recent study in which the effect of cell aspect ratio (the ratio of cell length to cell width), achieved by elongating (long rectangle) and expanding (square) micro-islands, on bone marrow–derived mesenchymal stem cells on their differentiation into osteogenic and adipogenic lineages (Yao et al., 2013). The higher aspect ratio of cells on long adhesive micro-islands with or without differentiation media stimulated them into osteogenic differentiation. This outcome was positively related with the cell cytoskeleton tension while it was negatively related with adipogenesis.

On the other hand, micro- and nanofabrication techniques have not been integrated into tissue engineering applications of many tissue types adequately because of the limitations of micro-/nanofabrication tools to 2D and smooth surfaces. Nevertheless, for thin and layered tissues that require anisotropic alignment of cells and ECM, nanopatterned biomaterials have become an important tool. Koo et al. cultured primary human keratocytes on chitosan and polydimethylsiloxane (PDMS) that were surface patterned with anisotropic topography of gratings in various widths and pitches varying between 350 nm and 20 µm for the construction of cornea tissue that requires orthogonal layers or lamellae of collagen for optimum light penetration (Koo et al., 2011). Experimental results showed that the sub-micrometer size gratings encouraged keratocyte alignment, cellular elongation capacity, and the alignment of deposited collagen I in the direction of patterns. Teo et al. (2012) also showed that the micro- and nanoscale pillars and wells guided bovine corneal endothelial cell (BCEC) by creating a monolayer, which resembled the natural corneal endothelium.

The nanogrove type of patterns is especially shown to be an ideal substrate for orienting dermal cells and their deposited matrix molecules. With this strategy, micron and submicron grooves (nearly 0.5 µm) with variable spacing ratios (1:1, 1:2, and 1:5) were tested for their effect on dermal wound healing process. The researchers measured fibroblast migration and alignment of synthesized extracellular matrix proteins, which is an essential function for fibroblasts in order to mimic well-organized natural dermis (Kim et al., 2012). The grooves of 1:2 spacing ratio resulted in higher wound healing efficiency in respect to migration speed and the grooves were detected to affect the orientation of ECM fibers and the length of the formed fiber bundles. Similarly, Clement et al. (2013) investigated human keratinocytes' proliferation capacity and laminin deposition in microgrooves (50 to 400 µm), which were made of collagen gel matrix for dermal–epidermal regeneration. It was demonstrated that the keratinocytes proliferated more in narrower channels (50 µm) while laminin secretion, which is an indication of wound re-epithelialization, was profound in larger channels.

In terms of cell alignment and contact guidance, nerve is one of the most potent tissues that can harness the potentials of micro- and nano-printed surfaces. Chan et al. (2013) used groove-ridge patterns to differentiate hPSC. Neural differentiation by the use of 2 µm gratings suggested that the physical topography was the dominant factor in neural differentiation of hPSC even without biochemical differentiation agents provided in culture medium. Pan et al. (2013) found similar results when various micro- and nano groove-ridge surfaces (width: 350 nm/2 µm/5 µm, height: 300 nm) were used for the differentiation of human-induced pluripotent stem cells. Very distinctly, the 350 nm width grating surface induced more enhanced upregulation of neuronal markers with or without pre-neuronal induction by using growth factor–enriched media.

3.4 CONCLUSIONS AND FUTURE TRENDS

The cascade events of mechanotransduction that starts by the interaction of transmembrane adhesion receptors with natural ECM is the physiological mechanism that needs to be imitated for the application of nanoscale biomaterials in tissue engineering through guiding cells by physical and biochemical cues. Therefore, this mechanism becomes a key strategy to guide cells and changes the lineage by either spontaneously formed nanoscale entities or by custom-designed nanoscale motifs on biomaterials. Although the potentials of nanofabrication techniques have not been fully harnessed yet to engineer the complex and hierarchical structure of ECM for engineering tissues, there are big advancements in ECM-like scaffolds toward engineering tissues. Bottom-up techniques that use small engineered materials to build complex structures is a promising strategy to solve the limitations associated with micro/nanofabrication techniques. Mineralized nanofibrous scaffolds also hold promise for engineering different tissues. In future, we can envision that mimicking microarchitecture of natural tissues by fine processing and creation of ECM-like nanofibrous structures can solve some of the challenges encountered in engineering of constructs with the aim of achieving complex organization of different cell types in customized 3D environment.

REFERENCES

Abdal-hay, A., Vanegas, P., Hamdy, A.S., Engel, F.B., Lim, J.H. Preparation and characterization of vertically arrayed hydroxyapatite nanoplates on electrospun nanofibers for bone tissue engineering applications. *Chem Eng J* 254 (2014): 612–622.

Aida, T., Meijer, E.W., Stupp, S.I. Functional supramolecular polymers. *Science* 335 (2012): 813–817.

Akasaka, T., Yokoyama, A., Matsuoka, M., Hashimoto, T., Watari, F. Thin films of single-walled carbon nanotubes promote human osteoblastic cells (Saos-2) proliferation in low serum concentrations. *Mater Sci Eng C* 30 (2010): 391–399.

Aminzare, M., Eskandari, A., Baroonian, M.H., Berenov, A., Razai, H.Z., Taheri, M., Sadrnezhaad, S.K. Hydroxyapatite nanocomposites: Synthesis, sintering, and mechanical properties. *Ceram Int* 39 (2013): 2197–2206.

Andric, T., Sampson, A.C., Freeman, J.W. Fabrication and characterization of electrospun osteon mimicking scaffolds for bone tissue engineering. *Mater Sci Eng C* 31 (2011): 2–8.

Bai, S., Liu, S., Zhang, C. et al. Controllable transition of silk fibroin nanostructures: An insight into in vitro silk self-assembly process. *Acta Biomater* 9 (2013): 7806–7813.

Bernstein, A., Niemeyer, P., Salzmann, G.M. Microporous calcium phosphate ceramics as tissue engineering scaffolds for the repair of osteochondral defects: Histological results. *Acta Biomater* 9 (2013): 7490–7505.

Biggs, M.J., Richards, R.G., Gadegaard, N. et al. The use of nanoscale topography to modulate the dynamics of adhesion formation in primary osteoblasts and ERK/MAPK signaling in STRO-1+ enriched skeletal stem cells. *Biomaterials* 28 (2009): 5094–5103.

Blit, P.H., Battiston, K.G., Yang, M., Santerre, J.P., Woodhouse, K.A. Electrospun elastin-like polypeptide enriched polyurethanes and their interactions with vascular smooth muscle cells. *Acta Biomater* 8 (2012): 2493–2503.

Bose, S., Tarafder, S. Calcium phosphate ceramic systems in growth factor and drug delivery for bone tissue engineering: A review. *Acta Biomater* 8 (2012): 1401–1420.

Braghirolli, D.I., Steffens, D., Pranke, P. Electrospinning for regenerative medicine: A review of the main topics. *Drug Discov Today* 19 (2014): 743–753.

Brown, A.C., Barker, T.H. Fibrin-based biomaterials: Modulation of macroscopic properties through rational design at the molecular level. *Acta Biomater* 10 (2014): 1502–1514.

Brown, B.N., Badylak, S.F. Extracellular matrix as an inductive scaffold for functional tissue reconstruction. *Transl Res* 163 (2014): 268–285.

Cai, L., and Heilshorn, S.C. Designing ECM-mimetic materials using protein engineering. *Acta Biomater* 10 (2014): 1751–1760.

Cao, L., Wang, J., Hou, J., Xing, W., Liu, C. Vascularization and bone regeneration in a critical sized defect using 2-N,6-O-sulfated chitosan nanoparticles incorporating BMP-2. *Biomaterials* 35 (2014): 684–698.

Carrick, L.M., Aggeli, A., Boden, N., Fisher, J., Ingham, E., Waigh, T.A. Effect of ionic strength on the self-assembly, morphology and gelation of pH responsive β-sheet tape-forming peptides. *Tetrahedron* 63 (2007): 7457–7467.

Castelletto, V., Gouveia, R.J., Connon, C.J., Hamley, I.W. Self-assembly and bioactivity of a polymer/peptide conjugate containing the RGD cell adhesion motif and PEG. *Eur Polymer J* 49 (2013): 2961–2967.

Cebotari, S., Tudorache, I., Ciubotaru, A. et al. Use of fresh decellularized allografts for pulmonary valve replacement may reduce the reoperation rate in children and young adults: Early report. *Circulation* 124(11) (2011): S115–S123.

Chai, Y., Carlier, A., Bolander, J. et al. Current views on calcium phosphate osteogenicity and the translation into effective bone regeneration strategies. *Acta Biomater* 8 (2012): 3876–3887.

Chan, L.Y., Birch, W.R., Yim, E.K.F, Choo, A.B.H. Temporal application of topography to increase the rate of neural differentiation from human pluripotent stem cells. *Biomaterials* 34 (2013): 382–392.

Chaudhry, A.A., Yan, H., Gong, K. High-strength nanograined and translucent hydroxyapatite monoliths via continuous hydrothermal synthesis and optimized spark plasma sintering. *Acta Biomater* 7 (2011): 791–799.

Chen, F.-M., Wu, L.-A., Zhang, M., Zhang, R., Sun, H.-H. Homing of endogenous stem/progenitor cells for in situ tissue regeneration: Promises, strategies, and translational perspectives. *Biomaterials* 32 (2011): 3189–3209.

Chen, W., Liu, J., Manuchehrabadi, N., Weir, M.D., Zhu, Z., Xu, H.H.K. Umbilical cord and bone marrow mesenchymal stem cell seeding on macroporous calcium phosphate for bone regeneration in rat cranial defects. *Biomaterials* 34(38) (2013): 9917–9925.

Cheng, C.W., Solorio, L.D., Alsberg, E. Decellularized tissue and cell-derived extracellular matrices as scaffolds for orthopaedic tissue engineering. *Biotechnol Adv* 32 (2014): 462–484.

Clement, A.L., Moutinho, T.J., Pins, G.D. Micropatterned dermal–epidermal regeneration matrices create functional niches that enhance epidermal morphogenesis. *Acta Biomater* 9 (2013): 9474–9484.

Combes, C.G., Rey, C.C. Amorphous calcium phosphates: Synthesis, properties and uses in biomaterials. *Acta Biomater* 6 (2010): 3362–3378.

Crapo, P.M., Gilbert, T.W., Badylak, S.F. An overview of tissue and whole organ decellularization processes. *Biomaterials* 32 (2011): 3233–3243.

Daamen, W.F., Veerkamp, J.H., Van Hest, J.C.M., Van Kuppevelt, T.H. Elastin as a biomaterial for tissue engineering. *Biomaterials* 28 (2007): 4378–4398.

Daley, W.P, Yamada, K.M. ECM-modulated cellular dynamics as a driving force tissue morphogenesis. *Curr Opin Genet Dev* 23 (2013): 408–414.

Davis, G.E. Matricryptic sites control tissue injury responses in the cardiovascular system: Relationships to pattern recognition receptor regulated events. *J Mol Cell Cardiol* 48 (2010): 454–460.

Davis, G.E., Bayless, K.J., Davis, M.J., Meininger, G.A. Regulation of tissue injury responses by the exposure of matricryptic sites within extracellular matrix molecules. *Am J Pathol* 156 (2000): 1489–1498.

De Volder, M.F.L., Tawfick, S.H., Baughman, R.H., Hart, A.J. Carbon nanotubes: Present and future commercial applications. *Science* 339 (2013): 535.

Deng, W., Fu, M., Cao, Y., Cao, X., Wang, M., Yang, Y., Qu, R., Li, J., Xu, X., Yu, J. Angelica sinensis polysaccharide nanoparticles as novel non-viral carriers for gene delivery to mesenchymal stem cells. *Nanomed Nanotechnol Biol Med* 9 (2013): 181–191.

Elgersma, R.C., Kroon-Batenburg, L.M.J., Posthuma, G., Meeldijk, J.D., Rijkers, D.T.S, Liskamp, R.M.J. pH-controlled aggregation polymorphism of amyloidogenic Ab(16–22): Insights for obtaining peptide tapes and peptide nanotubes, as function of the N-terminal capping moiety. *Eur J Med Chem* (2014): 1–11. doi.org/10.1016/j.ejmech.2014.07.089.

el-Kassaby, A., AbouShwareb, T., Atala, A. Randomized comparative study between buccal mucosal and acellular bladder matrix grafts in complex anterior urethral strictures. *J Urol* 179(4) (2008): 1432–1436.

Engler, A.J., Sen, S., Sweeney, H.I., Discher, D.E. Matrix elasticity directs stem cell lineage specification. *Cell* 126 (2006): 677–689.

Fricain, J.C., Schlaubitz, S., Le Visage, C. et al. A nano-hydroxyapatite-pullulan/dextran polysaccharide composite macroporous material for bone tissue engineering applications. *Biomaterials* 34 (2013): 2947–2959.

Gainza, G., Pastor, M., Aguirre, J.J. et al. A novel strategy for the treatment of chronic wounds based on the topical administration of rhEGF-loaded lipid nanoparticles: In vitro bioactivity and in vivo effectiveness in healing-impaired db/db mice. *J Contr Rel* 185 (2014): 51–61.

Gattazzo, F., Urciuolo, A., Bonaldo, P. Extracellular matrix: A dynamic microenvironment for stem cell niche. *Biochim Biophys Acta* 1840 (2014): 2505–2519.

Gershlak, J.R., Resnikoff, J.I.N., Sullivan, K.E., Williams, C., Wanga, R.M., Black III, L.D. Mesenchymal stem cells ability to generate traction stress in response to substrate stiffness is modulated by the changing extracellular matrix composition of the heart during development. *Biochem Biophys Res Commun* 439 (2013): 161–166.

Guex, A.G., Kocher, F.M., Fortunato, G. et al. Fine-tuning of substrate architecture and surface chemistry promotes muscle tissue development. *Acta Biomater* 8 (2012): 1481–1489.

Guvendiren, M., Burdick, J.A. Stem cell response to spatially and temporally displayed and reversible surface topography. *Adv Healthcare Mater* 2 (2013): 155–164.

Han, Y., Li, S., Wang, X., Chen, X. Synthesis and sintering of nanocrystalline hydroxyapatite powders by citric acid sol–gel combustion method. *Mater Res Bull* 39 (2004): 25–32.

Han, Y.L., Wang, S., Zhang, X. et al. Engineering physical microenvironment for stem cell based regenerative medicine. *Drug Discov Today* 19(6) (2014): 763–773.

Hansson, A., Francesco, T.D., Falson, F., Rousselle, P., Jordan, O., Borchard, G. Preparation and evaluation of nanoparticles for directed tissue engineering. *Int J Pharmaceut* 439 (2012): 73–80.

Hild, N., Fuhrer, R., Mohn, D. et al. Nanocomposites of high-density polyethylene with amorphous calcium phosphate: In vitro biomineralization and cytocompatibility of human mesenchymal stem cells. *Biomed Mater* 7 (2012): 054103.

Hopley, E.L., Salmasi, S., Kalaskar, D.M., Seifalian, A.M. Carbon nanotubes leading the way forward in new generation 3-D tissue engineering. *Biotechnol Adv* 32 (2014): 1000–1014.

Huang, C.-C., Ravindran, S., Yin, Z., George, A. 3-D self-assembling leucine zipper hydrogel with tunable properties for tissue engineering. *Biomaterials* 35 (2014): 5316–5326.

Hynes, R.O. The extracellular matrix: Not just pretty fibrils. *Science* 326 (2009): 1216–1219.

Jeon, O., Alt, D.S., Linderman, S.W., Alsberg, E. Biochemical and physical signal gradients in hydrogels to control stem cell behavior. *Adv Mater* 25(44) (2013): 6366–6372.

Jose, M.V., Thomas, V., Xu, Y., Bellis, S., Nyairo, E., Dean, D. Aligned bioactive multicomponent nanofibrous nanocomposite scaffolds for bone tissue engineering. *Macromol Biosci* 10(4) (2010): 433–444.

Kang, H., Shih, Y.-R.V., Hwang, Y., Wen, C., Rao, V., Seo, T., Varghese, S. Mineralized gelatin methacrylate-based matrices induce osteogenic differentiation of human induced pluripotent stem cells. *Acta Biomat* 10 (2014): 4961–4970.

Kim, D., Kim, J., Hyun, H., Kim, K., Roh, S. A nanoscale ridge/groove pattern arrayed surface enhances adipogenic differentiation of human supernumerary tooth-derived dental pulp stem cells in vitro. *Arch Oral Biol* 59 (2014): 765–774.

Kim, H.N., Hong, Y., Kim, M.S., Kim, S.M., Suh, K.-Y. Effect of orientation and density of nanotopography in dermal wound healing. *Biomaterials* 33 (2012): 8782–8792.

Kim, J.H., Jung, Y., Kim, B.S., Kim, S.H. Stem cell recruitment and angiogenesis of neuro-peptide substance P coupled with self-assembling peptide nano fiber in a mouse hind limb ischemia model. *Biomaterials* 34 (2013): 1657–1668.

Koo, S., Ahn, S.J., Zhang, H., Wang, J.C., Yim, E.K.F. Human corneal keratocyte response to micro- and nano-gratings on chitosan and PDMS. *Cell Mol Bioeng* 4 (2011): 399–410.

Koutsopoulos, S., Zhang, S. Long-term three-dimensional neural tissue cultures in functionalized self-assembling peptide hydrogels, Matrigel and Collagen I. *Acta Biomater* 9 (2013): 5162–5169.

Kurtz, A., Oh, S.-J. Age related changes of the extracellular matrix and stem cell maintenance. *Prev Med (Baltim)* 54 (2012): S50–S56 (Suppl).

Lander, A.D., Kimble, J., Clevers, H. et al. What does the concept of the stem cell niche really mean today? *BMC Biol* 10 (2012): 19.

Lee, J., Jang, J., Oh, H., Jeong, Y.H., Cho, D.-W. Fabrication of a three-dimensional nanofibrous scaffold with lattice pores using direct-write electrospinning. *Mater Lett* 93 (2013): 397–400.

Levett, P.A., Melchels, F.P.W., Schrobback, K., Hutmacher, D.W., Malda, J., Travis, J.K. A biomimetic extracellular matrix for cartilage tissue engineering centered on photocurable gelatin, hyaluronic acid and chondroitin sulfate. *Acta Biomater* 10 (2014): 214–223.

Li, J., Sun, H., Sun, D., Yao, Y., Yao, F., Yao, K. Biomimetic multicomponent polysaccharide/nano-hydroxyapatite composites for bone tissue engineering *Carbohydr Polym* 85 (2011): 885–894.

Lodish, H., Berk, A., Zipursky, S.L., Matsudaira, P., Baltimore, D., Darnell, J. *Molecular Cell Biology*, 4th edn., W.H. Freeman, New York, 2000.

Maeda, N., Miao, J., Simmons, T.J., Dordick, J.S., Linhardt, R.J.. Composite polysaccharide fibers prepared by electrospinning and coating. *Carbohyd Polym* 102 (2014): 950–955.

Matsugaki, A., Fujiwara, N., Nakano, T.. Continuous cyclic stretch induces osteoblast alignment and formation of anisotropic collagen matrix. *Acta Biomater* 8 (2013): 7227–7235.

McBeath, R., Pirone, D.M, Nelson, C.M, Bhadriraju, K., Chen, C.S. Cell shape, cytoskeletal tension, and rhoA regulate stem cell lineage commitment. *Dev Cell* 6 (2004): 483–495.

Mendes, A.C., Baran, E.T., Reis, R.L., Azevedo, H.S. Fabrication of phospholipid–xanthan microcapsules by combining microfluidics with self-assembly. *Acta Biomater* 9 (2013a): 6675–6685.

Mendes, A.C., Baran, E.T., Reis, R.L., Azevedo, H.S. Self-assembly in nature: Using the principles nature to create artificial versions of life forms. *Nanomed Nanobiotechnol* 5 (2013b): 582–612.

Morrison, S.J., Spradling, A.C. Stem cells and niches: Mechanisms that promote stem cell maintenance throughout life. *Cell* 132(4) (2008): 598–611.

Mu, Q., Du, G., Chen, T., Zhang, B., Yan, B. Suppression of human bone morphogenetic protein signaling by carboxylated single-walled carbon nanotubes. *ACS Nano* 26 (2009): 1139–1144.

Mujeeb, A., Miller, A.F., Saiani, A., Gough, J.E. Self-assembled octapeptide scaffolds for in vitro chondrocyte culture. *Acta Biomater* 9 (2013): 4609–4617.

Mwenifumbo, S., Shaffer, M.S., Stevens, M.M. Exploring cellular behaviour with multi-walled carbon nanotube constructs. *J Mater Chem* 17 (2007): 1894–1902.

Nagai, Y., Yokoi, H., Kaihara, K., Naruse, K. The mechanical stimulation of cells in 3D culture within a self-assembling peptide hydrogel. *Biomaterials* 33 (2012): 1044–1051.

Nagatomi, J., Arulanandam, B.P., Metzger, D.W., Meunier, A., Bizios, R. Frequency- and duration-dependent effects of cyclic pressure on select bone cell functions. *Tissue Eng* 7(6) (2001): 717–728.

Nagatomi, J., Arulanandam, B.P., Metzger, D.W., Meunier, A., Bizios, R. Cyclic pressure affects osteoblast functions pertinent to osteogenesis. *Ann Biomed Eng* 31(8) (2003): 917–922.

Namgung, S., Baik, K.Y., Park, J., Hong, S. Controlling the growth and differentiation of human mesenchymal stem cells by the arrangement of individual carbon nanotubes. *ACS Nano* 5(9) (2011): 7383–7390.

Newman, P., Minett, A., Ellis-Behnke, R., Zreiqat, H. Carbon nanotubes: Their potential and pitfalls for bone tissue regeneration and engineering. *Nanomed Nanotechnol* 9 (2013): 1139–1158. http://www.ncbi.nlm.nih.gov/pubmed/23770067.

Nivison-Smith, L., Rnjak, J., Weiss, A.S. Synthetic human elastin microfibers: Stable cross-linked tropoelastin and cell interactive constructs for tissue engineering applications. *Acta Biomater* 6 (2010): 354–359.

Orlando, G., Baptista, P., Birchall, M. et al. Regenerative medicine as applied to solid organ transplantation: Current status and future challenges. *Transpl Int* 24 (2011): 223–232.

Osathanon, T., Linnes, M.L., Rajachar, R.M., Ratner, B.D., Somerman, C.M., Giachelli, C.M. Microporous nanofibrous fibrin-based scaffolds for bone tissue engineering. *Biomaterials* 29(30) (2008): 4091–4099.

Pal, J., Sharma, S., Sanwaria, S., Kulshreshth, R., Nandan, B., Srivastava, R.K. Conducive 3D porous mesh of poly(ε-caprolactone) made via emulsion electrospinning. *Polymer* 55 (2014): 3970–3979.

Pan, F., Zhang, M., Wu, G., Lai, Y., Greber, B., Schöler, H.R., Chi, L. Topographic effect on human induced pluripotent stem cells differentiation towards neuronal lineage. *Biomaterials* 34 (2013): 8131–8139.

Park, J., Bauer, S., von der Mark, K., Schmuki, P. Nanosize and vitality: TiO$_2$ nanotube diameter directs cell fate. *Nano Lett* 7(6) (2007): 1686–1691.

Petreaca, M. and Martins-Green, M. The dynamics of cell–ECM interactions. In *Principles of Tissue Engineering*, Lanza, R., Langer, R., and Vacanti, J.P. (eds.), Elsevier Academic Press, Burlington, VA, 2007, pp. 81–100.

Pittrof, A., Park, J., Bauer, S., Schmuki, P. ECM spreading behavior on micropatterned TiO$_2$ nanotube surfaces. *Acta Biomater* 8 (2012): 2639–2647.

Prodanov, L., Semeins, C.M., Van Loon, J.J.W.A. et al. Influence of nanostructural environment and fluid flow on osteoblast-like cell behavior: A model for cell-mechanics studies. *Acta Biomater* 9 (2013): 6653–6662.

Pryzhkova, M.V., Aria, I., Cheng, Q., Harris, G.M., Zan, X., Gharib, M., Jabbarzadeh, E. Carbon nanotube-based substrates for modulation of human pluripotent stem cell fate. *Biomaterials* 35 (2014): 5098–5109

Quinn, R.W., Hilbert, S.L., Bert, A.A. et al. Performance and morphology of decellularized pulmonary valves implanted in juvenile sheep. *Ann Thoracic Surg* 92(1) (2011): 131–137.

Remlinger, N.T., Czajka, C.A., Juhas, M.E. et al. Hydrated xenogeneic decellularized tracheal matrix as a scaffold for tracheal reconstruction. *Biomaterials* 31(13) (2010): 3520–3526.

Saito, N., Usui, Y., Aoki, K. et al. Carbon nanotubes: Biomaterial applications. *Chem Soc Rev* 38(7) (2009): 1897–1903.

Salasznyk, R.M., Klees, R.F., Boskey, A., Plopper, G.E. Activation of FAK is necessary for the osteogenic differentiation of human mesenchymal stem cells on laminin-5. *J Cell Biochem* 100(2) (2007a): 499–514.

Salasznyk, R.M., Klees, R.F., Williams, W.A., Boskey, A., Plopper, G.E. Focal adhesion kinase signaling pathways regulate the osteogenic differentiation of human mesenchymal stem cells. *Exp Cell Res* 313(1) (2007b): 22–37.

Salvi, J.D., Lim, J.Y., Donahue, H.J. Increased mechanosensitivity of cells cultures on nanotopographies. *J Biomech* 43 (2010): 3058–3062.

Sargeant, T.D., Aparicio, C., Goldberger, J.E., Cui, H., Stupp, S.I. Mineralization of peptide amphiphile nanofibers and its effect on the differentiation of human mesenchymal stem cells. *Acta Biomater* 8 (2012): 2456–2465.

Schwartz, M.A., Chen, C.S. Deconstructing dimensionality. *Science* 339 (2013): 402–404.

Seabraa, A.B., Durán, N. Biological applications of peptides nanotubes: An overview. *Peptides* 39 (2013): 47–54.

Seo, C.H., Furukawa, K., Montagne, K., Jeong, H., Ushida, T. The effect of substrate microtopography on focal adhesion maturation and actin organization via the RhoA/ROCK pathway. *Biomaterials* 32 (2011): 9568–9575.

Shi, X., Sitharaman, B., Pham, Q.P. et al. In vitro cytotoxicity of single-walled carbon nanotube/biodegradable polymer nanocomposites. *J Biomed Mater Res A* 86 (2008): 813–823.

Sinthuvanich, C., Haines-Butterick, L.A., Nagy, K.J., Schneider, J.P. Iterative design of peptide-based hydrogels and the effect of network electrostatics on primary chondrocyte behaviour. *Biomaterials* 33 (2012): 7478–7488.

Sitharaman, B.., Shi, X., Walboomers, F.X. et al. In vivo biocompatibility of ultra-short single-walled carbon nanotube/biodegradable polymer nanocomposites for bone tissue engineering. *Bone* 43 (2008): 362–370.

Song, J.J., Ott, H.C. Organ engineering based on decellularized matrix scaffolds. *Trends Mol Med* 17(8) (2011): 424–432.

Stevenson, M.D., Piristine, H., Hogrebe, N.J. et al. A self-assembling peptide matrix used to control stiffness and binding site density supports the formation of microvascular networks in three dimensions. *Acta Biomater* 9 (2013): 7651–7661.

Suh, J.S., Lee, J.Y., Choi, Y.S., Chong, P.C., Park, Y.J. Peptide-mediated intracellular delivery of miRNA-29b for osteogenic stem cell differentiation. *Biomaterials* 34 (2013): 4347–4359.

Sun, B., Long, Y.Z., Zhang, H.D. et al. Advances in three dimensional nanofibrous macrostructures via electrospinning. *Prog Polym Sci* 39 (2014): 862–890.

Sur, S., Newcomb, C.J., Webber, M.J., Stupp, S.I. Tuning supramolecular mechanics to guide neuron development. *Biomaterials* 34 (2013): 4749–4757.

Takabayashi, Y., Ishihara, M., Sumi, Y., Takikawa, M., Nakamura, S., Kiyosawa, T. Platelet-rich plasma-containing fragmin/protamine micro/nanoparticles promote epithelialization and angiogenesis in split-thickness skin graft donor sites. *J Surg Res* 193 (2015): 3–91.

Teo, B.K.K., Goh, K.J., Ng, Z.J., Koo, S., Yim, E.K.F. Functional reconstruction of corneal endothelium using nanotopography for tissue-engineering applications. *Acta Biomater* 8 (2012): 2941–2952.

Toksoz, S., Mammadov, R., Tekinay, A.B., Guler, M.O. Electrostatic effects on nanofiber formation of self-assembling peptide amphiphiles. *J Colloid Interf Sci* 356 (2011): 131–137.

Tse, J.R., Engler, A.J. Stiffness gradients mimicking in vivo tissue variation regulate mesenchymal stem cell fate. *PLoS ONE* 6(1) (2011): e15978.

Uzunalli, G., Soran, Z., Erkal, T.S., Dagdas, Y.S., Dinc, E., Hondur, A.M., Bilgihan, K., Aydin, B., Guler, M.O., Tekinay, A.B. Bioactive self-assembled peptide nanofibers for corneal stroma Regeneration. *Acta Biomater* 10 (2014): 1156–1166.

Vines, J.B., Lim, D.-J., Anderson, J.M., Jun, H.-W. Hydroxyapatite nanoparticle reinforced peptide amphiphile nanomatrix enhances the osteogenic differentiation of mesenchymal stem cells by compositional ratios. *Acta Biomater* 8 (2012): 4053–4063.

Wan, A.C.A., Ying, J.Y. Nanomaterials for in situ cell delivery and tissue regeneration. *Adv Drug Deliver Rev* 62 (2010): 731–740.

Wang, P.-Y., Tsai, W.-B., Voelcker, N.H. Screening of rat mesenchymal stem cell behaviour on polydimethylsiloxane stiffness gradients. *Acta Biomater* 8 (2012): 519–530.

Wang, Y., Yao, M., Zhou, J. The promotion of neural progenitor cells proliferation by aligned and randomly oriented collagen nanofibers through $\beta 1$ integrin/MAPK signaling pathway. *Biomaterials* 32(28) (2011): 6737–6744.

Weissman, I. Stem cells: Units of development, units of regeneration, and units in evolution. *Cell* 100 (2000): 157–168.

Whited, B.M., Rylander, M.N. The influence of electrospun scaffold topography on endothelial cell morphology, alignment, and adhesion in response to fluid flow. *Biotechnol Bioeng* 111 (2014): 184–195.

Xiao, J.-q., Shi, X.-l., Ma, H.-c., Tan, J.-j., Zhang, L., Xu, Q., Ding, Y.-t. Administration of IL-1Ra chitosan nanoparticles enhances the therapeutic efficacy of mesenchymal stem cell transplantation in acute liver failure. *Arch Med Res* 44 (2013): 370–379.

Yao, X., Peng, R., Ding, J. Effects of aspect ratios of stem cells on lineage commitments with and without induction media. *Biomaterials* 34 (2013): 930–939.

You, L. and Jacobs, C.R. Cellular mechanotransduction. In *Nanoscale Technology in Biological Systems*, R.S. Greco, F.B. Prinz, R.L. Smith, eds., CRC Press, New York, 2005.

Zhang, L., Rakotondradany, F., Myles, A.J., Fenniri, H., Webster, T.J. Arginine-glycine-aspartic acid modified rosette nanotube–hydrogel composites for bone tissue engineering. *Biomaterials* 30 (2009): 1309–1320.

Zhu, K., Guo, C., Xia, Y., Lai, H., Yang, W., Wang, Y., Song, D., Wang, C. Transplantation of novel vascular endothelial growth factor gene delivery system manipulated skeletal myoblasts promote myocardial repair. *Int J Cardiol* 168 (2013): 2622–2631.

Zimenkov, Y., Conticello, V.P., Guo, L., Thiyagarajan, P. Rational design of a nanoscale helical scaffold derived from self-assembly of a dimeric coiled coil motif. *Tetrahedron* 60 (2004): 7237–7246.

4 Engineering Cell Surface Interface in Tissue Engineering and Cell-Based Biosensors

Anwarul Hasan, Amir Sanati-Nezhad,
Mohammad Ariful Islam, John Saliba,
Hani A. Alhadrami, Adnan Memic, and Ayad Jaffa

CONTENTS

Abstract: Tissue engineering has emerged to create artificial tissues and organs to mimic, repair, or replace damaged or injured tissues and organs using combinations of cells, biomaterials, and biologically active molecules. A major challenge in tissue engineering is to control the cellular behavior in the scaffold in three dimensions in a desired fashion to help the growth of engineered tissues in a controlled way. The ability to control the cell–surface interface spatially and temporally will make it possible to direct and control cell behavior. The recent advances in micro- and nano-fabrication technologies and microfluidic platforms have made it easy to engineer, manipulate, and modify cell–surface interface within the complex cell microenvironmental architecture. However, our understanding and ability to dynamically control the cellular behavior and the cell–surface interface has still remained limited. In this chapter, we focus on the recent advances in engineering cell–surface interface for cell-based biosensors and tissue engineering, with particular attention on surface chemistry, microfabrication, microfluidics, and dynamic aspects of the cell–surface interface modifications.

Keywords: Micro-nanopatterning, focal adhesions, cell–surface interface, biosensors, tissue engineering, extracellular matrix

4.1   INTRODUCTION

Tissue engineering is a rapidly growing field with enormous prospects for the development of artificial tissues and organs for the replacement or repair of diseased or damaged body parts (Barthes et al. 2014). These engineered tissues are usually developed by seeding or encapsulating living cells in three dimensional (3D) scaffolds and growing such constructs in controlled environments that mimic native body environments (Arghya et al. 2014; Hasan et al. 2014a,c). In tissue-engineered constructs of bone, tooth, heart, lungs, bladder, blood vessels, nerves, or any other organ, the cells need to reside in and perform essential cellular functions of adhesion, spreading, growth, differentiation, and proliferation in a 3D environment while averting acute immune responses (Goodman et al. 2009). The interaction between cells and the surrounding surfaces is therefore fundamental in tissue engineering as the biophysical and biochemical cues surrounding the cell–surface interface play a significant role in adhesion, spreading, proliferation, and migration of cells in the scaffolds. In addition, the extracellular cues including growth factors, cytokines, physical and chemical interactions with extracellular matrix (ECM), and chemokines significantly influence cellular processes such as morphogenesis, differentiation, tissue development, functionality, apoptosis, and regeneration (Ahmed et al. 2006). A major challenge in tissue engineering is understanding and controlling the cellular interaction with the ECM. The ability to engineer the cellular microenvironment will help in complete understanding of how cells receive and perceive information from their microenvironment (Ventre et al. 2012).

The cell-based biosensors, on the other hand, are another growing field that have found a plethora of applications, including early disease detection, environmental monitoring, pathogen testing, drug monitoring, and cell–biomaterial interactions monitoring (Edmondson et al. 2014). However, its success largely depends on a deeper understanding of cell–surface interaction for different chemical composition, mechanical properties, and topography of materials. Cell-based biosensors are becoming increasingly popular as they can sense specific biological molecules within the cellular microenvironment in real time at very low concentration levels through ultrasensitive optical, electrochemical, or acoustic sensing systems (Hasan et al. 2014b). Their advantages include miniature size, high specificity, high sensitivity, and shorter response time.

Fundamental questions such as how cells respond to the nanotexture of their surroundings or the dynamically changing microenvironment, and the mechanisms involved in such processes need to be better understood (Mendes 2013). Microfabrication technology and, particularly, microfluidic platforms have arisen as effective tools to overcome these challenges and help in answering the fundamental questions in tissue engineering and cell-based biosensors. Because these technologies offer precise control over multiple characteristics of the cellular microenvironment such as fluid flow, chemical gradients, and localized ECM. Microenvironmental cues such as the mechanical properties, chemical properties, and topographic features can also be easily manipulated using microfluidic platforms (Paul et al. 2014).

Numerous research groups are currently working on understanding the cell–material interface, with the aim to clarify the principles that govern cell–material cross-talk. In this chapter, we review the recent progress in understanding cell–surface interfaces for applications in tissue engineering and cell-based biosensors.

4.2 BIOLOGY OF CELL–SURFACE INTERFACE

Cells are naturally responsive to their surroundings. They respond to environmental features at a range of scale from macro down to molecular. Cell–surface interactions are mediated by the cell membrane. They interact with the surrounding surface interface primarily through proteins by the formation of focal points that join the cells' actin cytoskeleton to extracellular binding sites. Thus, it is critical to understand the structure and composition of cell membranes and recognize which molecules comprise the membrane and how they are arranged (Boyle 2008).

4.2.1 STRUCTURE AND COMPOSITION OF CELL MEMBRANE

Cell membrane separates the cells' interior from the extracellular environment. The basic functions of cell membranes are the protection of cells from the surroundings and controlling permeability of ions and organic molecules in and out of the cells (Boyle 2008). In various cellular processes such as cell signaling, ion conductivity, cell adhesion and attachment with extracellular structures, cell membrane plays an active role to give shape to the cells by anchoring the cytoskeleton and to form tissues by attaching the extracellular matrix and other cells. Cell membrane is basically composed of a double layer of phospholipids (glycolipids in plants) to which proteins are bound in various forms. A number of models have been proposed for the structure of the cell membrane including the "fluid mosaic" model (Singer and Nicolson 1972), where the double layer of phospholipids is described as the basic structure of cell membrane.

The three classes of amphipathic lipids such as phospholipids, glycolipids, and sterols are the major lipids in the composition of cell membrane. The protein receptors or the exposed proteins (such as antigens) on the cell membrane are the important components for cell–cell interactions and communications since they are exposed to the extracellular environment. The types and amounts of proteins in the membrane are highly variable. For example, the amount of protein in myelin membrane is less than 25% of the total membrane mass, whereas that in membranes that are involved in ATP production (i.e., the membrane of mitochondria or chloroplast) is about 75%. The cell membrane proteins have wide varieties in their structures and lipid bilayer association, which potentially influence their diverse properties and functionality (Boyle 2008), Figure 4.1a. The relative dimensions, concentrations, and chemistries of all membrane components can vary drastically depending on cell type, disease state, and life cycle.

4.2.2 STRUCTURE AND COMPOSITION OF EXTRACELLULAR MATRIX

The ECM is the noncellular component within all tissues and organs that provides important physical scaffolding and initiates essential biochemical processes and provides biomechanical strength as required for tissue differentiation, morphogenesis, and homeostasis (Järvcläinen et al. 2009). The modeling of the ECM governs the cell movement within the organism and guides them as the body grows, develops, and repairs itself (Boyle 2008). Fundamentally, ECM comprises of proteins, polysaccharides, and water (Frantz et al. 2010). It is constantly remodeled, thereby altering the mechanical and biochemical properties such as tensile strength, compressive strength, growth factor gradients, and local composition. It also maintains water retention and extracellular homeostasis by buffering action, binds with growth factors, thereby interacting with cell–surface receptors, and offers the essential morphological arrangement and physiological functions to induce signal transduction and control gene transcription in cells. These properties of ECM can vary greatly from one tissue to another, one physiological state to other state, or even within one tissue (Frantz et al. 2010).

The ECM proteins include collagen, elastin, fibronectin, laminin, and tenascin (Järveläinen et al. 2009; Schaefer and Schaefer 2010). Nevertheless, among all of the fibrous proteins, collagen is the most abundant one consisting about 30% of the total protein mass. Collagens regulate cell adhesion, provide tensile strength, support migration, and direct tissue development (Rozario and DeSimone 2010). Elastin is another essential protein that gives recoiling ability to the tissues that go through a repeated stretch (Lucero and Kagan 2006). Fibronectin is also an essential protein that plays a crucial role in cell attachment, functioning, and migration during cellular development, tumor metastasis, and cardiovascular diseases (Rozario and DeSimone 2010). Elastins are covered by fibrillins (which are glycoprotein microfibrils) that provide integrity to elastin (Wise and Weiss 2009).

Similar to fibronectin, the tenascin (another ECM protein) exhibits pleiotropic behavior of the cells such as fibroblast migration in wound healing (Tucker and Chiquet-Ehrismann 2009). The expression pattern of tenascins are complex, especially the rate of tenascin production significantly changes during embryogenesis. The tenascins also have the ability to modulate cell proliferation, differentiation, migration, and apoptosis. Tenascins are expressed in a variety of pathological conditions but not in normal adult tissues. For example, the tenascin-C appears in wound healing and in a condition such as tumor growth (Bosman and Stamenkovic 2003).

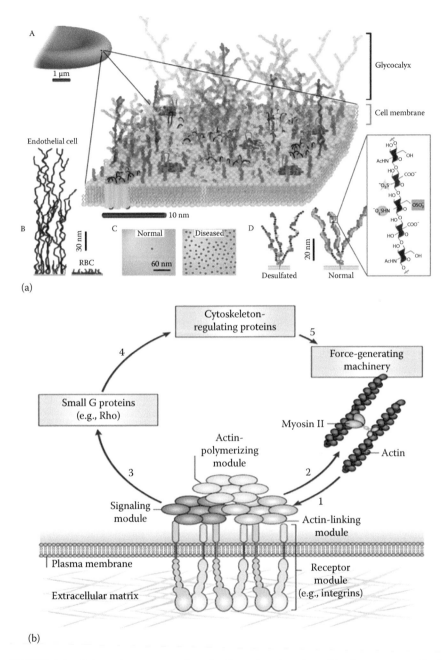

FIGURE 4.1 Schematic representation of cell–surface interface: (a) microstructure of cell surface and (b) an interplay between actin cytoskeleton and focal adhesion. (Reproduced by permission from Macmillan Publishers Ltd. *Nat. Chem.*, Mager, M.D., LaPointe, V., and Stevens, M.M., Exploring and exploiting chemistry at the cell surface, 3(8), 582–589, Copyright 2011; *Nat. Rev. Mol. Cell Biol.*, Geiger, B., Spatz, J.P., and Bershadsky, A.D., Environmental sensing through focal adhesions, 10(1), 21–33, Copyright 2009.)

Laminins, a major component of basement membrane, controls cell adhesion, migration, and differentiation. They exert their effects mainly through integrins and act as a mediator of the interaction between the ECM and the cells (Tucker and Chiquet-Ehrismann 2009). Some of the interactions provide specific functions, for example, laminin 1 induces differentiation in epithelial cells, whereas laminin 2 promotes neurite outgrowth from neural cells. On the other hand, laminin 5 is involved in cell adhesion and migration, although this function is dependent on the proteolytic processing of laminins by matrix metalloproteinases (MMPs) or plasmin. Moreover, vascular membrane represents laminin 5 and 10 both of which mediate adhesion of leukocytes, platelets, and endothelial cells.

The integrins are the type of proteins that provide mechanical stability and continuity in between inside and outside of the cells by linking the actin-filament system of cytoskeleton through a variety of linker proteins such as α-actinin, talin, paxillin, and vinculin. Moreover, integrin–ligand interaction triggers profound effects on cell proliferation, survival, and maintains the structure and functional properties of cytoskelcton. It also has direct influence on gene transcription (Bosman and Stamenkovic 2003).

Proteoglycans are mainly composed of glycosaminoglycans (GAG) that covalently link to a specific protein (except hyaluronic acid) (Iozzo and Murdoch 1996; Schaefer and Schaefer 2010). Proteoglycans can be divided into three main families such as cell–surface proteoglycans, small leucine-rich proteoglycans, and molecular proteoglycans. In composition of proteoglycans, GAGs are unbranched polysaccharide chains on the protein core composed of repeating units of disaccharide (such as D-glucuronic or L-iduronic acid, sulfated N-acetylglucosamine or N-acetylgalactosamine and galactose [-4-N-acetylglucosamine-β1, 3-galactose-β1]) that can be divided further into sulfated (heparan sulfate, chondroitin sulfate, and keratan sulfate) and nonsulfated (hyaluronic acid) GAGs (Schaefer and Schaefer 2010). These components have extended conformation with hydrophilic properties that is essentially required for hydrogel formation and provides ECM the ability to perform potential functions such as cushioning in joint movements.

4.2.3 FOCAL ADHESION

Focal adhesions are the integrin-based adhesion complexes on cell surfaces (Geiger et al. 2009), that can be defined as junctions of integrin adhesion, cellular signaling, and actin cytoskeleton. They are modified by the cells as required based on the alterations in physical force and chemical composition of their ECM environment (Wozniak et al. 2004), Figure 4.1b. At first, the polymerization of actin fibers and the contractility of myosin II fibers exert a force (step 1) that affect the vinculin, tali, etc., mechanoresponsive proteins, the integrin receptor modules, the P130CAS, and focal adhesion kinase signaling modules, and the coreceptor module such as sydecan 4. These mechanoresponsive components and modules form a mechanoresponsive network that retrospectively affect the actin cytoskeleton (step 2) based on the exerted force and its interaction with the extracellular matrix as well as with the entire system. Eventually, the GTPase-activating proteins and guanine nucleotide-exchange factors that govern the activation or inhibition of small G proteins, such as Rho and

Rac (step 3), will be activated. Actin polymerization and actomyosin contractility are affected by the G proteins through proteins that regulate the cytoskeleton (step 4), thus fine-tuning the force-generating machinery (step 5).

Focal adhesions involve diverse components such as scaffolding molecules, enzymes (e.g., kinases, phosphatases, proteases, and lipases) and GTPase. Depending on the size, composition, and subcellular localization, focal adhesions can be of different types such as focal complexes, focal adhesions, fibrillar adhesions, and 3D matrix adhesions. Focal complexes are also known as small focal adhesions, which converge at the periphery of migrating or spreading cells and are controlled by Rac and cell division control protein 42 (CDC42) (Nobes and Hall 1995). The focal adhesions are localized both at central region and cellular periphery in association with the stress fibers' ends in cells cultured on 2D rigid surfaces. The entirely defined subsets of focal complexes and focal adhesions have not been determined yet, and hence, it is unclear which components of focal adhesions differentiate them from the focal complexes (Zamir and Geiger 2001). On the other hand, the fibrillar adhesions initiate as an extended part of focal adhesions and contain tensin and $\alpha5\beta1$ integrin, whereas the 3D matrix adhesions are observed in fibroblasts when adhering to the fibronectin matrices (Cukierman et al. 2001, 2002) and collagen gels (Tamariz and Grinnell 2002) derived from 3D cell surface.

Focal adhesion kinase (FAK) is the key signaling component of focal adhesions that can be activated by numerous stimuli and acts as an integrator or biosensor to regulate cell motility. FAK localizes through the C-terminal focal adhesion targeting (FAT) domain to the focal adhesions (Hildebrand et al. 1993). This localization is very sensitive because FAK mutants that fail to localize to focal adhesions can cause disability of FAK substrates phosphorylation in response to cellular adhesion (Shen and Schaller 1999). Therefore, FAK is a crucial cellular signaling component that can influence cytoskeleton, regulation of cell movement, and structures of cell adhesion sites (Mitra et al. 2005).

4.3   ENGINEERING THE CELL–SURFACE INTERFACE

Given the importance of cell–surface interface for modulating cells function within their associated ECM, engineering surfaces for regulating the cell–material interactions are important in the success of engineered artificial tissues and implants. The roughness and topographical cues, the surface chemistry and wettability, and the bioactive coating are among the most important approaches to regulate protein adsorption, cell interaction, and the host responses (Han et al. 2014). Similar to the *in vivo* environment, the increasing evidence within *in vitro* environment indicates that cell–surface interactions, maintaining cell phenotype and behavior, may occur at multiple length scales; from nano to micro (Hasan and Lange 2007; Hasan et al. 2010). Hence, the substrates with engineered structures at nano-micro scales is beneficial for understanding the principles of cellular response to different matrix cues (Ranella et al. 2010). The engineered micro and nano patterns have been attractive for diverse applications such as engineering the artificial tissues, designing biosensors, and investigating the cellular biology as discussed later (Lim and Donahue 2007) (see Section 4.4).

4.3.1 NANO-MICROPATTERNING ON SURFACE FOR CONTROLLED CELL–SURFACE INTERACTION

In nano-micropatterning, cell surface interaction is regulated mainly by textured surfaces without considerable involvement of biomolecules (Figure 4.2a through d). A commonly used technique for nano-micropatterning of the proteins on resistant materials' surfaces is soft lithography (Kane et al. 1999), which can be grouped mainly to families of photolithography (Figure 4.2a), micromolding (Figure 4.2b), and microcontact printing (mCP) (Figure 4.2c). Photolithography usually involves a photomask while in micromolding a transfer mold is prepared by casting and curing an elastomeric polymer solution on a master mold, which is then used as a template for imprinting the micropatterns to a flat hydrogel. In microcontact printing, an elastomeric stamp formed by microstructured molding is inked with the desired proteins or molecules and transferred to the substrate. The nonstamped areas are then back filled with second biomolecule.

Using the indirect approach of cell attachment, the cells are attached to patterns on gold (and silver)-coated samples where the samples are coated with hydrophobic

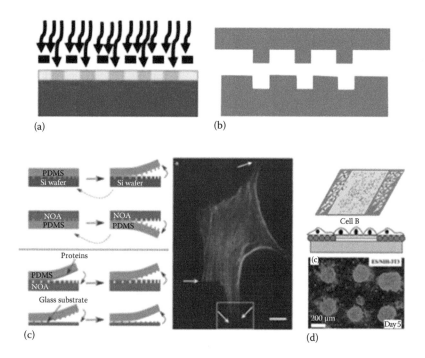

(a)   (b)

(c)   (d)

FIGURE 4.2 Techniques for engineering cell–surface interface: (a) photolithography (From Paul, A. et al., *Adv. Drug Deliv.*, 2014, in press), (b) micromolding techniques as basic methods for micro-nanotopographical patterning (From Paul, A. et al., *Adv. Drug Deliv.*, 71, 115, 2014), (c) microcontact printing involving double replication and subsequent lift-off is used for making protein gradient. C2C12 myoblast cell on a DNG of RGD peptides (From Ricoult, S.G. et al., *Small*, 9(19), 3308, 2013), (d) patterned cocultures of ES cells with fibroblasts (From Khademhosseini, A. et al., *Biomaterials*, 25(17), 3583, 2004).   *(Continued)*

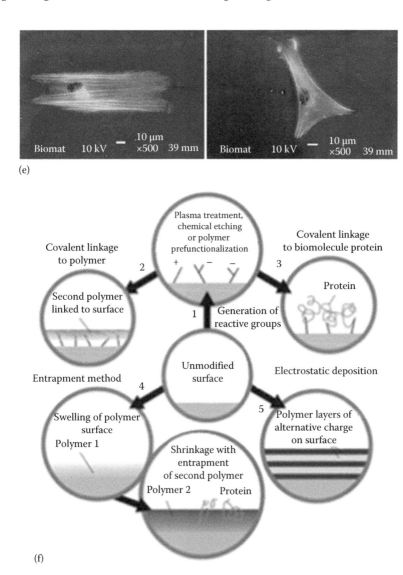

(e)

(f)

FIGURE 4.2 (*Continued*)   Techniques for engineering cell–surface interface: (e) cell response to the topographical patterns (From *Biomaterials*, 31(12), Lamers, E., Walboomers, X.F., Domanski, M., te Riet, J., van Delft, F.C.M.J.M., Luttge, R., Winnubst, L.A.J.A., Gardeniers, H.J.G.E., and Jansen, J.A., The influence of nanoscale grooved substrates on osteoblast behavior and extracellular matrix deposition, 3307–3316, Copyright 2010, with permission from Elsevier). Alignment of osteoblast-like cells to grooved substrates with different widths and depths. (f) Chemistry-based surface modification techniques including the generation of reactive groups at the surface by covalent linkage, noncovalent deposition of multilayer polymers, or anchoring polymers and/or proteins into a shrinking polymer surface through solvent exchange. (Place, E.S., George, J.H., Williams, C.K., and Stevens, M.M., Synthetic polymer scaffolds for tissue engineering, *Chem. Soc. Rev.*, 38(4), 1139–1151, 2009. Reproduced by permission of The Royal Society of Chemistry.)   (*Continued*)

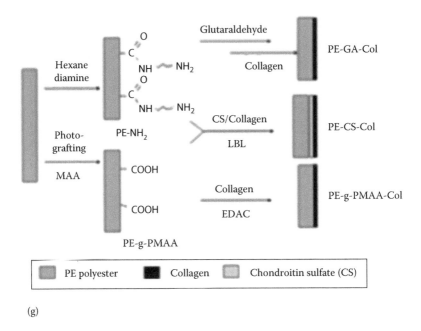

(g)

FIGURE 4.2 (*Continued*)   Techniques for engineering cell–surface interface: (g) Schematic of methods used to immobilize biomacromolecules on surface. (From *Coll. Surf. B: Biointerf.*, 60(2), Ma, Z., Mao, Z., and Gao, C., Surface modification and property analysis of biomedical polymers used for tissue engineering, 137–157, Copyright 2007, with permission from Elsevier.)

alkanethiolates and functionalized through the adsorption of fibronectin. In this case, the nonstamped area is passivated with the adsorption of molecules like ethylene–glycol-terminated thiols (Singhvi et al. 1994; Mrksich et al. 1997). The cell–surface interface can also be modulated by deposition of biotinylated poly-lactid–poly(ethylene glycol) (PLA–PEG) films on substrates followed by printing avidin on the substrates as the stable technique of patterning different biotinylated proteins such as carboxylic acid derivatized poly(ethylene terephthalate) (Yang et al. 2000). For direct approach of cell patterning, the molecules such as ECM proteins or synthetic peptides such as bovine serum albumin (BSA), immunoglobulins, and NgCAM are physiosorbed on the surface followed by attaching cells using fibronec-tin as the ink. Negative patterning is the other strategy of regulating cell–surface interface used for cell-based sensors where cell-repellent materials such as octa-decyltricholorosilane (OTS) are stamped onto substrate (Bernard et al. 1998) to form cell-free areas. The attachment of cells on cell-friendly areas is used to identify toxic agents or drugs in a high-throughput platform. This method provides a biomimetic environment with sufficient stimulating cues. Due to the nanoscale (4–250 nm) size of basement membranes of various tissues comprising pores and fibers (Flemming et al. 1999), nanopatterning is more biomimetic than micropatterning to affect the cell behavior for some specific cell types.

Various nanopatterning techniques have been developed including electron-beam lithography, nanoimprinting, nanoshaving, and dip-pen lithography to form patterns as small as tens of nanometer (Yasin et al. 2001). While some of these techniques, such as nanoimprint lithography technique, indirectly transfers patterns from a nano-structured mold to the substrate (Hoff et al. 2004), others such as dip-pen lithography using functionalized atomic force microscope tips or nanopipetting using a focused ion beam microscope directly create nanoscale patterns on the surface (Lee et al. 2002; Bruckbauer et al. 2004). For proper cell adhesion, a spacing of 10 nm to several hundred nanometers between adjacent nano/microfeatures has been suggested (Massia and Hubbell 1991). The scale of topography also affects the cell alignment. Generally, increasing the groove depth increases the cell alignment, while the groove width or pitch behaves inversely (Walboomers et al. 1999). The effect of groove depth on cell alignment was more pronounced compared to that of pattern pitch in case of human corneal epithelial cells patterned on ridges (Teixeira et al. 2003).

In addition to the scale of topography, the pattern of topography, for example, isotropic or anisotropic has also been shown to be important for cell function. In anisotropic topographies such as grooves and ridges, the alignment of cells and their associated cytoskeleton along the anisotropic direction has been observed for different cell types, irrespective of the micro- or nanoscale size of the pattern (Den Braber et al. 1998; Walboomers et al. 2000). It is still not clear how the cells respond to the pure topographic modifications as the only sites for integrin binding in these systems is the serum proteins adsorbed from the cell-culture media. While it is a nonspecific binding of proteins, it is hypothesized that the surface topography may induce directional protein adsorption through wetting of hydrophobic surfaces causing local protein concentration (Taborelli et al. 1997; Lim et al. 2005; Wilson et al. 2005). The enhanced cell response to high-surface energy of rough substrates may also be due to the advantage of the high surface-area-to-volume ratio of the structured substrates that increases the contact area of the cell membrane with the substrate (Ranella et al. 2010).

4.3.2 EXPLOITING SURFACE CHEMISTRY FOR CONTROLLED CELL–SURFACE INTERACTION

Despite the significant progress in engineering the surfaces using topography alteration, it is challenging to accurately guide the cell growth, differentiation, and fate only by these physical cues. Cell adhesion to different adsorbed proteins also mediates the adhesion to the surface. The aim is to selectively modify the surface and decorate it with integrin-binding peptides (RGD), purified ECM protein (fibronectin or collagen), a mixture of purified ECM proteins (matrigel or geltrex), and nonfouling agents to control cell adhesion or repulsion over the surface. However, these proteins may show nonspecific binding to the surface that may change the protein conformation and make the interaction nonfunctional due to hiding the active sites of proteins (Keselowsky et al. 2003). Surface chemistry can, therefore, play a critical role to control the protein patterning through adjusting the protein orientation and conformation on the surface.

Prior to patterning the proteins, surface chemistry using several different techniques has been exploited to selectively pattern self-assembled monolayers (SAMs) on the surface using different terminating functional groups (e.g., OH, CH_3, Br, COOH, $CH=CH_2$, NH_2, SH) that can further result in different compositions of absorbed proteins (Figure 4.2f and g) (Scotchford et al. 2002; Faucheux et al. 2004; Humphries et al. 2006). The methyl- and hydroxyl-surfaces, for instance, are the hydrophobic and hydrophilic functional groups, respectively, while amino- and carboxyl-groups represent negatively and positively charged surfaces, respectively. These surface chemistries control focal adhesion composition and signaling of cells and hence modulate the phenotype and function of cells (Neff et al. 1999), whereby the level of adhesion influences cell proliferation and differentiation.

In addition to the regular techniques of surface chemistry control, several advanced surface modification strategies have been developed that respond to external stimuli and induce direct cell adhesion or repulsion. For instance, poly(N-isopropyl acrylamide) (pNIPAAm) is a thermo-responsive polymer that dehydrates at temperatures above its lower critical solution temperature (LCST) in water and turns to its hydrated mode at temperatures below its LCST. This property was exploited to regulate on/off adhesion of cells to the surface proteins (Kanazawa et al. 1997). The patterns of pNIPAAm were grafted on polystyrene, while seeding cells over this surface before and after the temperature alteration provides a coculture of two cell types (Yamato et al. 2001). These types of advanced surface chemistries assist to answer basic biological questions in the context of cell–cell interaction and enable the development of novel artificial tissues made of multiple configurations of cell sheets, including single or multiple cell layers, for the purpose of tissue regeneration or repair (Yang et al. 2005). For instance, a layer of endothelial cells recovered from thermo-responsive surface, without using enzymatic detachment, has been grown on a sheet of hepatocytes in order to investigate the interaction of hepatocytes with nonparenchymal cells (Harimoto et al. 2002). Similar surface chemistry was applied to the surface of implanted material to induce cell adhesion and spreading (Roach et al. 2007), though for implanted materials, the surface modification has been carried out using a variety of different methods such as self-assembled layers, surface chemical gradients, and surface-active bulk additives (Ratner et al. 2004; Stevens and George 2005).

4.3.3 BIOACTIVE COATING

Although surface chemistry indirectly regulates protein absorption onto substrates, a more direct method to influence cell attachment is to functionalize the surfaces with bioactive peptides and amino acids that bind specifically to integrin receptors. These bioactive molecules present a different functionality with respect to the underlying monolayer. These molecules are either patterned on a SAM or deposited on a patterned SAM of nonfouling materials through grafting or physisorption (Ferretti et al. 2000). In this regard, PEG layers, phospholipid surfaces, and saccharide surfaces are known as the inert materials, resistant to cell attachment, and therefore, have been used as nonfouling materials to investigate cell-specific interaction of different peptides and their spacing patterns on cell function (Holland et al. 1998). Then, the fouling materials have been themselves covalently immobilized to different substrates under various surface chemistry protocols.

A large list of bioactive amino acid sequences have been used to promote and mediate cell-specific adhesion and function, as reviewed by Shin et al. (2003). Arginine–glycine–aspartic acid (or RGD) is the most widely used bioactive oligopeptide with great flexibility for surface modification. It has been applied to a number of different surfaces to activate the surface for improved cell adhesion. The combination of nonfouling PEG (99%)–RGD (1%) minimized protein adsorption, while promoting cell-specific adhesion via the RGD sites (VandeVondele et al. 2003). Other known bioactive molecules are derived from natural ECM proteins (e.g., collagen, fibronectin, laminin) or synthetic peptides (e.g., REDV, and KRSR), which can also play significant roles to surface bioactivities and modulate cell behavior.

In advanced forms of bioactive coating, the interplay of several bioactive cues on cell response was investigated by the development of strip assays, where the line patterns of different biomolecules are juxtaposed. These patterns can be created with vacuum pumps, microfluidics, microcontact printing, or by combining microcontact printing and microfluidics (Weschenfelder et al. 2013). Moreover, the microfluidic gradient generators, novel microcontact printing, laser-assisted adsorption by photobleaching, and colloid lithography have been employed to form nanodot gradients of proteins and biomolecules in order to investigate how cells respond to the gradient of bioactive materials (Toetsch et al. 2009; Li and Lin 2011; Dupin et al. 2013; Roy et al. 2013). A more advanced application of these patterns is to sequentially pattern biofouling and nonfouling molecules to enable coculturing of different cell types such as hepatocytes, fibroblasts, and stem cells with the aim of making more physiologically relevant organs patterned on 2D platforms (Yousaf et al. 2001; Khademhosseini et al. 2004).

4.4 IMPLICATIONS OF REGULATING THE CELL–SURFACE INTERFACE

All the aforementioned developments of engineered surfaces have implications in characterizing the cell response to a combination of various micro–nano patterns of surface topography, proteins, and biomolecules. However, the long-term goals are to fabricate engineered surfaces for artificial tissues mimicking the features of natural tissue environment as well as cell-based biosensors in such a way that cells perform their functions as in their physiological milieu (Lim and Donahue 2007).

4.4.1 CELL-BASED BIOSENSORS

In cell-based biosensors, the living cells are used as sensing elements to detect the function of biologically active analytes and biochemical compounds with high sensitivity and rapid response. These sensors have been applied in diverse fields such as environmental and pharmaceutical screening, and biomedicine. Cell-based assays are promising in drug discovery screening (Bhadriraju and Chen 2002; Schwenk et al. 2002). These sensors have also been tested to investigate basic biological aspects of cell functions such as effects of different combinatorial matrices of ECMs on the growth, differentiation, and fate of stem cells (Flaim et al. 2005). The advantage of these biosensors in comparison with other biosensors is that these sensors react to the changes in different environmental conditions such as toxins, pathogens, and pH (Figure 4.3a) (Falconnet et al. 2006).

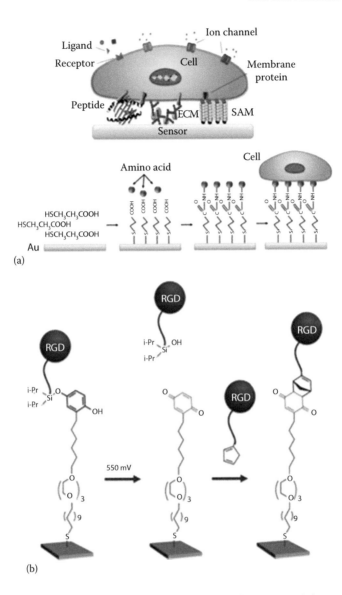

FIGURE 4.3 Implication of cell–surface interaction for biosensors and tissue engineering applications: (a) chemical modifications for cell-based biosensors. (Reproduced with permission from Liu, Q., Wu, C., Cai, H., Hu. N., Zhou, J., and Wang, P., Cell-based biosensors and their application in biomedicine, *Chem. Rev.*, 114, 6423–6461. Copyright 2014 American Chemical Society.) (b) Selective release of the RGD peptide from a monolayer presenting the O-silyl hydroquinone by electrochemical oxidation. (Mendes, P.M., Stimuli-responsive surfaces for bio-applications, *Chem. Soc. Rev.*, 37(11), 2512–2529. Reproduced by permission of The Royal Society of Chemistry.) 　　　　　　　　　　　　　　　　　　　*(Continued)*

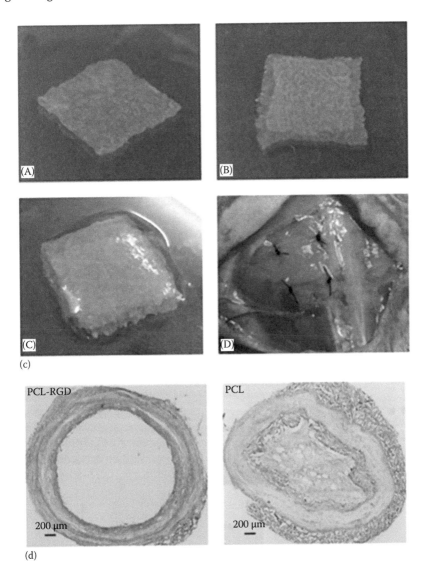

FIGURE 4.3 (*Continued*)   Implication of cell–surface interaction for biosensors and tissue engineering applications: (c) Smooth muscle cells cultured on temperature-responsive dishes harvested as intact sheets by simple temperature reduction. Multiple cell sheets were stacked and transplanted using homotypic layering of cell sheets, for creating 3D myocardial tissues. (From *Biomaterials*, 26(33), Yang, J., Yamato, M., Kohno, C., Nishimoto, A., Sekine, H., Fukai, F., and Okano, T., Cell sheet engineering: Recreating tissues without biodegradable scaffolds, 6415–6422, Copyright 2005, with permission from Elsevier.) (d) PCL-RGD grafts explanted after 4 weeks exhibited minimal thrombocity effect compared to the PCL grafts. The PCL graft was occluded and caused severe thrombocity filling up the entire lumen. (From *Biomaterials*, 33(10), Zheng, W., Wang, Z., Song, L., Zhao, Q., Zhang, J., Li, D., Wang, S., Han, J., Zheng, X.-L., and Yang, Z., Endothelialization and patency of RGD-functionalized vascular grafts in a rabbit carotid artery model, 2880–2891, Copyright 2012, with permission from Elsevier.)

To present high-performance cell-based assays, nonspecific interactions between the cell media and surface should be first suppressed in order to assure unbiased experimental results. This can be achieved by nonfouling surfaces reducing non-specific adsorption of biomolecules from cell media. Several native molecules such as agarose, mannitol, and albumin have been used to treat the substrate in order to reduce the protein adsorption (Luk et al. 2000), but because of limited stability, synthetic materials are now playing an essential role to modulate cell–surface interface in biosensors (Ratner and Bryant 2004). Poly(ethylene glycol) (PEG) as the known material has been widely used mainly in the form of copolymers such as (PLL-g-PEG) or PEG–PPO–PEG for making protein-resistant surfaces (Neff et al. 1999; Kenausis et al. 2000). The other types of protein/repellent surfaces are the lipid bilayers, polyelectrolyte multilayers deposited by the layer-by-layer (LbL) technique and smart polymers (Andersson et al. 2003; Shaikh Mohammed et al. 2004). Smart polymers such as poly(N-isopropylacrylamide) (p-NIPAAm), a thermo-responsive polymer, were used in cell-sheet engineering to enable controlled cell removal under a desired temperature (Yamato et al. 2003). The type of protein resistant material is also dependent on the substrate material underneath in a way that some are versatile to be used with different substrates such as PLL-g-PEG, while the others may need specific substrates such as in gold–thiol–based chemistry.

In recent models of cell-based biosensors, advanced cell-surface interfacing has been utilized. RGD peptide ligands and ECM protein fibronectin are copatterned over a nonfouling monolayer. The cells attached and evenly distributed across the regions of fibronectin and RGD peptide. Once the underlying gold layer is stimulated by an environmental factor, the cells are selectively released from the RGD region. In its advanced form, by using patterned surfaces with two different electroactive stimulator of RGD ligands in response to either reductive or oxidative potentials, the sensor has been used to selectively trigger the release of cells from the surface. The detection of cell release from specific region of surface due to the environmental reductive or oxidative agents can be used as a smart cell-based biosensor (Figure 4.3b) (Mendes 2008).

4.4.2 TISSUE ENGINEERING

Tissue engineering, mentioned earlier as the major application of engineered surfaces, is the science of regenerating functional tissues or organs using a combination of live cells, biomimetic matrices, specific growth factors, and external biophysical cues. Tissue engineering will enormously benefit from the ability to engineer and precisely control the cell–surface interface around the cellular microenvironment (Goodman et al. 2009). The regulation of cell–surface interface will specifically render biomaterials suitable for specific target tissues and organs and will influence the cellular behavior such as differentiation of stem cells into desired specific cell phenotypes in the engineered tissue. The nanotexture and composition of the scaffolds as well as the incorporation of macromolecules and ligands direct cellular processes and thus, carefully engineering the cell–surface interface of scaffolds yields a biomimetic engineered tissue or organ. Regulating the spatial arrangement of ligands, their density and conformation, the ligand solubility, and surface adsorption of peptides are the widely used techniques to modulate the cells' response in biomaterials within engineered tissues (Krijgsman et al. 2002; Hsu et al. 2004; Wilson et al. 2005). For instance, the vascular grafts and heart valves need to demonstrate sufficient shape recovery during the pulsation movement and also require to withstand against the high flow rate and pressure generated from blood flow. Cell–surface regulation in this regard can be achieved by coating the scaffold materials with suitable proteins such as laminin, vitronectin, and fibronectin (Santhosh Kumar and Krishnan 2001; Xue and Greisler 2003). This surface enhancement assisted in controlling the composition of the graft (made of collagen, elastin [Boland et al. 2004] and PLGA [Stitzel et al. 2006]) to match the construction of the tissue engineered artery (Williamson et al. 2006). In addition, these tissues need the ability to resist thrombosis. In native vascular and valvular tissues, the endothelium has the ability to prevent thrombosis. To do so in a engineered tissue, the cell–surface interface can be controlled by incorporating endothelium derived RGD macromolecules to induce endothelialization of vascular graft surfaces prior to implantations (Figure 4.3c and d) (Zheng et al. 2012). For synthesis of cartilage tissue, the incorporation of chitosan into electrospun nanofibrous scaffolds assisted in seeding of chondrocytes cells and formation of cartilage (chondrogenesis) (Subramanian et al. 2003). For an innervated tissue made of nanofibrous scaffolds, the modulation of cell–surface interface was used to covalently bind neuroactive D5 peptides to the surface and enhance the neuron adhesion, promote neurites extension, and further assist in the formation of innervated tissue (Ahmed et al. 2006).

4.5 DYNAMIC ASPECTS OF CELL MICROENVIRONMENT

The three-dimensional native ECM microenvironments continuously undergo remodeling. This dynamic process continues within the highly heterogeneous cellular space (Tibbitt and Anseth 2012; Paul et al. 2014). Components that make up the ECM have unique biophysical and biochemical properties (Liu 2010; Lu et al. 2012). The interactions between each of these ECM components with cells as well as cell–cell interactions are responsible for the formation of dynamic cellular microenvironment. For example, cells continually create, rearrange, realign, and break down ECM components that affect its properties. However, this is not a one-way progression but rather a two-way process where changes in the ECM can ultimately regulate cell behavior (Lu et al. 2012). Currently, there is still a large void in the detailed understanding of the highly dynamic and complex ECM microenvironment. Building platforms with the ability to dissect each of the biochemical and biophysical cues and to assess in-depth the role of spatial and temporal control signal is paramount to the development of biosensors or tissue engineered scaffolds. In addition, other processes such as cancer progression and cell migration in general can play vital roles in the spatiotemporal control of cell microenvironment and go beyond where their understanding can lend a hand in how disease treatment is sought. Therefore, taken all together, it is important to discuss the dynamic nature of the cell microenvironment, which is covered in this section.

4.5.1 CELL SIGNALING

Within the cellular microenvironment, numerous signaling molecules are present; these include biomolecules like cytokines and neurotransmitters to smaller molecules like hormones and ligands. All of these signaling processes are prone to dynamic changes and variations (Ashton-Beaucage and Therrien 2010). One example is integrin-mediated signaling, which represents a major cell signaling process often responsible for cell adhesion that induces the formation of focal adhesion structures consisting of a complex protein assembly. Predominately in the early stages of development, tissue morphogenesis is guided by integrins as they are responsible for determining the binding sites of ECM to cell. Additionally, integrin signaling can modulate, through transmembrane protein kinases pathways, cell migration and binding. One common receptor family that plays an important role in this pathway is receptor tyrosine kinases (RTK), which are high-affinity receptors for growth factors, cytokines, and hormones. These RTKs have been identified as underlying players in the signal transduction from an extracellular signal to the nucleus inducing a transcriptional activity (Ashton-Beaucage and Therrien 2010). However, there are other nonintegrin-based processes that are responsible for cellular dynamics that include other adhesion receptor families including cadherins, selectins, and others like laminin-binding proteins. Therefore, many of the biochemical and spatial data that cells convey are mediated through cytoskeletal interactions, determining what specific activity a cell should undertake as well as when and how.

Other biochemistry-based cell dynamics are controlled through the presence or absence of growth factors. Within the ECM, several soluble and matrix bound growth factors are present, controlling cell behavior through their cell–surface receptor interactions. These growth factor/receptor pair interactions are often present as concentration gradients that can play a significant role in the developmental processes, providing a pattern and direction for cell motion and activity (Kim et al. 2011).

Several studies have addressed the question of what role signaling molecules play when either integrated at the cell–surface interface or functionalized within biomaterials. It is evident that cell behavior significantly changes in response to the presence of bioactive compounds such as cytokines and growth factors as mentioned in Section 4.3.3. These biochemical cues are becoming a standard addition in the design of successful disease models and treatment strategies. For example, vascularization, represented in the formation of new blood vessels, is effected by the presence of growth factors such as FGF-2 and VEGF. Alternatively, others have reported on the regeneration of neurons by the addition of nerve growth factor within biomaterials. Another group has utilized cytokine immobilization on thin films to retain certain stem cell phenotypes in order to improve overall cell proliferation and viability. For increasing mesenchymal stem cell attachment and spreading, EGF was utilized on functionalized biomaterials (Mieszawska and Kaplan 2010). Others have shown that integration of bone bioactive motifs such as bone morphogenetic protein 2- or the osteopontin integration into scaffolds can result in *in vitro* bone formation after 28 days of culture (Mitchell et al. 2010). These examples clearly show that the cell–surface interface modified through biomolecular interactions is an important factor with which signaling and ultimately cell fate and behavior can be controlled.

4.5.2 TISSUE REMODELING

As a result of interactions between different cells and tissues, either among each other or with their neighboring microenvironment, reshaping of the ECM composition takes place. This process leads to alteration in ECM configuration followed by formation of functional tissues and organs, giving rise to specific architectures and characteristics. Several architectural and structural components of the ECM organization that defines the position of the cells within the ECM such as intercellular spacing, cell shape, and 3D cell position can also have regulatory effects on cell behavior. For instance, cell cycle of mammary epithelial cells has been correlated to ECM organization, where it was suggested as a suppressor of apoptosis, indicating there is a cellular response to ECM signaling. Similarly, tyrosine kinases have been implicated in structure-dependent ECM signaling, again indicating several ECM organizational factors could determine cellular response (Lukashev and Werb 1998).

Therefore, examining the interplay of the cell–surface interface and stimuli response within tissue-engineered models could provide a better understanding of the many processes at hand. The bioactive polymer network of engineered constructs should allow cell response to secreted signals providing insight into proteolytic remodeling, spatiotemporal control of cell adhesion, proteolytic matrix degradation, and guiding stem cell fate decision for generation of specific cell types and more complex architectures such as within innervated tissues (Figure 4.4a) (Lutolf and Hubbell 2005). One example of such an approach is the development of photocrosslinkable cell-laden hydrogels based on biomaterials such as PEG derivatives, gelatin and alginate. These and similar biomaterials have been shown to be able to mimic aspects of the natural ECM, particularly in their ability to be functionalized with bioactive signals to allow controlling cell phenotype and tissue formation (Mann et al. 2001).

4.5.3 DYNAMIC CONTROL OF CELL MICROENVIRONMENT USING MICROFLUIDICS

Microfluidics has the advantage of making microarrays containing multiple cell types (Paguirigan and Beebe 2009). The cells can also be cultured in both 2D and 3D microenvironment through droplet-based microfluidic systems (Meyvantsson et al. 2008). In their 3D model of cell-based arrays, the biomaterial surface needs to be modulated to provide biomimetic microenvironment for controlling cell functioning (Xu et al. 2011). The integration of the arrays with microfluidic platforms also makes the system automated, under controlled loading, by significantly reducing the cost of reagents and cells (Xu et al. 2011). Researchers have designed cell-laden matrices in 3D space, many of which also change over time (so-called 4D biology) and mimic functions of human tissues and organs *in vitro* (Tibbitt and Anseth 2012). Pioneering work by Bissell and colleagues revealed that mammary epithelial cells formed a normal acinus structure when encapsulated in a 3D material but aberrantly displayed cancerous phenotypes when cultured on a 2D substrate (Petersen et al. 1992). Other examples revealed that the materials-based presentation and timed removal of the peptide RGDs can enhance differentiation of mesenchymal stem cells

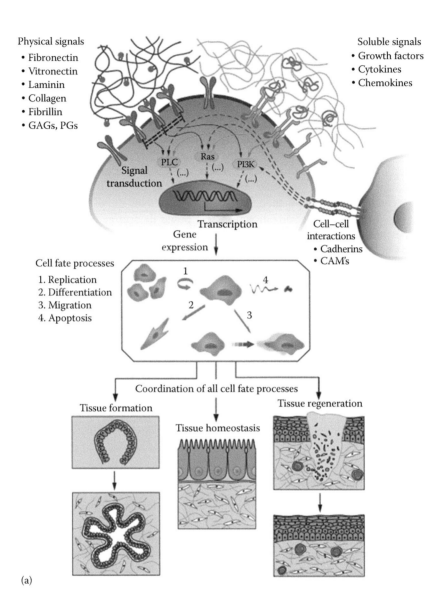

(a)

FIGURE 4.4 Dynamic aspects of cell microenvironment for cell signaling and tissue remodeling: (a) intricate reciprocal molecular interactions between cells and the surroundings to regulate the dynamic behavior of individual cells in multicellular tissues. Specific binding of signaling cues with cell–surface receptors induces intracellular signaling pathways regulating gene expression, cell phenotype, tissue formation, homeostasis, and tissue regeneration. (Reprinted with permission from Macmillan Publishers Ltd. *Nat. Biotechnol.*, Lutolf, M.P. and Hubbell, J.A., Synthetic biomaterials as instructive extracellular microenvironments for morphogenesis in tissue engineering, 23(1), 47–55, Copyright 2005.)   (*Continued*)

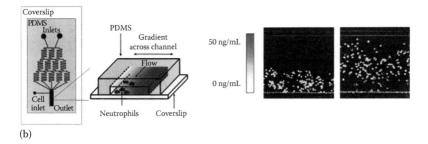

(b)

FIGURE 4.4 (*Continued*) Dynamic aspects of cell microenvironment for cell signaling and tissue remodeling: (b) A microfluidic device for dynamically controlling concentration gradient of the desired biomolecular cues such as IL-8. Top view and isometric view of the microfluidic device representing the gradient generating portion and observation portion, respectively. Neutrophil is exposed to IL-8 gradient, resulting in cell migration in response to linear increase in IL-8 concentration (0–50 ng/mL). (Reprinted with permission from Macmillan Publishers Ltd. *Nat. Biotechnol.*, Li Jeon, N., Baskaran, H., Dertinger, S.K.W., Whitesides, G.M., De Water, L.V., and Toner, M., Neutrophil chemotaxis in linear and complex gradients of interleukin-8 formed in a microfabricated device, 20(8), 826–830, Copyright 2002.)

into chondrocytes (Salinas and Anseth 2008; Kloxin et al. 2009). Thus the spatial and temporal control of microenvironment is crucial to engineer the tissue construct. The synergistic effects of chemical factor gradients, cell–cell interactions, mechanical sensing, and coordinated cell movements in tissue formation can be achieved through various microscale and microfluidic technologies as briefly discussed in Section 4.3. Adding to the advantages of microfluidics for making patterns of multiple cell types, they offer novel platforms for precise control and variation of cellular microenvironments in dynamic, automated, and reproducible ways. In the context of tissue engineering, they enable specific manipulation of environmental cues (attachment matrices containing self-assembling proteins, and gel-based substances), and cell shape and density. Moreover, mechanical (Jiang et al. 2013), chemical, and topographic properties can be precisely controlled (Unger et al. 2000; Gu et al. 2004; Mosadegh et al. 2007; Moon et al. 2009; Shamloo and Heilshorn 2010; Nikkhah et al. 2012) in a high throughput and automated platforms (Paul et al. 2014). Gilmore et al. captured and preserved rotavirus double-layered particles (DLPs) in a liquid environment contained in a microfluidic chamber by implementing an affinity capture technique (Gilmore et al. 2013). In another study, a gradient-based microfluidic device was used to infect the cells at many different concentrations of a virus simultaneously within a single channel by controlling laminar flow and diffusion (Walker et al. 2004). Gradients established by laminar flow and diffusion have been used in many other studies as well (Jeon et al. 2000; Dertinger et al. 2001, 2002; Li Jeon et al. 2002) (Figure 4.4b).

Xu et al. (2012) monitored the infection process of cells by a recombinant virus *in situ* in real time using a trilayer microfluidic device. They also used gradient-based microfluidic chip with a tree-like concentration gradient to conduct drug screening assays. Na et al. (2006) used micropatterning to create patterns of cell adhesive and repellant areas, thereby forming plaques with controlled number of cells with defined shapes and sizes. Microfluidic platforms have also been used as bioreactors containing independent chambers for production, preservation, and transduction of viruses or biomolecular compounds on a single device (Vu et al. 2007).

Challenges remain in long-term dynamical cell culturing and controlling the cell microenvironments temporally and spatially despite the substantial advances made. Future research might focus on the use of biochemical and mechanical stimuli for investigating cryptic biomolecular signals in various biomaterials similar to that in native ECM molecules (Lutolf and Hubbell 2005). The use of stimuli-responsive linkers, protecting groups and exposing mechanisms, will also be helpful in dynamically modulating the cellular microenvironment.

4.6 CONCLUSION AND FUTURE DIRECTIONS

Nanotechnology and nanofabrication techniques are allowing a better insight of cell–surface interface and parameters driving cell behavior and biology. The continuous advancements in the field of cell–surface interface will not only enhance fundamental biological studies, but it will have significant improvements in the field of tissue engineering via manufacturing synthetic substrates with controlled properties. There has been a growing interest in continuous real-time monitoring of cellular interaction with the ECM, which will give better insight on how to design and fabricate biomaterial interfaces for optimal results.

There are already biosensors showing great sensitivity and sensibility, but their challenge is in their size and integration with microfluidic platforms. Miniaturizing of these biosensors will increase their sensitivity and sensibility capabilities and allow the observation of interactions that may not have been noticed or fully understood yet. This knowledge when gained will offer great opportunities in the enhancement of engineering bioactive interfaces of biomaterials. The use of the few protein receptors in binding schemes on biomaterial surfaces has yielded significant results in cell adhesion already, and the incorporation of many other receptor classes offers great prospects for cell-specific surface design.

Further studies are required to increase our understanding of how artificial ECM interface topography influences the behavior among different cell types and how signal transduction pathways that initiated from cell attachment sites govern gene expression and ultimately cell behavior. Such studies will enhance our knowledge on how cells incorporate biophysical signals from their surrounding microenvironment. Without doubt, future research will move toward detailed analysis of cell functions and surface interface using expanding sets of more advanced and precise nanoscale tools. It is expected that the advancements in nanotechnology techniques will take us to the ultimate goal of fully understanding and mapping the structure and function of a living cell. Proper understanding and ability to control the cell–surface interface have massive applications in tissue engineering, cell-based biosensors, and regenerative medicine.

Although numerous studies have provided valuable insights of cell–surface interface in tissue engineering and cell-based biosensors, there are still many challenges and questions that need to be addressed. Further studies are required to enhance our understanding of how substrate topography influences the behavior among different cell types, and how signal transduction pathways initiated from cell attachment sites govern gene expression and ultimately cell behavior. Such studies will enhance our knowledge on how cells incorporate biophysical signals from their surrounding microenvironment.

REFERENCES

Ahmed, I., H.-Y. Liu, P.C. Mamiya, A.S. Ponery, A.N. Babu, T. Weik, M. Schindler, and S. Meiners. 2006. Three-dimensional nanofibrillar surfaces covalently modified with tenascin-C-derived peptides enhance neuronal growth in vitro. *Journal of Biomedical Materials Research Part A* 76(4):851–860.

Andersson, A.-S., K. Glasmästar, D. Sutherland, U. Lidberg, and B. Kasemo. 2003. Cell adhesion on supported lipid bilayers. *Journal of Biomedical Materials Research Part A* 64(4):622–629.

Arghya, P., A. Hasan, H. Al Kindi, A.K. Gaharwar, V.T.S. Rao, M. Nikkhah, S.R. Shin, D. Krafft, M.R. Dokmeci, D. Shum-Tim, and A. Khademhosseini. 2014. Injectable graphene oxide/hydrogel-based angiogenic gene delivery system for vasculogenesis and cardiac repair. *ACS Nano* 8(8):8050–8062.

Ashton-Beaucage, D. and M. Therrien. 2010. The greater RTK/RAS/ERK signalling pathway: How genetics has helped piece together a signalling network. *Medicine/Sciences* 26(12):1067–1073.

Barthes, J., H. Ozcelik, M. Hindie, A. Ndreu-Halili, A. Hasan, and N.E. Vrana. 2014. Cell microenvironment engineering and monitoring for tissue engineering and regenerative medicine: The recent advances. *BioMed Research International* 2014:921905.

Bernard, A., E. Delamarche, H. Schmid, B. Michel, H.R. Bosshard, and H. Biebuyck. 1998. Printing patterns of proteins. *Langmuir* 14(9):2225–2229.

Bhadriraju, K. and C.S. Chen. 2002. Engineering cellular microenvironments to improve cell-based drug testing. *Drug Discovery Today* 7(11):612–620.

Boland, E.D., J.A. Matthews, K.J. Pawlowski, D.G. Simpson, G.E. Wnek, and G.L. Bowlin. 2004. Electrospinning collagen and elastin: Preliminary vascular tissue engineering. *Frontiers in Bioscience: A Journal and Virtual Library* 9:1422–1432.

Bosman, F.T. and I. Stamenkovic. 2003. Functional structure and composition of the extracellular matrix. *Journal of Pathology* 200(4):423–428.

Boyle, J. 2008. Molecular biology of the cell. In B. Alberts, A. Johnson, J. Lewis, M. Raff, K. Roberts, and P. Walter, eds. *Biochemistry and Molecular Biology Education*, 5th edn., Garland Science, New York, Vol. 36(4), pp. 317–318.

Bruckbauer, A., D. Zhou, D.-J. Kang, Y.E. Korchev, C. Abell, and D. Klenerman. 2004. An addressable antibody nanoarray produced on a nanostructured surface. *Journal of the American Chemical Society* 126(21):6508–6509.

Cukierman, E., R. Pankov, D.R. Stevens, and K.M. Yamada. 2001. Taking cell-matrix adhesions to the third dimension. *Science* 294(5547):1708–1712.

Cukierman, E., R. Pankov, and K.M. Yamada. 2002. Cell interactions with three-dimensional matrices. *Current Opinion in Cell Biology* 14(5):633–639.

Den Braber, E.T., J.E. De Ruijter, L.A. Ginsel, A.F. Von Recum, and J.A. Jansen. 1998. Orientation of ECM protein deposition, fibroblast cytoskeleton, and attachment complex components on silicone microgrooved surfaces. *Journal of Biomedical Materials Research* 40(2):291–300.

Dertinger, S.K.W., D.T. Chiu, J.N. Li, and G.M. Whitesides. 2001. Generation of gradients having complex shapes using microfluidic networks. *Analytical Chemistry* 73(6):1240–1246.

Dertinger, S.K.W., X. Jiang, Z. Li, V.N. Murthy, and G.M. Whitesides. 2002. Gradients of substrate-bound laminin orient axonal specification of neurons. *Proceedings of the National Academy of Sciences of United States of America* 99(20):12542–12547.

Dupin, I., M. Dahan, and V. Studer. 2013. Investigating axonal guidance with microdevice-based approaches. *Journal of Neuroscience* 33(45):17647–17655.

Edmondson, R., J.J. Broglie, A.F. Adcock, and L. Yang. 2014. Three-dimensional cell culture systems and their applications in drug discovery and cell-based biosensors. *Assay and Drug Development Technologies* 12(4):207–218.

Falconnet, D., G. Csucs, H.M. Grandin, and M. Textor. 2006. Surface engineering approaches to micropattern surfaces for cell-based assays. *Biomaterials* 27(16):3044–3063.

Faucheux, N., R. Schweiss, K. Lutzow, C. Werner, and T. Groth. 2004. Self-assembled mono-layers with different terminating groups as model substrates for cell adhesion studies. *Biomaterials* 25(14):2721–30.

Ferretti, S., S. Paynter, D.A. Russell, K.E. Sapsford, and D.J. Richardson. 2000. Self-assembled monolayers: A versatile tool for the formulation of bio-surfaces. *TrAC: Trends in Analytical Chemistry* 19(9):530–540.

Flaim, C.J., S. Chien, and S.N. Bhatia. 2005. An extracellular matrix microarray for probing cellular differentiation. *Nature Methods* 2(2):119–125.

Flemming, R.G., C.J. Murphy, G.A. Abrams, S.L. Goodman, and P.F. Nealey. 1999. Effects of synthetic micro-and nano-structured surfaces on cell behavior. *Biomaterials* 20(6):573–588.

Frantz, C., K.M. Stewart, and V.M. Weaver. 2010. The extracellular matrix at a glance. *Journal of Cell Science* 123(24):4195–4200.

Geiger, B., J.P. Spatz, and A.D. Bershadsky. 2009. Environmental sensing through focal adhesions. *Nature Reviews Molecular Cell Biology* 10(1):21–33.

Gilmore, B.L., S.P. Showalter, M.J. Dukes, J.R. Tanner, A.C. Demmert, S.M. McDonald, and D.F. Kelly. 2013. Visualizing viral assemblies in a nanoscale biosphere. *Lab on a Chip—Miniaturisation for Chemistry and Biology* 13(2):216–219.

Goodman, S.B., E.G. Barrena, M. Takagi, and Y.T. Konttinen. 2009. Biocompatibility of total joint replacements: A review. *Journal of Biomedical Materials Research Part A* 90(2):603–618.

Gu, W., X. Zhu, N. Futai, B.S. Cho, and S. Takayama. 2004. Computerized microfluidic cell culture using elastomeric channels and Braille displays. *Proceedings of the National Academy of Sciences of the United States of America* 101(45):15861–15866.

Han, Y.L., S. Wang, X. Zhang, Y. Li, G. Huang, H. Qi, B. Pingguan-Murphy, Y. Li, T.J. Lu, and F. Xu. 2014. Engineering physical microenvironment for stem cell based regenerative medicine. *Drug Discovery Today* 19:763–773.

Harimoto, M., M. Yamato, M. Hirose, C. Takahashi, Y. Isoi, A. Kikuchi, and T. Okano. 2002. Novel approach for achieving double-layered cell sheets co-culture: Overlaying endothelial cell sheets onto monolayer hepatocytes utilizing temperature-responsive culture dishes. *Journal of Biomedical Materials Research* 62(3):464–470.

Hasan, A., C.F. Lange, and M.L. King. 2010. Effect of artificial mucus properties on the characteristics of airborne bioaerosol droplets generated during simulated coughing. *Journal of Non-Newtonian Fluid Mechanics* 165(21):1431–1441.

Hasan, A., A. Memic, N. Annabi, M. Hossain, A. Paul, M.R. Dokmeci, F. Dehghani, and A. Khademhosseini. 2014. Electrospun scaffolds for tissue engineering of vascular grafts. *Acta Biomaterialia* 10(1):11–25.

Hasan, A., Md. Nurunnabi, M. Morshed, A. Paul, A. Polini, T. Kuila, M. Al Hariri, Y.-K. Lee, and A.A. Jaffa. 2014. Recent advances in application of biosensors in tissue engineering. *BioMed Research International* 2014:307519.

Hasan, A., A. Paul, N.E. Vrana, X. Zhao, A. Memic, Y.-S. Hwang, M.R. Dokmeci, and A. Khademhosseini. 2014. Microfluidic techniques for development of 3D vascularized tissue. *Biomaterials* 35(26):7308–7325.

Hasan, M.A. and C.F. Lange. 2007. Estimating in vivo airway surface liquid concentration in trials of inhaled antibiotics. *Journal of Aerosol Medicine* 20(3):282–293.

Hildebrand, J.D., M.D. Schaller, and J.T. Parsons. 1993. Identification of sequences required for the efficient localization of the Focal Adhesion Kinase, pp125FAK, to cellular focal adhesions. *Journal of Cell Biology* 123(4):993–1005.

Hoff, J.D., L.-J. Cheng, E. Meyhöfer, L.J. Guo, and A.J. Hunt. 2004. Nanoscale protein patterning by imprint lithography. *Nano Letters* 4(5):853–857.

Holland, N.B., Y. Qiu, M. Ruegsegger, and R.E. Marchant. 1998. Biomimetic engineering of non-adhesive glycocalyx-like surfaces using oligosaccharide surfactant polymers. *Nature* 392(6678):799–801.

Hsu, S.-H., W.-P. Chu, Y.-S. Lin, Y.-L. Chiang, D.C. Chen, and C.-L. Tsai. 2004. The effect of an RGD-containing fusion protein CBD-RGD in promoting cellular adhesion. *Journal of Biotechnology* 111(2):143–154.

Humphries, J.D., A. Byron, and M.J. Humphries. 2006. Integrin ligands at a glance. *Journal of Cell Science* 119(Pt 19):3901–3903.

Iozzo, R.V. and A.D. Murdoch. 1996. Proteoglycans of the extracellular environment: Clues from the gene and protein side offer novel perspectives in molecular diversity and function. *FASEB Journal* 10(5):598–614.

Järveläinen, H., A. Sainio, M. Koulu, T.N. Wight, and R. Penttinen. 2009. Extracellular matrix molecules: Potential targets in pharmacotherapy. *Pharmacological Reviews* 61(2):198–223.

Jeon, N.L., S.K.W. Dertinger, D.T. Chiu, I.S. Choi, A.D. Stroock, and G.M. Whitesides. 2000. Generation of solution and surface gradients using microfluidic systems. *Langmuir* 16(22):8311–8316.

Jiang, B., T.M. Waller, J.C. Larson, A.A. Appel, and E.M. Brey. 2013. Fibrin-loaded porous poly(ethylene glycol) hydrogels as scaffold materials for vascularized tissue formation. *Tissue Engineering Part A* 19(1–2):224–234.

Kanazawa, H., K. Yamamoto, Y. Kashiwase, Y. Matsushima, N. Takai, A. Kikuchi, Y. Sakurai, and T. Okano. 1997. Analysis of peptides and proteins by temperature-responsive chromatographic system using *N*-isopropylacrylamide polymer-modified columns. *Journal of Pharmaceutical and Biomedical Analysis* 15(9):1545–1550.

Kane, R.S., S. Takayama, E. Ostuni, D.E. Ingber, and G.M. Whitesides. 1999. Patterning proteins and cells using soft lithography. *Biomaterials* 20(23):2363–2376.

Kenausis, G.L., J. Vörös, D.L. Elbert, N. Huang, R. Hofer, L. Ruiz-Taylor, M. Textor, J.A. Hubbell, and N.D. Spencer. 2000. Poly(L-lysine)-g-poly(ethylene glycol) layers on metal oxide surfaces: Attachment mechanism and effects of polymer architecture on resistance to protein adsorption. *Journal of Physical Chemistry B* 104(14):3298–3309.

Keselowsky, B.G., D.M. Collard, and A.J. García. 2003. Surface chemistry modulates fibronectin conformation and directs integrin binding and specificity to control cell adhesion. *Journal of Biomedical Materials Research Part A* 66(2):247–259.

Khademhosseini, A., K.Y. Suh, J.M. Yang, G. Eng, J. Yeh, S. Levenberg, and R. Langer. 2004. Layer-by-layer deposition of hyaluronic acid and poly-L-lysine for patterned cell co-cultures. *Biomaterials* 25(17):3583–3592.

Kim, S.-H., J. Turnbull, and S. Guimond. 2011. Extracellular matrix and cell signalling: The dynamic cooperation of integrin, proteoglycan and growth factor receptor. *Journal of Endocrinology* 209(2):139–151.

Kloxin, A.M., A.M. Kasko, C.N. Salinas, and K.S. Anseth. 2009. Photodegradable hydrogels for dynamic tuning of physical and chemical properties. *Science* 324(5923):59–63.

Krijgsman, B., A.M. Seifalian, H.J. Salacinski, N.R. Tai, G. Punshon, B.J. Fuller, and G. Hamilton. 2002. An assessment of covalent grafting of RGD peptides to the surface of a compliant poly (carbonate-urea) urethane vascular conduit versus conventional biological coatings: Its role in enhancing cellular retention. *Tissue Engineering* 8(4):673–680.

Lamers, E., X.F. Walboomers, M. Domanski, J. te Riet, F.C.M.J.M. van Delft, R. Luttge, L.A.J.A. Winnubst, H.J.G.E. Gardeniers, and J.A. Jansen. 2010. The influence of nanoscale grooved substrates on osteoblast behavior and extracellular matrix deposition. *Biomaterials* 31(12):3307–3316.

Lee, K.-B., S.-J. Park, C.A. Mirkin, J.C. Smith, and M. Mrksich. 2002. Protein nanoarrays generated by dip-pen nanolithography. *Science* 295(5560):1702–1705.

Li, J. and F. Lin. 2011. Microfluidic devices for studying chemotaxis and electrotaxis. *Trends in Cell Biology* 21(8):489–497.

Li Jeon, N., H. Baskaran, S.K.W. Dertinger, G.M. Whitesides, L.V. De Water, and M. Toner. 2002. Neutrophil chemotaxis in linear and complex gradients of interleukin-8 formed in a microfabricated device. *Nature Biotechnology* 20(8):826–830.

Lim, J.Y. and H.J. Donahue. 2007. Cell sensing and response to micro- and nanostructured surfaces produced by chemical and topographic patterning. *Tissue Engineering* 13(8):1879–1891.

Lim, J.Y., A.F. Taylor, Z. Li, E.A. Vogler, and H.J. Donahue. 2005. Integrin expression and osteopontin regulation in human fetal osteoblastic cells mediated by substratum surface characteristics. *Tissue Engineering* 11(1–2):19–29.

Liu, J.R. 2010. ["Zhi bai dihuang" or "zhi bo dihuang"?—A discussion on English translation of Chinese medicine prescriptions]. *Zhongguo Zhong Xi Yi Jie He Za Zhi* 30(11):1221–1226.

Liu, Q., C. Wu, H. Cai, N. Hu, J. Zhou, and P. Wang. 2014. Cell-based biosensors and their application in biomedicine. *Chemical Reviews* 114:6423–6461.

Lu, P., V.M. Weaver, and Z. Werb. 2012. The extracellular matrix: A dynamic niche in cancer progression. *Journal of Cell Biology* 196(4):395–406.

Lucero, H.A. and H.M. Kagan. 2006. Lysyl oxidase: An oxidative enzyme and effector of cell function. *Cellular and Molecular Life Sciences* 63(19–20):2304–2316.

Luk, Y.-Y., M. Kato, and M. Mrksich. 2000. Self-assembled monolayers of alkanethiolates presenting mannitol groups are inert to protein adsorption and cell attachment. *Langmuir* 16(24):9604–9608.

Lukashev, M.E. and Z. Werb. 1998. ECM signalling: Orchestrating cell behaviour and misbehaviour. *Trends in Cell Biology* 8(11):437–441.

Lutolf, M.P. and J.A. Hubbell. 2005. Synthetic biomaterials as instructive extracellular microenvironments for morphogenesis in tissue engineering. *Nature Biotechnology* 23(1):47–55.

Ma, Z., Z. Mao, and C. Gao. 2007. Surface modification and property analysis of biomedical polymers used for tissue engineering. *Colloids and Surfaces B: Biointerfaces* 60(2):137–157.

Mager, M.D., V. LaPointe, and M.M. Stevens. 2011. Exploring and exploiting chemistry at the cell surface. *Nature Chemistry* 3(8):582–589.

Mann, B.K., A.S. Gobin, A.T. Tsai, R.H. Schmedlen, and J.L. West. 2001. Smooth muscle cell growth in photopolymerized hydrogels with cell adhesive and proteolytically degradable domains: Synthetic ECM analogs for tissue engineering. *Biomaterials* 22(22):3045–3051.

Massia, S.P. and J.A. Hubbell. 1991. An RGD spacing of 440 nm is sufficient for integrin alpha V beta 3-mediated fibroblast spreading and 140 nm for focal contact and stress fiber formation. *Journal of Cell Biology* 114(5):1089–1100.

Mendes, P.M. 2008. Stimuli-responsive surfaces for bio-applications. *Chemical Society Reviews* 37(11):2512–2529.

Mendes, P.M. 2013. Cellular nanotechnology: Making biological interfaces smarter. *Chemical Society Reviews* 42(24):9207–9218.

Meyvantsson, I., J.W. Warrick, S. Hayes, A. Skoien, and D.J. Beebe. 2008. Automated cell culture in high density tubeless microfluidic device arrays. *Lab on a Chip* 8(5):717–724.

Mieszawska, A.J. and D.L. Kaplan. 2010. Smart biomaterials—Regulating cell behavior through signaling molecules. *BMC Biology* 8:59.

Mitchell, E.A., B.T. Chaffey, A.W. McCaskie, J.H. Lakey, and M.A. Birch. 2010. Controlled spatial and conformational display of immobilised bone morphogenetic protein-2 and osteopontin signalling motifs regulates osteoblast adhesion and differentiation in vitro. *BMC Biology* 8:57.

Mitra, S.K., D.A. Hanson, and D.D. Schlaepfer. 2005. Focal adhesion kinase: In command and control of cell motility. *Nature Reviews Molecular Cell Biology* 6(1):56–68.

Moon, J.J., M.S. Hahn, I. Kim, B.A. Nsiah, and J.L. West. 2009. Micropatterning of poly(ethylene glycol) diacrylate hydrogels with biomolecules to regulate and guide endothelial morphogenesis. *Tissue Engineering Part A* 15(3):579–585.

Mosadegh, B., C. Huango, J.W. Park, H.S. Shin, B.G. Chung, S.K. Hwang, K.H. Lee, H.J. Kim, J. Brody, and N.L. Jeon. 2007. Generation of stable complex gradients across two-dimensional surfaces and three-dimensional gels. *Langmuir* 23(22):10910–10912.

Mrksich, M., L.E. Dike, J. Tien, D.E. Ingber, and G.M. Whitesides. 1997. Using microcontact printing to pattern the attachment of mammalian cells to self-assembled monolayers of alkanethiolates on transparent films of gold and silver. *Experimental Cell Research* 235(2):305–313.

Na, K., M. Lee, B. Shin, Y. Je, and J. Hyun. 2006. Polymer-templated microarrays for highly reliable plaque purification. *Biotechnology Progress* 22(1):285–287.

Neff, J.A., P.A. Tresco, and K.D. Caldwell. 1999. Surface modification for controlled studies of cell–ligand interactions. *Biomaterials* 20(23):2377–2393.

Nikkhah, M., F. Edalat, S. Manoucheri, and A. Khademhosseini. 2012. Engineering microscale topographies to control the cell-substrate interface. *Biomaterials* 33(21):5230–5246.

Nobes, C.D. and A. Hall. 1995. Rho, Rac, and Cdc42 GTPases regulate the assembly of multimolecular focal complexes associated with actin stress fibers, lamellipodia, and filopodia. *Cell* 81(1):53–62.

Paguirigan, A.L. and D.J. Beebe. 2009. From the cellular perspective: Exploring differences in the cellular baseline in macroscale and microfluidic cultures. *Integrative Biology* 1(2):182–195.

Paul, A., A. Hasan, L. Rodes, M. Sangaralingam, and S. Prakash. 2014. Bioengineered baculoviruses as new class of therapeutics using micro and nanotechnologies: Principles, prospects and challenges. *Advance Drug Delivery*, 71:115–130.

Petersen, O.W., L. Ronnov-Jessen, A.R. Howlett, and M.J. Bissell. 1992. Interaction with basement membrane serves to rapidly distinguish growth and differentiation pattern of normal and malignant human breast epithelial cells. *Proceedings of the National Academy of Sciences of the United States of America* 89(19):9064–9068.

Place, E.S., J.H. George, C.K. Williams, and M.M. Stevens. 2009. Synthetic polymer scaffolds for tissue engineering. *Chemical Society Reviews* 38(4):1139–1151.

Ranella, A., M. Barberoglou, S. Bakogianni, C. Fotakis, and E. Stratakis. 2010. Tuning cell adhesion by controlling the roughness and wettability of 3D micro/nano silicon structures. *Acta Biomaterialia* 6(7):2711–2720.

Ratner, B.D., Hoffman, A.S., Schoen, F.J., Lemons, J.E. 2004. *Biomaterials Science: An Introduction to Materials in Medicine.* Elsevier, Waltham, MA.

Ratner, B.D. and S.J. Bryant. 2004. Biomaterials: Where we have been and where we are going. *Annual Review of Biomedical Engineering* 6:41–75.

Ricoult, S.G., M. Pla-Roca, R. Safavieh, G.M. Lopez-Ayon, P. Grütter, T.E. Kennedy, and D. Juncker. 2013. Large dynamic range digital nanodot gradients of biomolecules made by low-cost nanocontact printing for cell haptotaxis. *Small* 9(19):3308–3313.

Roach, P., D. Eglin, K. Rohde, and C.C. Perry. 2007. Modern biomaterials: A review—Bulk properties and implications of surface modifications. *Journal of Materials Science: Materials in Medicine* 18(7):1263–1277.

Roy, J., T.E. Kennedy, and S. Costantino. 2013. Engineered cell culture substrates for axon guidance studies: Moving beyond proof of concept. *Lab on a Chip* 13(4):498–508.

Rozario, T. and D.W. DeSimone. 2010. The extracellular matrix in development and morphogenesis: A dynamic view. *Developmental Biology* 341(1):126–140.

Salinas, C.N. and K.S. Anseth. 2008. The enhancement of chondrogenic differentiation of human mesenchymal stem cells by enzymatically regulated RGD functionalities. *Biomaterials* 29(15):2370–2377.

Santhosh Kumar, T.R. and L.K. Krishnan. 2001. Endothelial cell growth factor(ECGF) enmeshed with fibrin matrix enhances proliferation of EC in vitro. *Biomaterials* 22(20):2769–2776.

Schaefer, L. and R.M. Schaefer. 2010. Proteoglycans: From structural compounds to signaling molecules. *Cell and Tissue Research* 339(1):237–246.

Schwenk, J.M., D. Stoll, M.F. Templin, and T.O. Joos. 2002. Cell microarrays: An emerging technology for the characterization of antibodies. *Biotechniques* 33:54–61.

Scotchford, C.A., C.P. Gilmore, E. Cooper, G.J. Leggett, and S. Downes. 2002. Protein adsorption and human osteoblast-like cell attachment and growth on alkylthiol on gold self-assembled monolayers. *Journal of Biomedical Materials Research* 59(1):84–99.

Shaikh, M.J., M.A. DeCoster, and M.J. McShane. 2004. Micropatterning of nanoengineered surfaces to study neuronal cell attachment in vitro. *Biomacromolecules* 5(5):1745–1755.

Shamloo, A. and S.C. Heilshorn. 2010. Matrix density mediates polarization and lumen formation of endothelial sprouts in VEGF gradients. *Lab on a Chip—Miniaturisation for Chemistry and Biology* 10(22):3061–3068.

Shen, Y. and M.D. Schaller. 1999. Focal adhesion targeting: The critical determinant of FAK regulation and substrate phosphorylation. *Molecular Biology of the Cell* 10(8):2507–2518.

Shin, H., S. Jo, and A.G. Mikos. 2003. Biomimetic materials for tissue engineering. *Biomaterials* 24(24):4353–4364.

Singer, S.J. and G.L. Nicolson. 1972. The fluid mosaic model of the structure of cell membranes. *Science* 175(4023):720–731.

Singhvi, R., A. Kumar, G.P. Lopez, G.N. Stephanopoulos, D.I. Wang, G.M. Whitesides, and D.E. Ingber. 1994. Engineering cell shape and function. *Science* 264(5159):696–698.

Stevens, M.M. and J.H. George. 2005. Exploring and engineering the cell surface interface. *Science* 310(5751):1135–1138.

Stitzel, J., J. Liu, S.J. Lee, M. Komura, J. Berry, S. Soker, G. Lim, M. Van Dyke, R. Czerw, and J.J. Yoo. 2006. Controlled fabrication of a biological vascular substitute. *Biomaterials* 27(7):1088–1094.

Subramanian, A., H.Y. Lin, D. Vu, and G. Larsen. 2003. Synthesis and evaluation of scaffolds prepared from chitosan fibers for potential use in cartilage tissue engineering. *Biomedical Sciences Instrumentation* 40:117–122.

Taborelli, M., M. Jobin, P. Francois, P. Vaudaux, M. Tonetti, S. Szmukler-Moncler, J.P. Simpson, and P. Descouts. 1997. Influence of surface treatments developed for oral implants on the physical and biological properties of titanium(I). Surface characterization. *Clinical Oral Implants Research* 8(3):208–216.

Tamariz, E. and F. Grinnell. 2002. Modulation of fibroblast morphology and adhesion during collagen matrix remodeling. *Molecular Biology of the Cell* 13(11):3915–3929.

Teixeira, A.I., G.A. Abrams, P.J. Bertics, C.J. Murphy, and P.F. Nealey. 2003. Epithelial contact guidance on well-defined micro- and nanostructured substrates. *Journal of Cell Science* 116(10):1881–1892.

Tibbitt, M.W. and K.S. Anseth. 2012. Dynamic microenvironments: The fourth dimension. *Science Translational Medicine* 4(160):160ps24.

Toetsch, S., P. Olwell, A. Prina-Mello, and Y. Volkov. 2009. The evolution of chemotaxis assays from static models to physiologically relevant platforms. *Integrative Biology* 1(2):170–181.

Tucker, R.P. and R. Chiquet-Ehrismann. 2009. The regulation of tenascin expression by tissue microenvironments. *Biochimica et Biophysica Acta—Molecular Cell Research* 1793(5):888–892.

Unger, M.A., H.P. Chou, T. Thorsen, A. Scherer, and S.R. Quake. 2000. Monolithic microfabricated valves and pumps by multilayer soft lithography. *Science* 288(5463):113–116.

VandeVondele, S., J. Vörös, and J.A. Hubbell. 2003. RGD-grafted poly-l-lysine-graft-(polyethylene glycol) copolymers block non-specific protein adsorption while promoting cell adhesion. *Biotechnology and Bioengineering* 82(7):784–790.

Ventre, M., F. Causa, and P.A. Netti. 2012. Determinants of cell-material crosstalk at the interface: Towards engineering of cell instructive materials. *Journal of the Royal Society Interface* 9(74):2017–2032.

Vu, H.N., Y. Li, M. Casali, D. Irimia, Z. Megeed, and M.L. Yarmush. 2007. A microfluidic bioreactor for increased active retrovirus output. *Lab on a Chip—Miniaturisation for Chemistry and Biology* 8(1):75–80.

Walboomers, X.F., L.A. Ginsel, and J.A. Jansen. 2000. Early spreading events of fibroblasts on microgrooved substrates. *Journal of Biomedical Materials Research* 51(3):529–534.

Walboomers, X.F., W. Monaghan, A.S.G. Curtis, and J.A. Jansen. 1999. Attachment of fibroblasts on smooth and microgrooved polystyrene. *Journal of Biomedical Materials Research* 46(2):212–220.

Walker, G.M., M.S. Ozers, and D.J. Beebe. 2004. Cell infection within a microfluidic device using virus gradients. *Sensors and Actuators B: Chemical* 98(2–3):347–355.

Weschenfelder, M., F. Weth, B. Knoll, and M. Bastmeyer. 2013. The stripe assay: Studying growth preference and axon guidance on binary choice substrates in vitro. *Methods in Molecular Biology* 1018:229–246.

Williamson, M.R., R. Black, and C. Kielty. 2006. PCL–PU composite vascular scaffold production for vascular tissue engineering: Attachment, proliferation and bioactivity of human vascular endothelial cells. *Biomaterials* 27(19):3608–3616.

Wilson, C.J., R.E. Clegg, D.I. Leavesley, and M.J. Pearcy. 2005. Mediation of biomaterial-cell interactions by adsorbed proteins: A review. *Tissue Engineering* 11(1–2):1–18.

Wise, S.G. and A.S. Weiss. 2009. Tropoelastin. *International Journal of Biochemistry and Cell Biology* 41(3):494–497.

Wozniak, M.A., K. Modzelewska, L. Kwong, and P.J. Keely. 2004. Focal adhesion regulation of cell behavior. *Biochimica et Biophysica Acta—Molecular Cell Research* 1692(2–3):103–119.

Xu, F., J. Wu, S. Wang, N.G. Durmus, U.A. Gurkan, and U. Demirci. 2011. Microengineering methods for cell-based microarrays and high-throughput drug-screening applications. *Biofabrication* 3(3):034101.

Xu, N., Z.F. Zhang, L. Wang, B. Gao, D.W. Pang, H.Z. Wang, and Z.L. Zhang. 2012. A microfluidic platform for real-time and in situ monitoring of virus infection process. *Biomicrofluidics* 6(3):Article number 034122.

Xue, L. and H.P. Greisler. 2003. Biomaterials in the development and future of vascular grafts. *Journal of Vascular Surgery* 37(2):472–480.

Yamato, M., C. Konno, S. Koike, Y. Isoi, T. Shimizu, A. Kikuchi, K. Makino, and T. Okano. 2003. Nanofabrication for micropatterned cell arrays by combining electron beam-irradiated polymer grafting and localized laser ablation. *Journal of Biomedical Materials Research Part A* 67(4):1065–1071.

Yamato, M., O.H. Kwon, M. Hirose, A. Kikuchi, and T. Okano. 2001. Novel patterned cell coculture utilizing thermally responsive grafted polymer surfaces. *Journal of Biomedical Materials Research* 55(1):137–140.

Yang, J., M. Yamato, C. Kohno, A. Nishimoto, H. Sekine, F. Fukai, and T. Okano. 2005. Cell sheet engineering: Recreating tissues without biodegradable scaffolds. *Biomaterials* 26(33):6415–6422.

Yang, Z., A.M. Belu, A. Liebmann-Vinson, H. Sugg, and A. Chilkoti. 2000. Molecular imaging of a micropatterned biological ligand on an activated polymer surface. *Langmuir* 16(19):7482–7492.

Yasin, S., D.G. Hasko, and H. Ahmed. 2001. Fabrication of <5 nm width lines in poly(methylmethacrylate) resist using a water: Isopropyl alcohol developer and ultrasonically-assisted development. *Applied Physics Letters* 78(18):2760–2762.

Yousaf, M.N., B.T. Houseman, and M. Mrksich. 2001. Using electroactive substrates to pattern the attachment of two different cell populations. *Proceedings of the National Academy of Sciences* 98(11):5992–5996.

Zamir, E. and B. Geiger. 2001. Molecular complexity and dynamics of cell-matrix adhesions. *Journal of Cell Science* 114(20):3583–3590.

Zheng, W., Z. Wang, L. Song, Q. Zhao, J. Zhang, D. Li, S. Wang, J. Han, X.-L. Zheng, and Z. Yang. 2012. Endothelialization and patency of RGD-functionalized vascular grafts in a rabbit carotid artery model. *Biomaterials* 33(10):2880–2891.

5 Protein Micropatterning Techniques for Tissue Engineering and Stem Cell Research

Neerajha Nagarajan, Kenneth Hung,
and Pinar Zorlutuna

CONTENTS

5.1   INTRODUCTION

Tissues are composed of multiple cell types that are organized into a functional group. In addition to the cellular composition of the tissue, the organization of these cells is vital for the tissues to perform their intended function [1,2]. For example, as with all other vital organs in the human system, the unique architecture of the liver tissue, which is composed of hepatocytes on the apical and basolateral surfaces, sinusoidal endothelial cells (ECs), and Kupffer cells within the perisinusoidal space, is key to its functionality [3]. Therefore, recapitulating the tissue architecture *in vitro* is very important for many biomedical applications, including tissue engineering and stem cell research [4,5].

Recapitulating cellular interactions and cell microenvironments have been shown to be effective in controlling the cell shape [6], organization [7,8], phenotype [9,10], and function [11,12]. Patterning of cell-interacting proteins is a widely used and versatile way of patterning cells and cell microenvironments *in vitro*. These cell-interacting proteins are often cell-adhesive proteins such as extracellular matrix (ECM) proteins, and recently, peptides derived from these proteins have been used toward the same end [13–16]. To micropattern cell-interacting proteins or peptides, microfabrication techniques that are mostly based on photolithography were borrowed from integrated circuit industry and have been modified and exploited extensively. When applying these fabrication techniques to cells and biomolecules, several additional factors such as denaturation of the proteins, toxicity to cells, biocompatibility, and robustness in cell culture conditions should be considered [16–18]. Over the years, photolithography techniques have paved the way to soft lithography techniques [19–21] and more recently computer-based direct write techniques [63–65].

In this chapter, we discuss various surface micropatterning techniques for generating protein patterns and their application in stem cell culture and tissue engineering. We also look into factors influencing patterned protein adhesion on substrates, surface–protein interactions, and their role in guiding target cell behavior. Micropatterning proteins and cells may not exactly create all the *in vivo* conditions, but the ability to manipulate the shape and distribution of the cells can give insight into cell responses under biomimetic conditions. As we will see in later sections, with broadening of the applications of microfabrication techniques, the techniques themselves have undergone vast modifications, to better adapt for bioengineering applications.

5.2   SURFACE ENGINEERING

In vitro cell culture can benefit from modifying the surfaces of the substrates or materials that the cells are grown on. These modifications can include tuning the substrate stiffness or addition of ECM proteins for controlling cell adhesion, proliferation, phenotype and function, and often achieved by altering the physical and biochemical characteristics of the surface to help improve the functionality and compatibility of the original material, usually by creating a biomimetic microenvironment [18]. Surface modifications not only can enhance biocompatibility of the materials but also can be used to promote cell recruitment to the site of repair, promoting regeneration. For example, a polymeric surface can be functionalized with a layer of ECM protein such as fibronectin or the functional peptide group derived from it (arginine–glycine–aspartate [RGD]), which can enhance cell attachment, viability, and function [22,23].

However, protein transfer onto substrates is not a straightforward process and a number of physical and chemical factors influence the degree of protein attachment to the surface.

5.2.1 SURFACE–PROTEIN INTERACTIONS

Protein molecules are polymeric chains consisting of amino acid subunits, which have pendant groups some of which are capable of gaining or losing charge depending on the surrounding environment [23–25]. The mechanism of protein adhesion can be categorized into physiosorption or noncovalent interaction, and chemisorption or covalent interactions. Conformation of protein molecule on adhesion determines its activity, that is, strength of interaction and stability. Physiosorption refers to attachment of proteins mainly through physical force of interactions, such as van der Waals interactions, electrostatic forces, and hydrophobic–hydrophobic interactions. It is characterized by a conformational change upon binding, which is influenced by the pH, ionic strength, and temperature of the environment. Therefore, the choice of substrate and buffer used for protein transfer are important criteria for maintenance of stability and activity of the protein after adsorption. Figure 5.1A through C illustrates the mechanism of protein physiosorption and the influence of physical characteristics of the substrate [25].

Proteins form aggregates in solution, which then deposit onto the surface and eventually spread out as seen in Figure 5.1A [25]. Rabe et al. used Forster resonance energy transfer (FRET) technique to investigate the cluster formation in protein solution and its interaction with the substrate on deposition. Protein adhesion requires specific surface affinity to surface charge and wettability (Figure 5.1B and C) [25]. Physiosorption-based protein transfer is convenient to employ, requiring no special equipment and can be used on most surfaces for cell culture such as glass slides, polystyrene dishes, or other polymeric surfaces.

At the same time, protein attachment through physiosorption can be unstable since no chemical bonds are formed but rather the adhesion strength is determined by the difference in surface charges and properties. Another characteristic of this technique is that adsorbed protein molecule can be easily displaced by another with higher affinity to the substrate. Removal of protein from surface is known as elution and can also be caused by thermodynamic fluctuations such as structurally disoriented protein attachment causing collision with other molecules in the solution [25]. This occurs due to breakage of hydrogen bonds and electrostatic interactions that had initially bond the protein to the surface. Over the years, various researchers have investigated protein–substrate interactions at their interface and developed kinetic models that predict the mechanism of protein adhesion [26].

On the other hand, chemisorption is driven by a chemical reaction between the adsorbate and the substrate at the interface [27]. Chemical linkages are established between the biomolecule and the substrate through ionic or covalent bonds, depending on the reactive species on the adsorbate and the surface. Hence, unlike

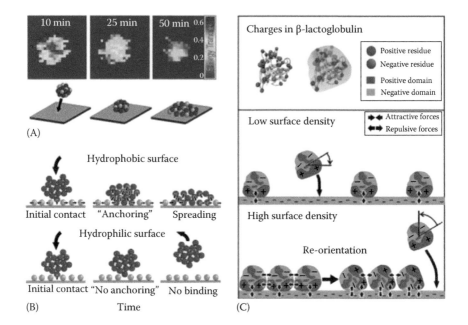

FIGURE 5.1 (A) FRET imaging illustrates protein adsorption and spreading on the surface. (B) Mechanism of protein adhesion relative to substrate hydrophobicity. (C) Schematic representation of change in orientation of surface-adsorbed protein. (From Rabe, M. et al., *Adv. Colloid Interf. Sci.*, 162(1–2), 87, 2011.)

physiosorption, usually additional substrate treatments are required to functionalize the substrate. The most common application of chemisorption for protein transfer is the formation of self-assembled monolayers (SAMs) (see Section 5.3.3). SAMs are formed by chemisorption of thiols (RS–H) onto gold surfaces forming Au–SR bonds [28,29]. The "R" side chain is then free to form covalent bonds with the desired protein molecule. Figure 5.2B shows chemisorption of bovine serum albumin (BSA) after functionalization of wafer [28]. The fabrication process is described in Section 5.3.

5.2.2 PROTEINS–CELL INTERACTIONS

Cell attachment onto the surface occurs in sequential steps that involve (1) cell attachment to ECM proteins, (2) formation of membrane or filopodial extensions, (3) extension of the cell membrane, and (4) formation and contraction of stress fibers that aids cell migration and proliferation [30]. Interaction of cells with respective ECM and surface attachment occurs through a specialized heterodimeric transmembrane unit called the integrin complex. The complex consists of two subunits α and β, which self-associates to form up to 24 distinct known αβ combinations with distinct binding characteristics (Figure 5.3A). The subunits bind to specific motifs within the surface proteins, for example, the receptor domain motif of the fibronectin complex composed of three amino acids arginine–glycine–aspartate (RGD) [30–33].

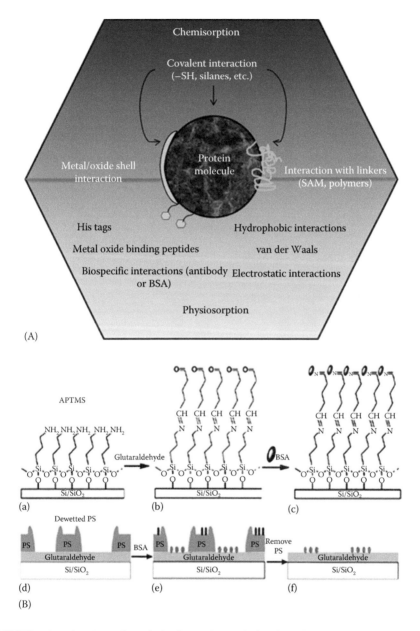

FIGURE 5.2 (A) Illustration of physiosorption and chemisorption interaction of proteins and surface. (Adapted from Aili, D. and Stevens, M.M., *Chem. Soc. Rev.*, 39, 3358, 2010.) (B) Schematic representation of the chemical functionalization of a silicon wafer and binding of bovine serum albumin. Top row: schematic representation of the chemical functionalization of a silicon wafer with (3-aminopropyltrimethoxy) silane (APTMS), followed by glutaraldehyde (b) and chemisorption of BSA from solution (c); bottom row: protein patterning was achieved by dewetting a polystyrene (PS) film (d). Gray ovals represent chemisorbed BSA molecules. (From Lim, S.K. et al., *Soft Matter*, 9(9), 2598, 2013.)

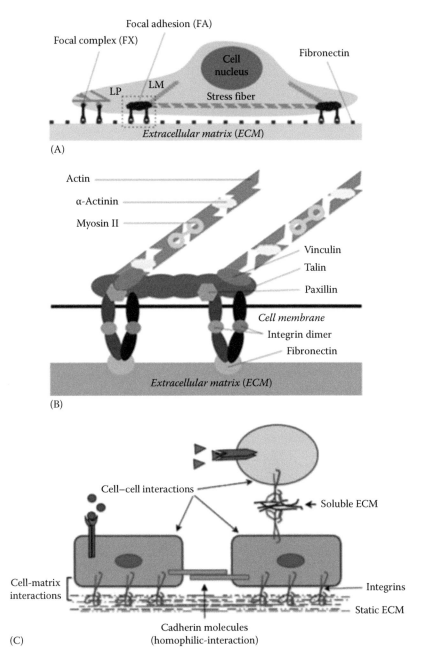

FIGURE 5.3 (A) Surface–protein interaction—binding of integrin complex with fibronectin. (From Hoffmann, M. and Schwarz, U.S., *BMC Syst. Biol.*, 7, 2, 2013.) (B) Cell interaction within microenvironment. (From Shekaran, A. and Garcia, A.J., *J. Biomed. Mater. Res. A*, 96(1), 261, 2011, Copyright Creative Commons Attribution License.) (C) Intercellular interactions.

The surface receptor–integrin interactions create dynamic changes at the basement membrane causing cytoskeletal rearrangement, tension, and formation of stress fibers [34,35]. This also causes an internal conformation change mediated by RhoGTPase and thereby activating RhoA-mediated GTPase signaling cascade. The adhesion molecules are intermediary participants in other signaling pathways within the cell, therefore, cell behavior including cell size [34,35], proliferation, and function [36,37] are greatly affected by them. While mechanical cues from the ECM are transmitted through this complex to the cell, another transmembrane receptor complex family called the cadherins mediates interactions with neighboring cells (Figure 5.3B) [33].

5.2.3 Nonfouling Treatment

For selective patterning of proteins and cells, surface modifications are necessary to avoid nonspecific interactions between the surface and the ambient proteins that cells usually reside in (i.e., cell culture media). Researchers have proposed a variety of techniques over the years for selective protein patterning. These include inactivation or complete removal of adsorbed proteins and inhibition of initial adsorption of proteins through use of masked regions [41,42,68].

The importance of nonfouling treatment can be best evidenced in immunohistochemistry; a technique that is based on highly specific, protein ligand–receptor interactions and where nonspecific attachments would yield false results. In immunochemistry, BSA has been commonly used to "block" nonspecific protein domains until a molecule with higher affinity for the substrate displaces it [38]. BSA readily sticks to surfaces and other molecules and moreover, it does not have an integrin-binding site, hence it is a favorable choice to deter unwanted, random cell attachment.

In protein patterning, commonly used molecules for creating nonfouling sites on a substrate are BSA, Pluronic-F127, poly(ethylene glycol) (PEG), and its derivatives. Pluronics are triblock copolymer surfactants with a polymer chain of poly(ethylene oxide)–poly(propylene oxide)–poly(ethylene oxide) repeating units. They adhere to highly hydrophobic surfaces and can deter further adsorption of proteins. Another commonly used blocking agent is PEG grafted poly-L-lysine (PLL-g-PEG), a cationic polymer. This requires treatment of target surface with plasma in the presence of pressurized air for 30–60 s, which results in deposition of a uniform layer of negative ions on the surface, making it highly adhesive to the cationic PLL-g-PEG polymers. The factors governing resistance to protein binding by PEG-grafted polymers and related quantitative information have been reviewed and investigated in detail [38,40]. It has been concluded that a combination of factors, namely, high hydrophilicity, large molecular size, and lack of protein binding sites prevent protein attachment.

5.3 TECHNIQUES FOR PROTEIN PATTERNING

In the following section, we will briefly cover some of the common techniques used for micropatterning proteins, especially those that have found wide application in stem cell and tissue engineering research. These include micropatterning SAMs, microcontact printing (µCP), microfluidic printing, stencil-based patterning, and direct-write techniques [39,43].

5.3.1   PHOTOLITHOGRAPHY: LIGHT-BASED PATTERNING

Photolithography is the classic technique for microfabricating structures and the starting point of most microfabrication technologies. This technique was initially used in the semiconductor industry for patterning microchips and eventually applied to biological problems. A schematic of this process is represented in Figure 5.4A [44,45]. The basic principle of photopatterning is to use focused light energy to pattern photosensitive polymers called photoresist. Subsequent chemical treatment called "developing," dissolves out the unreacted regions of the photoresist, leaving

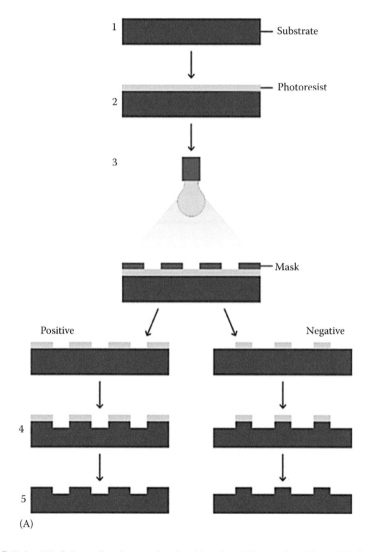

FIGURE 5.4   (A) Schematic of steps involved in photolithography. (From Khaleel, H.R. et al., Design, fabrication, and testing of flexible antennas, in: Kishk, A., ed., *Advancement in Microstrip Antennas with Recent Applications*, InTech, 2013, pp. 363–383.)   *(Continued)*

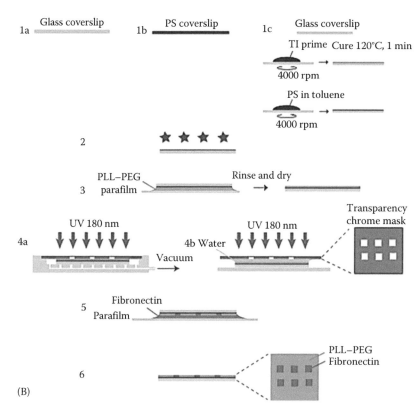

FIGURE 5.4 (*Continued*) (B) Schematic of protein patterning through deep UV (185 nm) exposure. (From Azioune, A. et al., *Microtubules: In Vivo*, 97, 133, 2010.)

behind defined patterns, which can be used directly or subjected to further treatment to attain the desired surface properties such as increased hydrophilicity through plasma treatment [45].

The chemicals used throughout this process are generally organic solvents or strong acids or bases, which can readily denature proteins. Therefore, this technique is not very suitable for direct protein patterning. But over the years, advancement in photolithography has enabled use of other approaches based on similar principles. For example, deep UV (185 nm) is a direct patterning technique suitable for patterning proteins on glass substrates. As seen in the work by Azioune et al., a PLL-g-PEG-coated glass coverslip is exposed to deep UV light through a chrome mask to selectively oxidize the surface, which is then followed by fibronectin treatment that results in selective attachment of fibronectin [46].

Alternatively, photosensitive proteins can be used directly, instead of a photoresist. David Tirrell's group synthesized photosensitive amino acid *para*-azido-phenylalanine (pN$_3$Phe) in a bacterium (*Escherichia coli*) and then incorporated it into ECM proteins such as elastin and fibronectin. Exposure to UV light (~365 nm) causes the formation of highly reactive nitrogen groups by the azidophenyl group, which then cross-links with the protein molecules [45]. The major drawback of this technique is that the generation of such genetically engineered proteins is costly and time consuming. This has given way to alternative techniques that is based on protein photobleaching, which is the photochemical depletion of a dye or fluoropore. In the study by Belisle et al., fluorescein-tagged biotin-coated substrate was illuminated through a spatial filter, which acts as a mask, using a high-energy laser, resulting in selective photobleaching of the protein. The photobleaching process resulted in the generation of free radicals, which then aided attachment of other organic molecules such as fluoropore-tagged streptavidin, which bound to unexposed biotin molecules and aided visualization of the patterned proteins. Additionally, biotinylated proteins or antibodies can interact and bind with streptavidin [47].

5.3.2 SOFT LITHOGRAPHY

Pioneered by George Whitesides' group, soft lithography is a group of patterning techniques that uses an elastomeric polymer such as poly(dimethylsiloxane) (PDMS) [19–24]. PDMS is used to generate a pattern or form a device through replica molding from a microfabricated master. The master is fabricated by photolithography as described in the previous section. Soft lithography can be divided into three major classes depending on the mode of material transfer onto the substrate: (1) microstamping or microcontact printing, (2) microfluidic patterning, and (3) PDMS stencils.

5.3.2.1 Microcontact Printing

μCP is one of the most common methods for protein and cell micropatterning. It is relatively inexpensive and does not require any complex equipment while producing submicron-level patterns with high fidelity. μCP is performed by "inking" a polymer stamp and placing it in contact with a surface such as glass or tissue culture plastic, and transferring the "ink" to the desired substrate with pressure application [48]. The polymer stamp will have a relief pattern consisting of the desired micropattern shape and layout. The process itself is similar to conventional stamping with ink and paper; however, the ink can be a variety of polymers including dendrimers, DNA, and proteins. For cellular micropatterning, an ECM protein such as fibronectin may be stamped to promote cell attachment only to the stamped region. In brief, a silicon wafer master is fabricated with a relief pattern through photolithography. The PDMS pre-polymer and curing agent are mixed, degassed, and poured into the master. After curing in an oven, the stamps are peeled off and cut into desired shapes. Stamps are inked by covering the patterned surface with the ink solution. After aspirating the ink solution and briefly rinsing, the stamp is quickly dried under a stream of nitrogen gas. Finally, a micropattern is obtained by pressing the stamp against a suitable surface. A schematic of the process is illustrated in Figure 5.5 [48].

Material selection plays an important role in the success of μCP. Materials that are too stiff will not make conformal contact with slightly rough surfaces, causing incomplete patterns or no pattern transfer at all. On the other hand, materials that

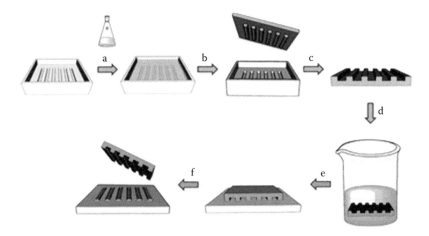

FIGURE 5.5 Schematic of microcontact printing. (From Kaufmann, T. and Ravoo, B.J., *Polym. Chem.*, 1, 371, 2010.)

are too compliant will cause the relief patterns to collapse or the spaces in between patterns will sag under load, preventing transfer of especially submicron-level patterns. Additionally, the surface wettability of the stamp plays an important role in ink physiosorption and transfer. The most common material used for µCP is PDMS, though other materials such as poly(ether-block-ester), polyolefin elastomers, or even hydrogels such as agarose have been used [48]. PDMS stamps offer easy fabrication through thermal curing, are relatively inexpensive and easy to acquire, and have a suitable stiffness range for µCP down to submicron-level patterns. One of the limitations of PDMS, is its highly hydrophobic nature. This makes it difficult to ink with polar molecules such as proteins and DNA, resulting in incomplete pattern transfer. To resolve this issue, the protein incubation period may be increased or the surface of the PDMS stamp can be modified to increase hydrophilicity using techniques such as surface oxidation of the stamp by oxygen plasma treatment or chemical treatments of the substrate surface with silanes [48–50].

Microcontact printing can also be used for selective patterning for various types of molecules using a single multilayered stamp (Figure 5.6) [49]. Fabrication of multilayered stamps follows a similar procedure as described earlier, except in this case the photoresist-coated silicon wafer has features of multiple heights. Curing PDMS prepolymer against the patterned photoresist master generates the PDMS stamp with varying layers of relief features. Each discrete layer could be brought into contact individually by adjusting the vertical pressure applied on the stamp [49]. The larger the applied pressure on the stamp, the greater is the area of contact between the stamp and the substrate, and proteins from deeper levels are transferred. One of the major drawbacks of this technique is that PDMS stamps deform over time, hence it becomes difficult to align the features on top of each other when several layers of patterns are required. To overcome this, the stamps could be made more rigid by either increasing the cross-linking or providing a solid glass support as base [48].

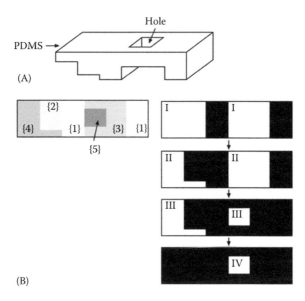

(A)

(B)

FIGURE 5.6 Multilayered stamp and formation of multiple protein patterns. (A) 3D structure of the pattern and (B) cross-sections of the patterns. (From Tien, J. et al., Fabrication of aligned microstructures with a single elastomeric stamp, *Proc. Natl. Acad. Sci. USA*, 99(4), 1758–1762, Copyright 2002 National Academy of Sciences, U.S.A.)

5.3.2.2 Microfluidics-Based Patterning

Another commonly used soft lithography technique that is rapidly gaining attention, especially for tissue engineering applications, is microfluidics-based protein patterning. Fabrication of these devices follows the same procedure as that of fabrication of PDMS molds for microcontact printing, except that the PDMS mold here is fabricated to form channels [51–54]. PDMS microchannels are attached to the desired surface, and filled with the desired polymer "ink," resulting in micropatterns on the substrate in the shape of the microchannels (Figure 5.7) [52,54]. The advantage of microfluidic patterning is that it allows proteins to be deposited while in the fluid phase, avoiding some of the issues encountered with drying of proteins during μCP. Since this technique circumvents the drying of proteins, the proteins remain more stable and are not denatured.

On the other hand, this technique is not devoid of any limitations; and protein loss to the walls of the device is a common occurrence. Owing to the micron-scale of these devices, they have high surface to volume ratio, and thus protein loss is a significant factor. Filling the channels made out of hydrophobic PDMS with hydrophilic solutions, such as a protein solution, through capillary action is the next challenge (Figure 5.8) [54]. This can be overcome by exposing the devices to oxygen plasma treatment, which makes it highly hydrophilic, resulting in smooth flow of hydrophilic solutions. Another major challenge is that protein flow within the channel is determined by the size of the channels. Channels with lower depths pose high resistance to fluid flow. In such cases,

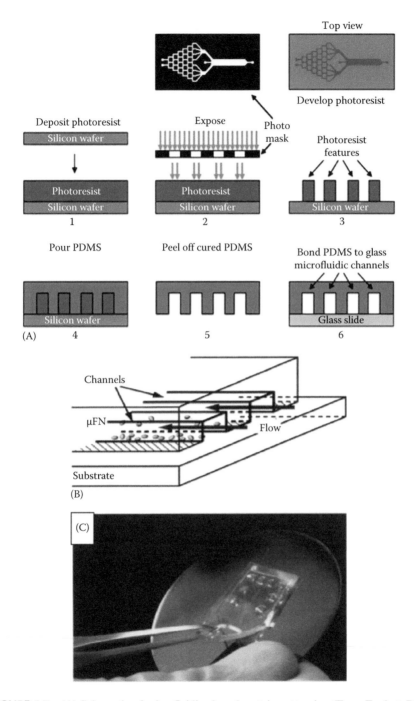

FIGURE 5.7 (A) Schematic of microfluidics-based protein patterning (From Englert, D.L. et al., *Nat. Protoc.*, 5, 864, 2010); (B) fluid flow in microfluidic channels; and (C) microfluidic device. (From Delamarche, E. et al., *J. Am. Chem. Soc.*, 120, 500, 1998.)

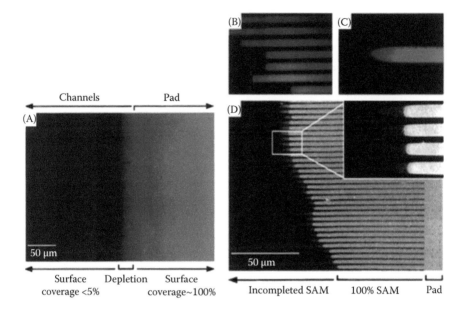

FIGURE 5.8 Depletion of proteins in small microchannels. (From Delamarche, E. et al., *J. Am. Chem. Soc.*, 120, 500, 1998.)

protein exchange proves difficult as the fluids cannot be flushed out easily for further treatments. At the same time, long exposure to solvent-based proteins causes swelling up of the channels. Since the polymer used for these devices is elastomeric PDMS, as in the case of fabricated stamps, the integrity of the microfabricated channels is less and is prone to buckling. Hence, careful consideration of these parameters are required during the design of these microfluidic channels.

5.3.2.3 Stencil-Based Patterning

This is one of the simplest ways of protein patterning. The idea behind this technique is borrowed from art making, where a thin material with holes (called "stencils") is used to deposit paint (or protein solution in this case) within predefined patterns. The stencils serve as masks during processes such as protein deposition, cell attachments, and other surface modifications [19,21]. In its earliest applications, stencils were made of thin metal foils and used to block nonspecific deposition of metal vapors such as palladium. However, metal foils cannot form perfect conformal contact with the substrate and are not suitable for depositing materials in solution such as proteins [58]. Alternative approaches have been developed that use more elastic materials as stencils such as thin PDMS [56] or Parylene membranes [57,59]. The PDMS thin films remain a popular choice owing to their ease and low cost of fabrication. They also form better seals on contact with the substrate, hence gaining an advantage over other techniques.

The fabrication of PDMS stencils is similar to the fabrication of features described in Section 5.3.2.1, except in this case we fabricate "holes" instead of features. A SU8 mold with features is fabricated initially and then casted using PDMS, which then has to be excluded from the top of the features to form holes. The exclusion process can be carried out in three ways. In one of the ways, the master wafer can be spin coated with PDMS where the thickness of the film is determined by the spin speed and spin time [56]. Though this method is the easiest one to fabricate stencils, it can sometimes result in uneven PDMS stencils, owing to the formation of menisci or bubbles between the feature walls and PDMS. Another convenient way of exclusion is to cover the PDMS-filled master with a hard flat surface that forms a bridge across the features. The cover has to be hard enough not to sag in between the features and yet soft enough to form good conformal contact. Thin Mylar film supported by hard glass substrate is a preferred choice for cover, as after curing treatment the PDMS film can be easily peeled off the Mylar sheet (Figure 5.9) [14,57].

The last method is referred to as "microfluidic molding" or "capillary filling," which basically involves tightly capping the master with a cover and then filling the microchamber with PDMS pre-polymer [14,56]. In a study by Master et al., a UV-curable pre-polymer was used as the stencil material and fabricated similarly as

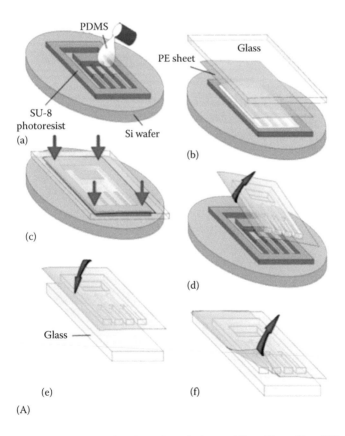

FIGURE 5.9 (A) Stencil fabrication through exclusion molding. (From Hsu, C.H. et al., *Lab Chip*, 4(5), 420, 2004.) *(Continued)*

(B)

FIGURE 5.9 (*Continued*)   (B) Phase-contrast imaging of thin PDMS stenciled, single cell within a stencil hole. (From Hardelauf, H. et al., *Lab Chip*, 11, 2763, 2011.)

shown in Figure 5.10. They also showed that stencil could be used for the deposition of both single and multiple proteins. To achieve patterning of multiple proteins, stencils are stacked up in layers and each protein deposition is followed by treatment with antifouling agent before the first stencil layer is removed. Incubation with antifouling agent ensures that secondary deposition of proteins in same area does not occur unless driven by specific interaction. As seen in Figure 5.10A and B, two stencil layers with 30 and 600 μm diameter holes were used to pattern BSA-AlexaFluor™ 488 and BSA-AlexaFluor™ 555, respectively. At the same time, Figure 5.10C shows patterning of varied concentration of BSA. This was also achieved as a two-step process but with a single layer. First, only the upper half was treated with a low-concentration BSA (1 μg/mL) for 20 min, followed by passivation with Pluronic, and then followed by treatments of the entire surface with a higher BSA concentration (10 μg/mL) until saturation [55].

Though stencils provide a convenient way to selectively mask regions during protein adsorption and cell attachment, they are prone to deformation during use. Moreover, fabrication of holes is not as easy as fabrication of features as they are subject to tear and failure of formation of holes. This has been replaced by the use of parylene-C stencils, which are more resistant to wear and can be reused multiple times [56,59].

5.3.3 SELF-ASSEMBLED MONOLAYERS

Some organic molecules undergo spontaneous chemisorption as a uniform monolayer of "self-assembled" molecules upon contact with a specific surface [12]. In order to minimize surface energy, the molecules adhere as densely packed, crystalline layer of single-molecule thickness, hence they are commonly referred to as SAMs. They are bifunctional organic molecules and bind to the surface through one end while the other remains free, enabling further functionalization. SAMs had been predominantly used for solid surface–liquid interactions, especially for studying molecular interaction at the interface and the effect of surface wettability.

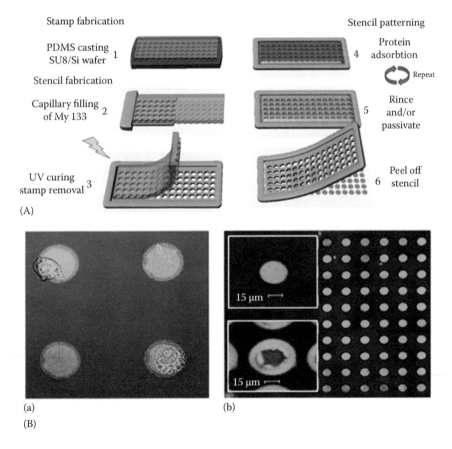

FIGURE 5.10 (A) Stencil fabrication through capillary filling and protein patterning; and (B) cells confined within pattern. *(Continued)*

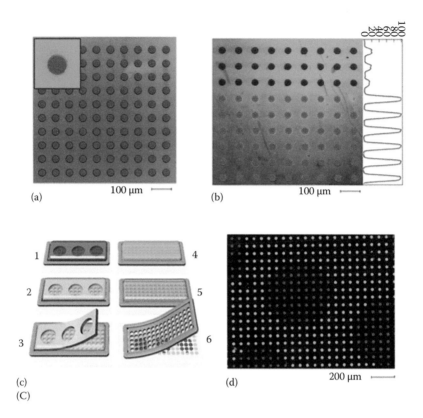

(a) 100 μm

(b) 100 μm

(c)
(C)

(d) 200 μm

FIGURE 5.10 (*Continued*) (C) gradient protein patterning. (From Masters, T. et al., *PLoS ONE*, 7(8), e44261, 2012, Copyright Creative Commons Attribution License.)

Based on ease of synthesis, two commonly used SAMs are alkanethiols (CH_3-$(CH_2)_n$-1-SH, $n > 9$) that bind to metals such as gold (Au) or silver (Ag) through their sulfur end, and the alkanesilanes, that bind to silicon or oxidized surfaces through reaction of silane/siloxane group with hydroxyl group of the surface (Figure 5.11) [12]. The use of SAMs for micropatterning proteins is appealing because it provides the ability to control the structure and interfacial properties of the surface at the molecular level. The functional end of the SAM molecule, for example, an alkanethiol, can be varied as needed to selectively control protein adhesiveness. For example, an oligo(ethylene glycol) end terminal group is highly protein resistant, leading to the formation of "inert surfaces." Depending on the end functional group, proteins can either directly bind to the SAM molecules or through a cross-linker. For example, glutaraldehyde is a commonly used cross-linking molecule that can bind to an amine group on both ends, so it acts as a linker, binding the amine group of aminosilane and the target protein [28,29].

Micropatterning SAMs can be carried out with a similar strategy as patterning proteins. Four major strategies commonly used are photolithography, microcontact printing, selective modification/removal, and finally selective formation of SAMs [12,28,29]. Selective removal or modification is carried out through ablation

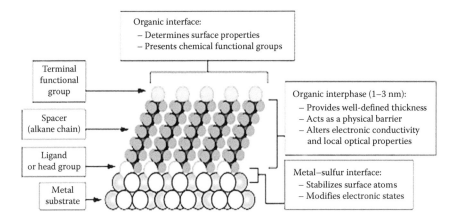

FIGURE 5.11 SAM binding on surface. (From Love, J. et al., *Chem. Rev.*, 105(4), 1103, 2005.)

with exposure to laser beam, degradation on exposure to electron beam, electrochemical desorption, mechanical shear with the aid of atomic force microscope (AFM), or through oxygen plasma etching through a patterned mask [13]. Selective formation of SAMs occurs on chemical interaction of SAM molecules with a preexisting pattern. For example, alkanethiol SAM formation on a gold micropattern substrate over a glass substrate is a commonly used procedure [60–62].

5.3.4 Direct Write Techniques

Direct write technology is one of the recent developments, a novel approach for fabrication of electronic and sensor devices of sizes ranging from micro to nano scales [63–67]. Recent advancements include the adaptation of this technique to biological applications, especially for stem cell and tissue engineering research. These fabrication methods employ a computer-controlled stage that translates the signals from the system. In response, a pattern generating stage (such as an inkjet printer) or the substrate itself is moved creating "prints" or "patterns" of controlled architecture and composition [63,64]. In the following section, three of the most commonly used direct write techniques—laser-based printing [65–68], inkjet printing [69–77], and the AFM-based lithography (or dip pen lithography) [77–83], are reviewed.

5.3.4.1 Laser-Based Patterning

Laser-based printing is a form of noncontact patterning technique where a laser is used to guide and deposit protein particles onto target solid surfaces [64,65]. The ability to capture and guide target proteins or particles depends on its refractive index relative to its surrounding and also the strength of the optical force. A higher refractive index of the particle coupled with strong applied optical force can easily guide it steadily along the light path for deposition onto the target surface. This technique can also be applied for a variety of organic and inorganic particles in both gas and liquid phases and sometimes used for directly patterning the cells in culture medium [65,67]. Odde et al. have used a laser-guided technique for manipulation of particles and cells, and found that in contrast to optical trapping, which uses high numerical aperture to axially trap the target particle in three dimension, low-numerical aperture laser provides a weak force that guides (or pushes) the target particle axially (Figure 5.12A and B) [65,66]. Hence, this way, multiple particles can be targeted simultaneously. Moreover, hollow optical fibers are sometimes coupled to the light source, allowing for transmission of high-intensity beam over a distance of 1 cm.

In addition to particle guidance, laser beams are also used to pattern proteins through selective removal of surface molecules. In a recent work by Heinz et al., exposing poly-L-lysine-poly(ethyleneglycol) (PLL-PEG) and polystyrene (PS)-coated glass coverslips to laser pulses of defined power causes selective ablation of

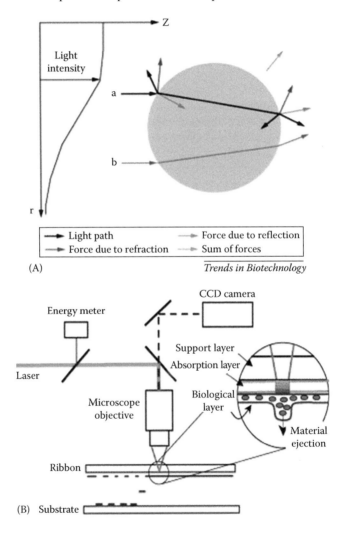

FIGURE 5.12 (A) Force generation via laser beam. (From Odde, D.J. and Renn, M.J., *Biotechnol. Bioeng.*, 67(3), 312, 2000.) (B) Laser force based protein patterning. (From Barron, J.A. et al., *Biosens. Bioelectron.*, 20, 246, 2004.)

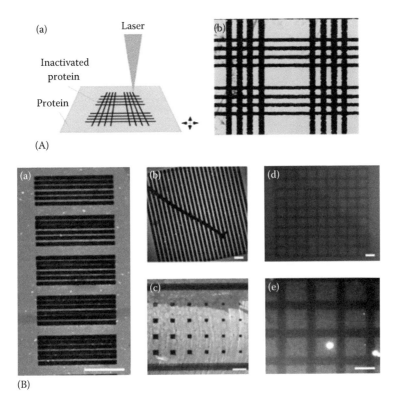

FIGURE 5.13 (A) Schematic of laser inactivation patterning of proteins and (B) examples of laser patterning on different substrate materials: (a) silicon, (b) quartz, (c) glass, (d) polystyrene, (e) poly(dimethylsiloxane) (PDMS). (From Heinz, W.F. et al., *Lab Chip*, 11(19), 3336, 2011.)

the molecules. As seen in the schematic representation, the substrate is coated with a protein to be patterned and is mounted on a xyz translation stage. The stage is then moved in a predetermined pattern and speed. Exposure to focused UV laser causes denaturation or removal of the protein and a desired pattern of functional protein is formed [68] (Figure 5.13).

5.3.4.2 Inkjet Printing

Inkjet printing is a form of noncontact patterning technique that is based on the working principle of a desktop printer, where numerical data from a computer are translated to printing patterns on substrates with ink drops, or in this case, protein solutions [69] or sometimes cell suspensions [73]. This idea was first devised for biological applications in the early 1990s for printing DNA arrays [71] and proteins, especially ECM proteins [69,72]. Since then, this process has grown as a versatile technique for direct writing on two-dimensional surfaces, owing to its ability to accurately predict the placement of predetermined quantity of the particle. The physical parameters that determine printability are viscosity, density, surface tension, and nozzle diameter of the printing system. Printing is possible only within a defined range of rheological and surface tension properties [69]. This process is classified into two broad categories,

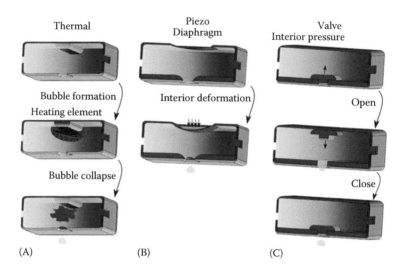

FIGURE 5.14   Schematic illustration of working of (A) thermal inkjet technology; (B) piezo inkjet technology; and (C) valve inkjet technology. (From Romanov, V. et al., *Analyst*, 139, 1303, 2014.)

continuous inkjet printing (CIJ) and drop on demand inkjet printing (DIJ). In the CIJ, a liquid is forced under pressure through a small orifice, resulting in a jet stream of droplets. These droplets are electrically charged on generation and can be conveniently steered in flight by an electric field. On the other hand, in DIJ ink is contained in a reservoir chamber and drops are generated only when required by propagation of a pressure pulse within the filled chamber. Inkjet emission can be brought about through three mechanism and are commonly referred to as (1) piezo actuation [75], valve jet [76], and thermal inkjet [77] (Figure 5.14A through C) [77]. As the namesake, in thermal inkjet, ejection of droplet occurs due to sudden change in chamber pressure brought about by rapid heating. While in the piezo-actuation-based system, a pressure change is induced by volumetric change in the ink reservoir. Finally, in the valve jet ejection system that may also be classified as "demand inkjet printing," a reservoir valve is opened and closed through a computer control [77].

Depending on the means of droplet generation, inkjet printers can be categorized into three other categories as piezoelectric, thermal, and electrostatic inkjet printers. The difference among these lie in their operating temperature range, throughput, reproducibility, precision, printable viscosity, and versatility [69,77].

5.3.4.3   Dip-Pen Lithography

Another direct write protein micropatterning technique is dip-pen lithography, a type of scanning probe lithography where an AFM tip is used for patterning proteins [78–83]. The AFM tip consists of a cantilever with a micron-sized conical probe at the end, which is used to "write" on the surface just like writing with an ink pen. As seen in the figure, the molecular protein solution (ink solution) is transferred from the tip of the probe onto the surface of the substrate via capillary forces

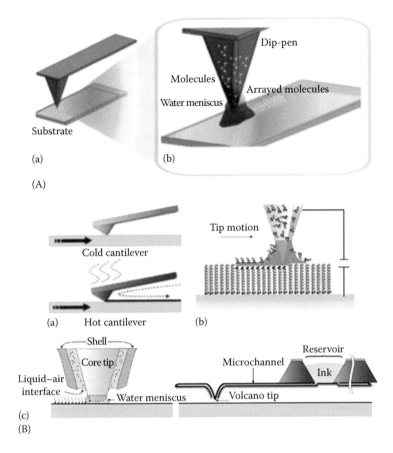

FIGURE 5.15 (A) Schematic illustration dip-pen lithography. (From Romanov, V. et al., *Analyst*, 139, 1303, 2014.) (B) Variants of dip-pen lithography. (From Salaita, K. et al., *Nat. Nanotechnol.*, 2(3), 145, 2007.)

(Figure 5.15A and B) [77,79]. This form of patterning technique can be used to fabricate up to nanoscale features on different substrates as long as the surfaces are smooth. Otherwise, an additional step of surface smoothing is required [78]. This is comparable to the texture of the paper influencing the resolution and convenience of conventional writing by hand with an ink pen.

DPL is a powerful patterning technique compared to traditional AFM lithography and can pattern molecules down to the scale of 5 nm [80,81]. However, this resolution is easily affected by environmental factors. For example, increasing the relative humidity disturbs the size of the water meniscus bridging the tip and the substrate and thereby reducing the effective resolution. Another factor to consider is the speed of tip movement, which has to be slow enough for uniform transfer of the ink [78]. Moreover, a single dot is transferred at a time sequentially, which causes the patterning process to be slow. A recent advancement of this technique is the development of parallel dip-pen nanolithography with multiple arrays of cantilevers [82,83].

5.4 NEED FOR MICROPATTERNING

5.4.1 BIOMIMICRY

As a part of a highly structured architecture within the organs and tissues, the cells are in constant physical and biochemical interaction with their microenvironment (Figure 5.16). The cell microenvironment composed of the ECM and neighboring cells impose boundary conditions, such as stress, that influence its shape, polarity, and function [84,85]. In addition, the biochemical composition along with the stiffness of the microenvironment guides the adhesion of the cells onto the matrix and thereby affecting intracellular signal cascades. These signaling pathways regulate cytoskeletal remodeling, subsequently affecting migration and differentiation behavior of the cells [86]. Therefore, the composition and properties of the microenvironment are crucial for cellular function.

When cultured in regular Petri dishes, the cells are exposed to macroscale, homogenous and static environment, which fails to simulate the *in vivo* conditions. Unlike traditional cell culture, micropatterning methods provide the ability to control and manipulate specific microenvironmental cues according to the desired conditions. This procedure allows construction of tissue-like conditions through fabrication of culture substrates with microscopic features that promote cell adhesion and function [22,93].

5.4.2 MODULATION OF CELL BEHAVIOR

In vivo, localized patterns arise due to variety of stimuli that includes gradients of diffusible factors, mechanical forces, and biochemical signals from neighboring cells and adhesion to ECM. Micropatterning methods allow for developing both homogeneous and heterogeneous microenvironments as needed [88,89]. For example, varying the size of the micropatterned proteins determines the amount of adhesive protein molecules available for interaction, which in turn influences the strength of signaling cascade and eventually the formation of stress fibers [90,91]. *In vivo*, ECM is not homogenous and large areas of the cells may remain unattached to any substrate.

Mimicking ECM heterogeneity, Therey et al. micropatterned irregular geometries such as V- and T-shaped micropatterns. Despite the noncontinuity, the cells were able to remodel their actin bundles and form large Rho-A-dependent stress fibers over the nonadhesive regions (Figure 5.17) [85]. This gave unique insight into migration behavior of the cells. Not all cells have similar shape, neither all tissues have similar architecture. Muscle cells and fibroblasts have slightly elongated shape as compared to other cells such as hepatocytes. The shape of the micropattern also determines the shape of the cells *in vitro* [87,88]. In a study carried out by Levina et al., it was shown that anisotropic cell adhesion within the pattern led to an anisotropic cellular organization of actin filaments. Stress fibers were found to align along the long axis of the cells for mesenchymal cells, fibroblasts, and muscle cells. Though cell spreading along the short axis was limited, some cells compensated for it through further elongation along the long axis, creating more focal adhesion [6]. This was especially true for epithelial cells while the fibroblasts retained a fixed cell length irrespective of the shape of the micropattern. Interestingly, other studies have

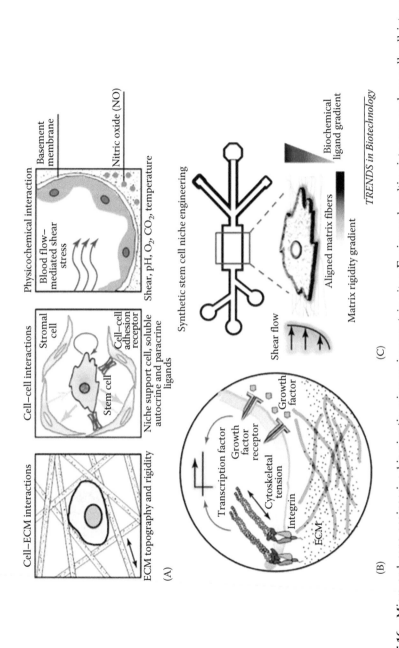

FIGURE 5.16 Micro and nano-engineering biomimetic microenvironment *in vitro:* Engineered multivariate cues such as cell–cell interactions, cell–ECM interaction, soluble factors and biophysical factors like substratum rigidity, topography, shear stress, and pH (A); Specialized microenvironment that facilitate both homotypic and heterotypic juxtracrine and paracrine interactions *in-vivo* (B); Techniques such as microfluidics and micro-nanopatterning allows precise control of parameters like shear stress, biochemical gradients, substrate rigidity, and cell positioning (C). (From Kshitiz et al., *Trends Biotechnol.,* 29(8), 399, August 2011.)

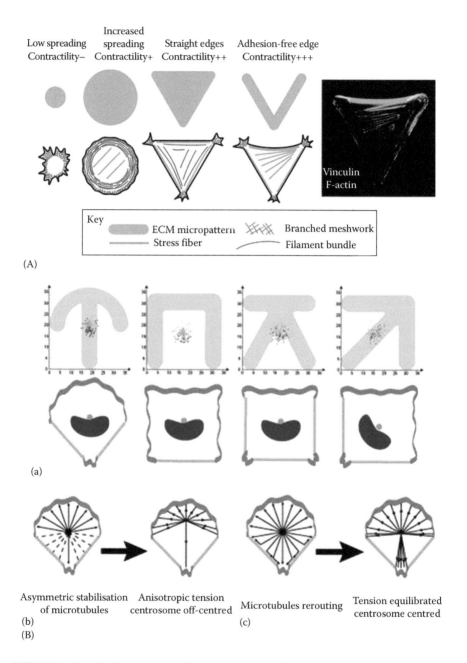

FIGURE 5.17 (A) Variation in cell and cell contractility with change in adhesion area. (Adapted from Thery, M. et al., *Cell Motil. Cytoskel.*, 63, 341, 2006.) (B) Cytoskeletal remodeling also results in orientation of microtubules within the cell. (From Thery, M. et al., *Proc. Natl. Acad. Sci. USA*, 103(52), 19771, 2006.)

found a correlation between shape of the cells and contractility. In elongated mesenchymal cells, high levels of myosin-II were observed on actin stress fibers, which resulted in increased cell contractility [90,99]. In contrast, elongation of vascular smooth muscle cells and endothelial cells resulted in reduction of F-actin content, cytoskeletal stiffness, and thus contractility [2].

5.4.3 GUIDING STEM CELL BEHAVIOR USING MICROPATTERNED PROTEINS

Many environmental cues within a stem cell niche play an important role in maintaining the stem cells in quiescent state, as well as directing their commitment to different lineages. In this section, we review how micropatterns provide cues that contribute to regulation of stem cell differentiation into various cell lineages. Culturing stem cells on micropatterned proteins and its effect on cell shape and function with respect to stem cell behavior has been extensively investigated. Various studies have demonstrated that aside from biochemical cues, the shape and size of micropattern regulates commitment of stem cells [90–96]. In one of the earlier studies by McBeath et al., influence of cell shape as a regulatory factor for determining commitment of human mesenchymal stem cells (hMSCs) was studied [97]. In their work, fibronectin islands of different sizes were microcontact-printed on PDMS substrates surrounded by regions of nonadhesive Pluronic-F108. They found that commitment to adipogenesis was favored in case of high-density cell population and in smaller patterns, while osteogenic commitment was preferred for cells seeded on larger areas. In high-density culture, cell adhesion and spreading on the substrate was limited, cell–cell interactions and paracrine signaling was observed to increase, and cells attained a nonelongated, rounded shape [97]. Therefore, by mimicking the *in vivo* microenvironment of the adipocytes, hMSCs showed adipogenic cell features such as a rounded, spherical shape, which is also functionally important as it allows for maximal lipid storage. On the other hand, hMSCs seeded on larger surfaces at lower densities attained a more elongated shape, and this is functionally important as elongated cell shape facilitates osteoblast matrix deposition during bone remodeling. RhoA GTPase is the central regulator of contractility in many cells and was found in higher concentration in well-spread osteogenic cells than unspread cells (Figure 5.18) [93,97]. Irrespective of the culture media used, higher RhoA and Rho-kinase (ROCK) activity was shown to induce osteogenic differentiation. RhoA activity effects ROCK-mediated cytoskeletal tension, thereby resulting in increased or decreased cell contractility. Thus, protein micropatterning enabled investigation of the effect of cell shape on cytoskeletal tension and RhoA signaling, and eventually on stem cell fate.

Micropatterned protein patterns of different size and shape generation of localized adhesion cues, which stimulates signaling cascades, that regulates cytoskeletal remodeling, gene expression, and cell differentiation, various studies have been conducted and leads to were reviewed recently [2]. Irrespective of the cell type, cytoskeletal remodeling, in turn, defines the shape of the cell, contractility, polarity, cell division, and migration pattern, and eventually the tissue architecture, which is usually crucial for its function [87,93].

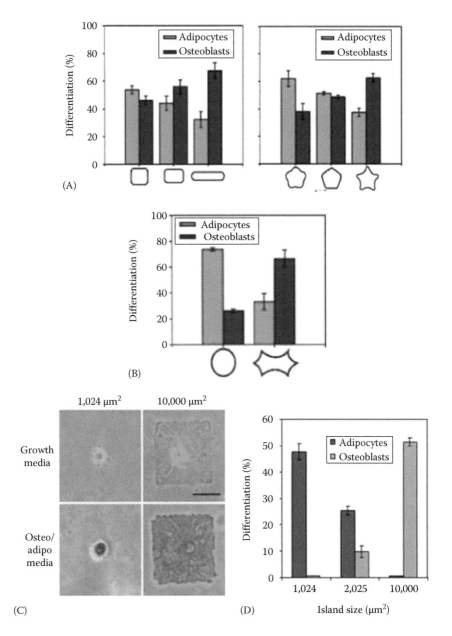

FIGURE 5.18 (A, B) Dependence of stem cell fate on shape of micropatterns (From Kilian, K.A. et al., Geometric cues for directing the differentiation of mesenchymal stem cells, *Proc. Natl. Acad. Sci. USA*, 107(11), 4872–4877, Copyright 2010 National Academy of Sciences, U.S.A.); (C) differentiation of stem cells into adipocytes or osteoblasts was independent of type of media used; and (D) effect of size of micropattern on degree of cell differentiation. (From McBeath, R. et al., *Develop. Cell*, 6(4), 483, 2004.)

Adhesion to ECM proteins defines the spatial distribution of the cell that acts as key mechanical stimulation defining not only the shape of the cell but also providing directional cues during asymmetric cell division. Spatial distribution of cell adhesion defines polarization of actin and microtubule network especially during cell division [84]. Asymmetric adhesion cues results in dynamic remodeling of actin network and microtubules, which plays a significant role during cell division, consequently resulting in asymmetric positioning of cytokinetic plane. This asymmetry generates daughter cells with distinct fates owing to their unequal sizes and adhesion-associated factors. Micropatterning asymmetric and symmetric geometries can also provide a controlled microenvironment to identify the underlying mechanisms during occurrence of biased DNA segregation (Figure 5.19A) [98]. Biased DNA segregation is a mitotic event in which the chromatid carrying the original template DNA strands and the template copies are segregated unequally (or in a biased manner). The factors influencing

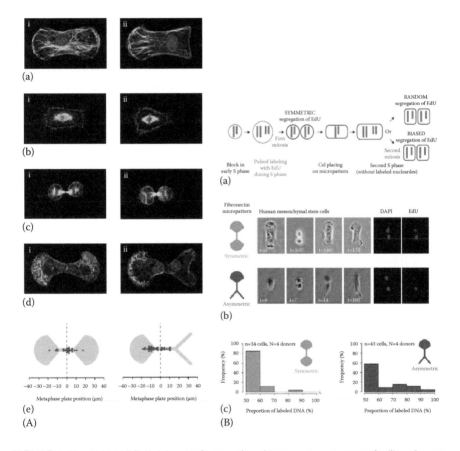

FIGURE 5.19 (A) hMSC division on fibronectin micropatterns—symmetric (i) and asymmetric (ii). Images show cells in interphase (a), metaphase (b), telophase (c), and cytokinesis (d). (B) Segregation of hMSCs. (From Freida, D. et al., *Cell Rep.*, 5(3), 601, 2013.)

biased segregation are still largely under investigation, and this behavior has been observed in several cells including tumor cells. In a recent study, Freida et al. took advantage of micropatterned proteins to investigate the role of geometric cues on biased DNA segregation in human bone marrow stem cells. Cell division is difficult to be monitored directly *in vivo* and through traditional *in vitro* culturing, thus it is difficult to distinguish between random and biased segregation. In this study, a singular chromatid was marked through EdU labeling for tracking segregation pattern [98]. On symmetric micropatterns, random segregation of sister chromatids to the daughter cells was observed while for cells cultured on asymmetric micropatterns the segregation was biased. It can be concluded that spatial distribution of cell adhesion regulates biased chromatid segregation. However, this, in turn, also depends on the cell type as the mesenchymal cells exhibited more sensitivity for this behavior unlike the more robust fibroblast cells [98].

5.4.4 PATTERNED CO-CULTURE SYSTEMS FOR TISSUE ENGINEERING

The focus of tissue engineering research is to produce artificial tissues or whole organs for clinical applications, drug testing, and disease models [4,15]. Many tissues have a heterogeneous cell composition, meaning they are composed of several different cell types with a specific phenotype. *In vivo*, cell–cell interactions occur through direct cell contact or indirectly through exchange of soluble factors, and these interactions play an important role in regulation of individual cell responses. During traditional *in vitro* culture, this interaction is lost through the isolation, digestion, and selection steps carried out to obtain a purified population of a target cell type. Initial *in vitro* co-cultures involved simpler culture systems where two or more cell types were co-cultured randomly. Though these experiments presented a brief insight into heterotypic cell–cell interactions, it was limited by lack of control over degree of cell–cell contact and spatial distribution of each cell type. Therefore, using micropatterned proteins to create co-cultures with more control over the cellular distribution has enabled better understanding of *in vitro* cell–cell interactions.

For example, co-culturing hepatocytes with fibroblasts was shown to greatly improve liver-specific function, that was influenced by the density of each cell type as well as the proximity of the hepatocytes to the fibroblasts [100–102]. In one of the earlier studies, patterned collagen strips were fabricated using photolithography and seeded with hepatocytes [101]. After washing, hepatocytes remained attached to these adhesive collagen strips and regions in between left unoccupied, where the second cell type, fibroblasts, was attached after a second cell seeding. Though this technique provided some control over the localization of the cells, the fidelity was not very high as adhesion of second cell type over the initially seeded cells was a problem. Furthermore, the system was not capable of any temporal control. In later studies, to overcome these shortcomings, a micromechanical co-culture system, which provided dynamic control over the cell–cell interaction, was developed (Figure 5.20) [102].

In this set-up, microfabricated silicon-based two interlocking parts were coated with fibronectin and seeded with two different cell types and manually manipulated to control the interlocking distance between these two parts temporally

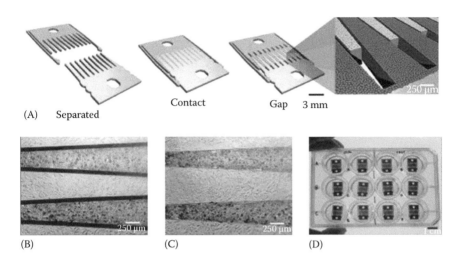

(A) Separated   Contact   Gap 3 mm

(B)   (C)   (D)

FIGURE 5.20 Micromechanical device for culturing two different cell types: hepatocytes and fibroblasts. Microfabricated silicon parts can be fully separated (left) or locked together in contact mode with comb fingers to mimic close proximity (center), or slightly separated to allow a controlled gap between neighboring cells (right) and cells are cultured on top surfaces of the comb fingers (inset) (A). (B and C) Bright-field images of hepatocytes (darker cells) and 3T3 fibroblasts cultured on the comb fingers. (D) Devices place within a standard 12-well plate for long term culture. (From Hui, E.E. and Bhatia, S.N., Micromechanical control of cell–cell interactions, *Proc. Natl. Acad. Sci. USA*, 104(14), 5722–5726, Copyright 2007 National Academy of Sciences, U.S.A.)

during the culture. This device design allowed for dynamic control of cell interaction through control of proximity of the two cell types with respect to each other.

As described in Section 5.3.2.2, a multilayered stamp can be used to sequentially microcontact print multiple proteins for creating co-culture systems (Figure 5.21A) [49]. Similarly, other soft lithography techniques such as microfluidics-based patterning and stencil-based patterning can be used to create patterned co-cultures. When applying microfluidic-based strategies, multiple isolated channels can be contained in a bi-layered device and hence allowing patterning of multiple types of proteins and cells (Figure 5.21B) [53]. As an alternative, microfabricated parylene membranes can be used as multilayered stencils to seed each layer with a different cell-type sequentially (Figure 5.21C) [59]. In another study, Fukuda et al. used a stencil-based approach and a combination of three major ECM components: hyaluronic acid (HA), fibronectin (FN), and collagen to create patterned co-cultures (Figure 5.22A through C) [103]. HA was spin coated on bare glass to form a thin layer over which a PDMS mold with holes was placed tightly. Due to the capillary action, spin-coated HA was receded from the void spaces exposing bare glass on which FN coating was carried out. The hepatocytes were seeded on these FN islands, followed by seeding of 3T3 cells on collagen, which was coated on the HA. Micropatterned protein systems developed for co-cultures hold significant potential for investigating cell–cell and cell–microenvironment interactions, and for engineering multicellular tissues.

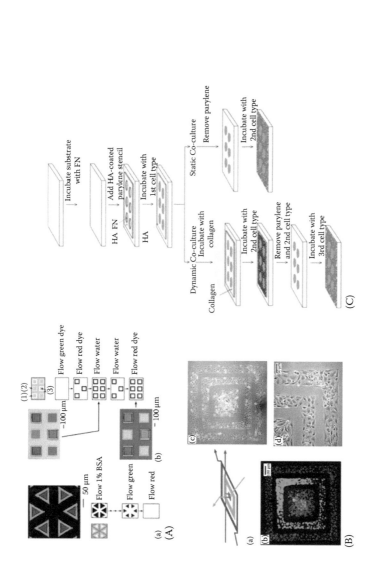

FIGURE 5.21 (A) Multilayer PDMS stamps for simultaneous patterning. (From Tien, J., et al., Fabrication of aligned microstructures with a single elastomeric stamp. *Proc. Natl. Acad. Sci. USA*, 99(4), 1758–1762, Copyright 2002 National Academy of Sciences, U.S.A.) (B) Microfluidic-based co-culture patterning. (From Chiu, D. et al., Patterned deposition of cells and proteins onto surfaces by using three-dimensional microfluidic systems, *Proc. Natl. Acad. Sci. USA*, 97(6), 2408–2413, Copyright 2000 National Academy of Sciences, U.S.A.) (C) Patterning using multilayered parylene-C stencils. (From Wright, D. et al., Reusable, reversibly sealable parylene membranes for cell and protein patterning. *J. Biomed. Mater. Res. A*, 85(2), 530–538, Copyright 2008 National Academy of Sciences, U.S.A.)

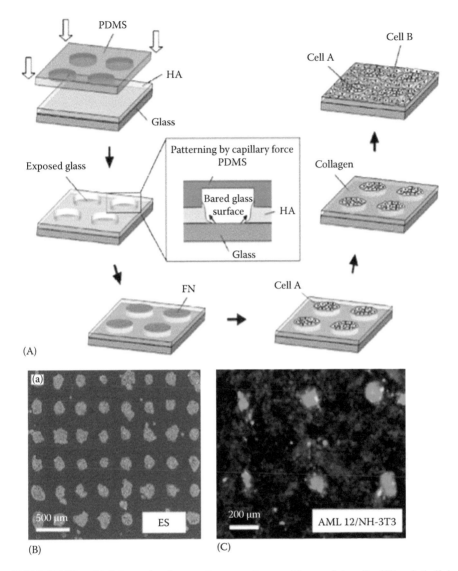

FIGURE 5.22 (A) Schematic of patterning co-cultures with use of stencils; (B) endothelial cells (cell type 1); (C) 3T3 cells (red) and AML cells. (From Fukuda, J. et al., *Biomaterials*, 27(8), 1479, 2006; Khademhosseini, A. and B.G. Chung, Microscale technologies for tissue engineering, In: *Life Science Systems and Applications Workshop, 2009 (LiSSA 2009)*, IEEE/NIH, 2009, pp. 56–57.)

5.5 CONCLUSION

Microfabrication is a general term that defines manufacturing miniature devices and control systems with micrometer resolution. With the advancements in biomedical engineering and the microfabrication technologies, these techniques have been adapted for biomedical applications ranging from tissue engineering to drug screening and toxicology assays. As we have seen throughout the chapter, the physical and biochemical cues from the cellular microenvironment that emerge the through interaction of the cells with other cells and the ECM result in dynamic regulation of the cell behavior. Using micropatterning techniques to mimic this microenvironmental factors for directing cell behavior has given important insights in cellular physiology and cell response with respect to various microenvironmental factors, as well as for creating biomimetic tissue constructs toward tissue replacement or regeneration and for directing stem cell differentiation.

ACKNOWLEDGMENTS

This material is based upon work supported by the National Science Foundation under Grant No. 1530884.

REFERENCES

1. Khademhosseini, A. and B.G. Chung, Microscale technologies for tissue engineering. In: *Life Science Systems and Applications Workshop, 2009 (LiSSA 2009)*. IEEE/NIH, 2009, pp. 56–57.
2. Thery, M., Micropatterning as a tool to decipher cell morphogenesis and functions. *J Cell Sci*, 2010. 123(Pt 24): 4201–4213.
3. Smedsrød, B. et al., Cell biology of liver endothelial and Kupffer cells. *Gut*, 1994. 35(11): 1509–1516.
4. William, D., Benefit and risk in tissue engineering. *Mater Today*, 2004. 7(5): 24–29.
5. Van Blitterswijk, C.A., Tissue assembly and organization: Developmental mechanisms in microfabricated tissues, *Biomaterials*, 2009. 30(28): 4851–4858.
6. Levina, E. et al., Cytoskeletal control of fibroblast length: Experiments with linear strips of substrate. *J Cell Sci*, 2001. 114(Pt 23): 4335–4341.
7. Gallego-Perez, D. et al., High throughput assembly of spatially controlled 3D cell clusters on a micro/nanoplatform. *Lab Chip*, 2010. 10(6): 775–782.
8. Liu, J. and Z. Gartner, Directing the assembly of spatially organized multicomponent tissues from the bottom up. *Trends Cell Biol*, 2012. 22(12): 683–691.
9. Liu, W., Mechanical regulation of cellular phenotype: Implications for vascular tissue regeneration. *Cardiovasc Res*, 2012. 95(2): 215–222.
10. Kamei, K. et al., Phenotypic and transcriptional modulation of human pluripotent stem cells induced by nano/microfabrication materials. *Adv Healthcare Mater*, 2013. 2(2): 287–291.
11. Li, N., A. Tourovskaia, and A. Folch, Biology on a chip: Microfabrication for studying the behavior of cultured cells. *Crit Rev Biomed Eng*, 2003. 31(5–6): 423–488.
12. Love, J. et al., Self-assembled monolayers of thiolates on metals as a form of nanotechnology. *Chem Rev*, 2005. 105(4): 1103–1169.
13. Folch, A. and M. Toner, Microengineering of cellular interactions. *Annu Rev Biomed Eng*, 2000. 2: 227–256.
14. Zhang, S., L. Yan, M. Altman, M. Lässle, H. Nugent, F. Frankel, D.A. Lauffenburger, G.M. Whitesides, and A. Rich. Biological surface engineering: a simple system for cell pattern formation. *Biomaterials*, 1999. 20(13): 1213–1220.
15. Chung, B., L. Kang, and A. Khademhosseini, Micro- and nanoscale technologies for tissue engineering and drug discovery applications. *Exp Opin Drug Discov*, 2007. 2(12): 1653–1668.

16. Fischer, T. and H. Hess, Materials chemistry challenges in the design of hybrid bionanodevices: Supporting protein function within artificial environments. *J Mater Chem*, 2007. 17(10): 943–951.

17. Cavalcanti-Adam, E.A. et al., Cell adhesion and response to synthetic nanopatterned environments by steering receptor clustering and spatial location. *HFSP J*, 2008. 2(5): 276–285.

18. Lim, J.Y. and H.J. Donahue, Cell sensing and response to micro- and nanostructured surfaces produced by chemical and topographic patterning. *Tissue Eng*, 2007. 13(8): 1879–1891.

19. Kane, R.S. et al., Patterning proteins and cells using soft lithography. *Biomaterials*, 1999. 20(23–24): 2363–2376.

20. Liu, W.W., Z.L. Chen, and X.Y. Jiang, Methods for cell micropatterning on two-dimensional surfaces and their applications in biology. *Chin J Anal Chem*, 2009. 37(7): 943–949.

21. Qin, D., Y. Xia, and G.M. Whitesides, Soft lithography for micro- and nanoscale patterning. *Nat Protoc*, 2010. 5(3): 491–502.

22. von der Mark, K. et al., Nanoscale engineering of biomimetic surfaces: Cues from the extracellular matrix. *Cell Tissue Res*, 2010. 339(1): 131–153.

23. Aili, D. and M.M. Stevens, Bioresponsive peptide–inorganic hybrid nanomaterials. *Chem Soc Rev*, 2010. 39: 3358–3370

24. Tan, J.L. et al., Simple approach to micropattern cells on common culture substrates by tuning substrate wettability. *Tissue Eng*, 2004. 10(5–6): 865–872.

25. Rabe, M., D. Verdes, and S. Seeger, Understanding protein adsorption phenomena at solid surfaces. *Adv Colloid Interf Sci*, 2011. 162(1–2): 87–106.

26. Brusatori, M.A. and P.R. Van Tassel, A kinetic model of protein adsorption/surface-induced transition kinetics evaluated by the scaled particle theory. *J Colloid Interf Sci*, 1999. 219(2): 333–338.

27. Lim, S.K., S. Perrier, and C. Neto, Patterned chemisorption of proteins by thin polymer film dewetting. *Soft Matter*, 2013. 9(9): 2598–2602.

28. Ostuni, E. et al., Using self-assembled monolayers to pattern ECM proteins and cells on substrates. *Methods Mol Biol*, 2009. 522: 183–194.

29. Ostuni, E., L. Yan, and G.M. Whitesides, The interaction of proteins and cells with self-assembled monolayers of alkanethiolates on gold and silver. *Colloids Surf B: Biointerf*, 1999. 15(1): 3–30.

30. Huttenlocher, A. and A.R. Horwitz, Integrins in cell migration. *Cold Spring Harb Perspect Biol*, 2011. 3(9): a005074.

31. Hoffmann, M. and U.S. Schwarz, A kinetic model for RNA-interference of focal adhesions. *BMC Syst Biol*, 2013. 7: 2

32. Condic, M.L. and P.C. Letourneau, Ligand-induced changes in integrin expression regulate neuronal adhesion and neurite outgrowth. *Nature*, 1997. 389(6653): 852–856.

33. Shekaran, A. and A.J. Garcia, Extracellular matrix-mimetic adhesive biomaterials for bone repair. *J Biomed Mater Res A*, 2011. 96(1): 261–272.

34. Wary, K.K., Recognizing scientific excellence in the biology of cell adhesion, *Cell Commun Signal*, 2005, 3: 7. doi:10.1186/1478-811X-3-7.

35. Cox, E.A., S.K. Sastry, and A. Huttenlocher, Integrin-mediated adhesion regulates cell polarity and membrane protrusion through the Rho family of GTPases. *Mol Biol Cell*, 2001. 12(2): 265–277.

36. Chen, C.S. et al., Geometric control of cell life and death. *Science*, 1997. 276(5317): 1425–1428.

37. Chen, C.S. et al., Micropatterned surfaces for control of cell shape, position, and function. *Biotechnol Prog*, 1998. 14(3): 356–363.

38. Alamdari, O.G. et al., Micropatterning of ECM proteins on glass substrates to regulate cell attachment and proliferation. *Avicenna J Med Biotechnol*, 2013. 5(4): 234–240.

39. Thakar, R.G. et al., Cell-shape regulation of smooth muscle cell proliferation. *Biophys J*, 2009. 96(8): 3423–3432.
40. Falconnet, D. et al., Surface engineering approaches to micropattern surfaces for cell-based assays. *Biomaterials*, 2006. 27(16): 3044–3063.
41. Mandal, K., M. Balland, and L. Bureau, Thermoresponsive micropatterned substrates for single cell studies. *PloS ONE*, 2012. 7(5): e37548.
42. Otsuka, H., Nanofabrication of nonfouling surfaces for micropatterning of cell and microtissue. *Molecules*, 2010. 15(8): 5525–5546.
43. Fink, J. et al., Comparative study and improvement of current cell micro-patterning techniques. *Lab Chip*, 2007. 7(6): 672–680.
44. H.R. Khaleel, Al-Rizzo, H.M., and Abbosh, A.I., Design, fabrication, and testing of flexible antennas. In: *Advancement in Microstrip Antennas with Recent Applications*, Kishk, A. (ed.), InTech, 2013, pp. 363–383.
45. Carrico, I.S. et al., Lithographic patterning of photoreactive cell-adhesive proteins. *J Am Chem Soc*, 2007. 129(16): 4874–4875.
46. Azioune, A. et al., Protein micropatterns: A direct printing protocol using deep UVs. *Microtub In Vivo*, 2010. 97: 133–146.
47. Belisle, J.M., D. Kunik, and S. Costantino, Rapid multicomponent optical protein patterning. *Lab Chip*, 2009. 9(24): 3580–3585.
48. Kaufmann, T. and B.J. Ravoo, Stamps, inks and substrates: Polymers in microcontact printing. *Polym Chem*, 2010. 1(4): 371–387.
49. Tien, J., C. Nelson, and C. Chen, Fabrication of aligned microstructures with a single elastomeric stamp. *Proc Natl Acad Sci USA*, 2002. 99(4): 1758–1762.
50. Tan, J.L., J. Tien, and C.S. Chen, Microcontact printing of proteins on mixed self-assembled monolayers. *Langmuir*, 2002. 18(2): 519–523.
51. Hsu, C.H., C. Chen, and A. Folch, "Microcanals" for micropipette access to single cells in microfluidic environments. *Lab Chip*, 2004. 4(5): 420–424.
52. Englert, D.L., M.D. Manson, and A. Jayaraman, Investigation of bacterial chemotaxis in flow-based microfluidic devices. *Nat Protoc*, 2010. 5: 864–872.
53. Chiu, D. et al., Patterned deposition of cells and proteins onto surfaces by using three-dimensional microfluidic systems. *Proc Natl Acad Sci USA*, 2000. 97(6): 2408–2413.
54. Delamarche, E. et al., Microfluidic networks for chemical patterning of substrates: Design and application to bioassays. *J Am Chem Soc*, 1998. 120: 500–508.
55. Masters, T. et al., Easy fabrication of thin membranes with through holes. Application to protein patterning. *PLoS ONE*, 2012. 7(8): e44261.
56. Folch, A. et al., Microfabricated elastomeric stencils for micropatterning cell cultures. *J Biomed Mater Res*, 2000. 52(2): 346–353.
57. Hardelauf, H. et al., High fidelity neuronal networks formed by plasma masking with a bilayer membrane: Analysis of neurodegenerative and neuroprotective processes. *Lab Chip*, 2011. 11: 2763–2771.
58. Carter, S., Haptotactic islands: A method of confining single cells to study individual cell reactions and clone formation. *Exp Cell Res*, 1967. 48: 189–193.
59. Wright, D. et al., Reusable, reversibly sealable parylene membranes for cell and protein patterning. *J Biomed Mater Res A*, 2008. 85(2): 530–538.
60. Jinno, S. et al., Microfabricated multilayer parylene-C stencils for the generation of patterned dynamic co-cultures. *J Biomed Mater Res A*, 2008. 86(1): 278–288.
61. Chen, C.S. et al., Using self-assembled monolayers to pattern ECM proteins and cells on substrates. *Methods Mol Biol*, 2000. 139: 209–219.
62. Mrksich, M. et al., Using microcontact printing to pattern the attachment of mammalian cells to self-assembled monolayers of alkanethiolates on transparent films of gold and silver. *Exp Cell Res*, 1997. 235(2): 305–313.

63. Hon, K.K.B., L. Li, and I.M. Hutchings, Direct writing technology—Advances and developments. *CIRP Ann Manuf Technol*, 2008. 57(2): 601–620.
64. Tasoglu, S. and U. Demirci, Bioprinting for stem cell research. *Trends Biotechnol*, 2013. 31(1): 10–19.
65. Barron, J.A. et al., Biological laser printing of genetically modified *Escherichia coli* for biosensor applications. *Biosens Bioelectron*, 2004. 20: 246–252.
66. Odde, D.J. and M.J. Renn, Laser-guided direct writing of living cells. *Biotechnol Bioeng*, 2000. 67(3): 312–318.
67. Schiele, N.R. et al., Laser-based direct-write techniques for cell printing. *Biofabrication*, 2010. 2(3): 032001.
68. Heinz, W.F., M. Hoh, and J.H. Hoh, Laser inactivation protein patterning of cell culture microenvironments. *Lab Chip*, 2011. 11(19): 3336–3346.
69. Delaney, J.T., P.J. Smith, and U.S. Schubert, Inkjet printing of proteins. *Soft Matter*, 2009. 5(24): 4866–4877.
70. Derby, B., Bioprinting: Inkjet printing proteins and hybrid cell-containing materials and structures. *J Mater Chem*, 2008. 18(47): 5717–5721
71. Goldmann, T. and J. Gonzalez, DNA-printing: Utilization of a standard inkjet printer for the transfer of nucleic acids to solid supports. *J Biochem Biophys Methods*, 2000. 42(3): 105–110.
72. McWillian, I., M.C. Kwan, and D. Hall, Inkjet printing for the production of protein microarrays. *Methods Mol Biol*, 2011. 785: 345–361.
73. Roth, E.A. et al., Inkjet printing for high-throughput cell patterning. *Biomaterials*, 2004. 25(17): 3707–3715.
74. Fujita, S. et al., Development of super-dense transfected cell microarrays generated by piezoelectric inkjet printing. *Lab Chip*, 2012. 13: 77–80.
75. Kim, J.D., J.S. Choi, B.S. Kim, Y.C. Choi, and Y.W. Cho. Piezoelectric inkjet printing of polymers: Stem cell patterning on polymer substrates. *Polymer*, 2010. 51(10): 2147–2154.
76. Setti, L. et al., Thermal inkjet technology for the microdeposition of biological molecules as a viable route for the realization of biosensors. *Anal Lett*, 2004. 37: 1559–1570.
77. Romanov, V. et al., A critical comparison of protein microarray fabrication technologies. *Analyst*, 2014. 139: 1303.
78. Piner, R.D. et al., "Dip-Pen" nanolithography. *Science*, 1999. 283(5402): 661–663.
79. Salaita, K., Y. Wang, and C.A. Mirkin, Applications of dip-pen nanolithography. *Nat Nanotechnol*, 2007. 2(3): 145–155.
80. Lee, K.B. et al., Protein nanoarrays generated by dip-pen nanolithography. *Science*, 2002. 295(5560): 1702–1705.
81. Wilson, D.L. et al., Surface organization and nanopatterning of collagen by dip-pen nanolithography. *Proc Natl Acad Sci USA*, 2001. 98(24): 13660–13664.
82. Bullen, D. et al., Parallel dip-pen nanolithography with arrays of individually addressable cantilevers. *Appl Phys Lett*, 2004. 84(5): 789–791.
83. Brown, K.A. et al., A cantilever-free approach to dot-matrix nanoprinting. *Proc Natl Acad Sci USA*, 2013. 110(32): 12921–12924.
84. Pouthas, F. et al., In migrating cells, the Golgi complex and the position of the centrosome depend on geometrical constraints of the substratum. *J Cell Sci*, 2008. 121(Pt 14): 2406–2414.
85. Thery, M. et al., Anisotropy of cell adhesive microenvironment governs cell internal organization and orientation of polarity. *Proc Natl Acad Sci USA*, 2006. 103(52): 19771–19776.
86. Gomez, E.W. et al., Tissue geometry patterns epithelial-mesenchymal transition via intercellular mechanotransduction. *J Cell Biochem*, 2010. 110(1): 44–51.

87. Roca-Cusachs, P. et al., Micropatterning of single endothelial cell shape reveals a tight coupling between nuclear volume in G1 and proliferation. *Biophys J*, 2008. 94(12): 4984–4995.
88. Gomes, S. et al., Natural and genetically engineered proteins for tissue engineering. *Prog Polym Sci*, 2012. 37(1): 1–17.
89. Engel, E. et al., Nanotechnology in regenerative medicine: The materials side. *Trends Biotechnol*, 2008. 26(1): 39–47.
90. Vallenius, T., Actin stress fibre subtypes in mesenchymal-migrating cells. *Open Biol*, 2013. 3: 130001.
91. Vignaud, T. et al., Reprogramming cell shape with laser nano-patterning. *J Cell Sci*, 2012. 125(Pt 9): 2134–2140.
92. Kshitiz, A.J. et al., Micro- and nanoengineering for stem cell biology: The promise with a caution. *Trends Biotechnol*, 29(8). August 2011: 399–408.
93. Kilian, K.A. et al., Geometric cues for directing the differentiation of mesenchymal stem cells. *Proc Natl Acad Sci USA*, 2010. 107(11): 4872–4877.
94. Ruiz, S.A. and C.S. Chen, Emergence of patterned stem cell differentiation within multicellular structures. *Stem Cells*, 2008. 26(11): 2921–2927.
95. Huang, N.F. and S. Li, Regulation of the matrix microenvironment for stem cell engineering and regenerative medicine. *Ann Biomed Eng*, 2011. 39(4): 1201–1214.
96. Watt, F.M., Stem cell fate and patterning in mammalian epidermis. *Curr Opin Genet Dev*, 2001. 11(4): 410–417.
97. McBeath, R. et al., Cell shape, cytoskeletal tension, and RhoA regulate stem cell lineage commitment. *Develop Cell*, 2004. 6(4): 483–495.
98. Freida, D. et al., Human bone marrow mesenchymal stem cells regulate biased DNA segregation in response to cell adhesion asymmetry. *Cell Rep*, 2013. 5(3): 601–610.
99. Thery, M. et al., Cell distribution of stress fibres in response to the geometry of the adhesive environment. *Cell Motil Cytoskel*, 2006. 63: 341–355.
100. Kaji, H. et al., Engineering systems for the generation of patterned co-cultures for controlling cell-cell interactions. *Biochim Biophys Acta*, 2011. 1810(3): 239–250.
101. Bhatia, S. et al., Microfabrication of hepatocyte/fibroblast co-cultures: Role of homotypic cell interactions. *Biotechnol Progr*, 1998. 14(3): 378–387.
102. Hui, E.E. and S.N. Bhatia, Micromechanical control of cell-cell interactions. *Proc Natl Acad Sci USA*, 2007. 104(14): 5722–5726.
103. Fukuda, J. et al., Micropatterned cell co-cultures using layer-by-layer deposition of extracellular matrix components. *Biomaterials*, 2006. 27(8): 1479–1486.

6 Cell–Scaffold Interfaces in Skin Tissue Engineering

Beste Kinikoglu

CONTENTS

6.1 INTRODUCTION

Skin tissue engineering is based on almost 40 years of research since the first culture of keratinocytes on feeder layers and their use as epidermal sheets for burn treatment was realized (Rheinwald and Green, 1975). Over the past 40 years, our knowledge of cell biology and wound healing increased significantly and great efforts have been made to create substitutes that mimic human skin. Skin is the largest organ in humans and serves as a protective barrier at the interface between the human body and the surrounding environment (Groeber et al., 2011). It protects the underlying organs against pathogenic microbial agents, mechanical disturbances, and UV radiation; it also prevents loss of body fluid and plays a very important role in immune defense and thermoregulation (Böttcher-Haberzeth et al., 2010). Skin is basically composed of two layers: a stratified epidermis and an underlying dense connective tissue, that is, dermis. The two are attached to each other at the basement membrane region. Skin comprises several different cell types. Keratinocytes are the most common cell type in the epidermis and form the surface barrier layer. Melanocytes are found in the lower layer of the epidermis and provide skin color. Fibroblasts form the lower dermal layer and provide strength and resilience (MacNeil, 2007).

The predominant function of tissue-engineered skin is to restore barrier function to patients in whom this has been severely compromised, as in the cases of burns, soft tissue traumas, skin necrosis, scars, congenital giant nevus, and skin tumors. Methods for tissue-engineering skin include

1. Cells delivered on their own
2. Cells delivered within 2D or 3D biomaterials
3. Biomaterials for replacement of the skin's dermal layer (both with and without cells)
4. Biomaterials/scaffolds to support the replacement of both the epidermis and dermis (MacNeil, 2008)

For the treatment of deep wounds such as full-thickness burns, where the epidermis and all of the dermis is lost, it is necessary to replace both epidermal and dermal layers of the skin. For such cases, the tissue-engineered skin substitute should be full thickness, comprising both layers. For the reconstruction of such full-thickness skin equivalents, a 3D dermal scaffold is required to support the growth of fibroblasts and synthesis of new extracellular matrix (ECM). The general approach in full-thickness skin tissue engineering is first to design a suitable biocompatible, porous 3D scaffold with good mechanical properties. This scaffold is then seeded with fibroblasts, where they synthesize several types of collagen, glycoproteins, glycosaminoglycans of human ECM, and thus induce a remodeling of the initial matrix (Berthod et al., 1993). The resulting living dermal equivalent could be used either to prepare the wound for epidermalization in the treatment of burns, or as a bioactive tissue releasing growth factors in the treatment of chronic wounds (Damour et al., 1994; Braye et al., 2001). This dermal equivalent is epidermalized by keratinocytes to obtain a full-thickness skin equivalent. The culture of keratinocytes on top of the dermal equivalent and at an air–liquid interface gives rise to a fully differentiated stratified epidermis. The air–liquid interface mimics the *in vivo* environment and is achieved by placing the skin equivalents on semipermeable membranes such that the keratinocytes are directly exposed to air and ambient oxygen concentration, while the underlying dermis is in contact with the nutrient medium absorbed by the membrane. This configuration promotes epidermal differentiation. The quality of the dermal equivalent determines the quality of the multistratified epidermis (Auxenfans et al., 2009), and the quality of the former is very much dependent on the scaffold. The scaffold aims to mimic the natural ECM by providing volume and sites for cell attachment, proliferation, migration, and synthesis of new ECM. Like the natural ECM itself, the scaffold modulates the phenotype of different cell types involved, their gene expression, changes at proteome of the seeded cells, and function of the seeded cells (Kamel et al., 2013).

At the cell–scaffold interface, both an appropriate physical and a chemical environment profoundly affect the overall behavior of the engineered tissue (Mata et al., 2007). Cellular behavior and subsequent tissue development at the cell–scaffold interface involve adhesion, motility, proliferation, differentiation, and functional maturity (Johnson, 2014). In skin tissue engineering, the physicochemical properties of the cell–scaffold interface, such as surface biochemistry, nature of the biopolymer, protein immobilization, hydrophilicity/hydrophobicity, surface charge, surface topography, porosity, three-dimensionality, and mechanical properties, all influence cellular response (Johnson, 2014). Therefore, it is crucial to control and engineer the cell–scaffold interface by investigating the physicochemical properties that would enhance specific and desirable cell behaviors. The anticipated outcome of this research would be the development of a bioactive soft tissue scaffold for skin tissue engineering. This chapter will discuss physicochemical parameters that influence cell behavior at the cell–scaffold interface and current approaches and technologies used to modify this interface to enhance the cellular response in tissue engineering of skin.

6.2 CHEMICAL PROPERTIES OF THE CELL–SCAFFOLD INTERFACE IN SKIN TISSUE ENGINEERING

6.2.1 BIOPOLYMER CHEMISTRY AT THE CELL–SCAFFOLD INTERFACE

The nature of a polymer surface has important consequences for cell function and significant implications for soft tissue engineering. Bulk chemistry of the scaffold is an important parameter in determining the biocompatibility of the scaffold. It can control cytotoxicity, as most skin tissue engineering scaffolds are made of biodegradable polymers and must eventually release the by-products of their degradation. For treatment of deep skin defects, autografts are still the gold standard (Priya et al., 2008). Therefore, skin tissue engineers aim to develop cultured skin substitutes that match the quality of auto skin grafts.

Scaffolds used in the construction of such skin substitutes can be broadly grouped as natural scaffolds and synthetic scaffolds according to the origin of the polymeric material used. Natural scaffolds are either cadaver or animal derived de-epithelialized acellular matrices or they are mostly constructed using natural polymers extracted from animals. Natural polymers have the advantage of responding to the environment via degradation and remodeling through the action of the enzymes. They are also generally nontoxic, even at high concentrations (Dang and Leong, 2006). The first artificial dermis was made using collagen (Yannas and Burke, 1980), and the majority of biomaterials in clinical use for skin regeneration today are based on natural or extracted collagen (MacNeil, 2008).

Collagen has been the material of choice for skin tissue engineering scaffolds due to its high biocompatibility and biodegradability. In addition, it is adhesive, fibrous, cohesive, and can be used in combination with other materials. On the other hand, it might be antigenic through telopeptides, though it is possible to remove these small telopeptides proteolytically before use (Glowacki and Mizuno, 2008). Collagen is the most abundant protein in all animals, being the predominant component of the ECM (Shoulders and Rainers, 2009). One-third of total protein in humans and three-quarters of the dry weight of skin is collagen. The source of collagen, bovine, avian, or porcine, used in dermal scaffold construction, does not affect the behavior of human dermal fibroblasts seeded in the scaffold or the mechanical properties of the resulting dermal equivalent (Parenteau-Bareil et al., 2011). Other natural polymers used to construct dermal scaffolds are hyaluronic acid, chondroitin sulfate, gelatin, elastin, chitosan, silk, and fibronectin. Plant extracts have recently emerged as a new class of biomaterials for skin tissue engineering scaffolds, promoting human dermal fibroblast growth and epidermal differentiation of adipose-derived stem cells (Jin et al., 2013). Synthetic scaffolds have also been widely investigated for skin tissue engineering, mainly to improve the mechanical stability of the scaffold against enzymatic degradation. Currently, nonwoven polygalactic acid scaffolds (Dermagraft and Transcyte) are in clinical use for skin replacement. Others, such as poly(ε-caprolactone) (PCL), poly(glycolic acid) (PGA), poly(ethyleneglycol-terephthalate) (PEGT), poly (butylene terephthalate) (PBT), poly(ethylene oxide) (PEO), and poly(ethylene glycol) (PEG), are being evaluated *in vitro* and *in vivo*. These scaffolds are mostly in the form of highly porous solid foams, which will be discussed later in this chapter.

Hydrogels are insoluble hydrophilic polymeric networks having high water content and soft tissue like mechanical properties that make them highly attractive scaffolds. For example, dimethylaminoethyl methacrylate:methacrylic acid hydrogels were shown to promote wound healing and reduce wound contraction, a significant complication in burn wound healing (Bullock et al., 2010). In another study, application of pullulan-collagen hydrogels to murine excisional wounds resulted in improved early cutaneous wound healing (Wong et al., 2011).

A relatively new and advanced class of polymers for tissue engineering applications is "recombinant polymers," which are proteins designed using recombinant DNA technology and contain desired peptide sequences. They can be shaped into 3D scaffolds, thin films, or fibers. Elastin-like recombinant polymers (ELRs) form a subclass of these biocompatible protein-based polymers, especially suitable for soft tissue engineering. They are composed of the pentapeptide repeat Val-Pro-Gly-Xaa-Gly (VPGXG), which is derived from the hydrophobic domain of tropoelastin and where X represents any natural or modified amino acid, except proline (Chilkoti et al., 2006). Elastin-like recombinant protein fiber mats were shown to support adhesion and proliferation of normal human skin fibroblasts (Rnjak et al., 2009; Machado et al., 2013). Moreover, it is possible to enrich these polypeptides with short peptides having specific bioactivity, which are easily inserted into the polymer sequence. The first active peptides inserted in the polymer chain were the well-known general-purpose cell adhesion tripeptide RGD (R = L-arginine, G = glycine and D = L-aspartic acid) and the REDV (E = L-glutamic acid and V = L-valine), which is specific to endothelial cells. The resulting bioactivated (VPGVG) derivatives, especially those based on RGD, showed a high capacity to promote cell attachment (Rodriguez-Cabello et al., 2007). Previous studies showed that the addition of ELR containing RGD sequences to a collagen-based scaffold significantly increased the proliferation of both oral fibroblasts and keratinocytes, and the thickness of the epithelium formed by the latter in a 3D tissue equivalent (Figure 6.1) (Kinikoglu et al., 2011a,b). In these studies, an ELR engineered to contain the cell adhesion sequence RGD was blended with collagen type I isolated from rat tails. The mixture was either freeze-dried or electrospun to obtain a 3D porous scaffold that was cross-linked by dehydrothermal treatment (DHT) to increase its mechanical stability. Control scaffolds were prepared in the same way except for the addition of ELR. The scaffolds were seeded with primary human

Oral mucosal equivalent based on
ELR-collagen scaffold

Oral mucosal equivalent based on
control collagen scaffold

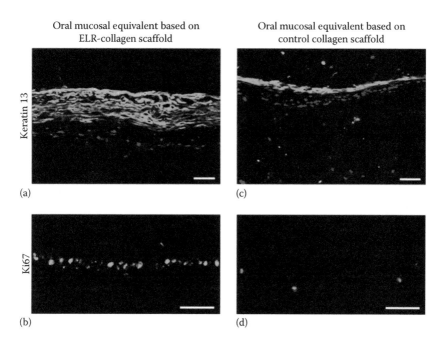

Keratin 13

Ki67

(a)   (c)

(b)   (d)

FIGURE 6.1 Influence of the ELR on the thickness of the reconstructed epithelium and on the expression of keratin 13. (a) Oral mucosal equivalent based on the ELR-collagen electrospun scaffold had a thick epithelium containing a high number of proliferative basal cells Ki67 positive (b), (c) in the control collagen electrospun scaffold the epithelium was thin, containing a few cell layers, and a few proliferative basal cells Ki67 positive (d). Keratin 13 was expressed strongly in both. Immunolabeling is shown in green, cell nuclei in red. Scale bars = 50 µm. (From Kinikoglu, B. et al., *J. Mater. Sci. Mater. Med.*, 22(6), 1541, June 2011a. With permission.)

oral fibroblasts and cultured for 3 weeks under submerged conditions. During this period, fibroblasts migrated through the thickness of the scaffold, proliferated, and filled the pores of the scaffold with newly synthesized ECM, such as collagen type I, which was detected by using transmission electron microscopy (TEM). At the end of 3 weeks, primary human oral epithelial cells were seeded on top, lifted to an air–liquid interface where they proliferated, formed a well-organized and continuous basement membrane expressing laminin 332 and a multistratified epithelium expressing keratin 13. The epithelium formed on scaffolds containing ELR was thicker compared to the controls and all of its basal cells were in proliferative stage as in native tissues, shown by immunostaining against Ki67, marker of proliferative cells. Normally, these basal epithelial cells lose their proliferative capacity during *in vitro* culture (Tomakidi et al., 1998). However, in the tissue equivalent based on the ELR-containing scaffold, the basal cells retained their proliferative capacity after 6 weeks of *in vitro* culture, suggesting that the epithelium would still be able to self-renew when transplanted *in vivo*. These studies have shown the promise of bioengineered polymers such as ELRs as scaffolds for soft tissue engineering.

6.2.2 Protein Immobilization at the Cell–Scaffold Interface

Polymeric scaffolds can be made more suitable for cell attachment and growth by surface modification. Various peptide sequences derived from ECM molecules such as vitronectin, laminin, collagen, and fibrinogen have also been used to mediate cell attachment (LeBaron and Athanisou, 2000). Among these, laminin 332 (formerly termed laminin 5) is of special interest for skin tissue engineering since it is the major component of the basement membrane of skin, synthesized by keratinocytes for attachment. Its interaction with integrins plays important roles in the adhesion, proliferation, and migration of skin cells (Tsuruta et al., 2008). Recently, it was shown that dermal scaffolds tethered with laminin 332 α3 promoted cell adhesion, migration, and proliferation, and they enhanced wound healing *in vivo* (Damodaran et al., 2013).

Peptides and proteins are not the only molecules to promote cell attachment on scaffold surfaces. Mammalian cells have considerable amount of carbohydrates on their surfaces, which can act as ligands for lectins on different mammalian cell types (Atala and Lanza, 2002). This interaction between carbohydrates and their lectin receptors regulates cell migration and attachment to each other. For example, heparin-coated polystyrene surfaces showed higher or comparable growth rate for fibroblasts than fibronectin-coated and gelatin-coated surfaces, retaining the bioactivity of molecules such as heparin-binding fibroblast growth factor 2 (FGF-2) (Ishihara et al., 2000).

Collagen is the major component of the skin dermis, and collagen type I is known to contain both RGD and DGEA (aspartic acid–glycine–glutamic acid–alanine) sequences known to interact with integrins. Therefore, it has been investigated as a coating material for dermal scaffolds. Gautam and colleagues modified the surface of PCL/gelatin dermal scaffolds by collagen type I immobilization on the surface of the scaffold (Gautam et al., 2014). Mouse fibroblasts seeded on these scaffolds showed good attachment, high proliferation and viability. Likewise, coating of PCL/collagen nanofibrous matrices with a thin layer of type I collagen gel significantly stimulated skin keratinocyte adhesion, proliferation, and migration by regulating the activation and distribution of integrin β1 and the downstream effectors of Rac1 and Cdc42, facilitating the deposition of laminin-332, and promoting the expression of active MMPs (MMP-2 and 9) (Fu et al., 2014). Gelatin coating of poly(3-hydroxybutyric acid) (PHB) scaffolds also supported the growth of human skin fibroblasts and keratinocytes (Nagiah et al., 2013).

Certain short amino acid sequences, identified by analysis of active fragments of ECM molecules, appear to bind to receptors on cell surfaces and mediate cell adhesion. A large number of ECM proteins (fibronectin, collagen, vitronectin, thrombospondin, tenascin, laminin, and entactin) contain the RGD sequence (Lanza et al., 2014). Since RGD sequence is very critical for the adhesion of many cell types to ECM, researchers have examined the addition of this peptide to tissue engineering scaffolds. RGD grafting enhances the adhesion of both human skin fibroblasts and skin keratinocytes to substrates *in vitro* (VandeVondele et al., 2003; Wang et al., 2006). *In vivo*, its addition into a polymeric scaffold was shown to enhance the rate and the quality of dermal wound healing with an increase in cellularity and ECM organization compared to controls (Waldeck et al., 2007). In another study, RGD-conjugated multifunctional peptide fibrils promoted human dermal fibroblast spreading with well-organized actin stress fibers and focal contacts (Ohga et al., 2009).

6.2.3 Hydrophobicity/Hydrophilicity at the Cell–Scaffold Interface

The surface hydrophobicity/hydrophilicity is well known as a key factor to govern cell response. On hydrophilic surfaces, cells generally show good spreading, proliferation, and differentiation (Chang and Wang, 2011). In skin tissue engineering, several approaches were tested in order to increase the hydrophilicity of the scaffold at the cell–scaffold interface. For example, the surface hydrophilicity of the poly(lactic acid) scaffolds was enhanced by the addition of 10% hydrophilic PEG (Hendrick and Frey, 2014). The structural morphology of the resulting PEGylated scaffolds was preserved. The surface hydrophobicity can be assessed by measuring contact angle through water spread of a sessile droplet on a surface. The lower the contact angle, the more hydrophilic the surface is. Fibroblasts were found to have maximum adhesion when contact angles were between 60° and 80° (Tamada and Ikada, 1993; Wei et al., 2009). An ideal dermal scaffold should possess moderately hydrophilic surface that would improve the initial adhesion and growth of keratinocytes and fibroblasts (Lee and Kim, 2013). To study the effect of hydrophilicity on cell adhesion, growth, and collagen synthesis, rat skin fibroblasts were seeded onto the surfaces of 13 different polymeric materials (Tamada and Ikada, 1994). These surfaces had varying surface energies ranging from very hydrophobic to very hydrophilic. Surface energy derives from the bonding potential of molecules at a surface and closely linked with surface hydrophobicity. The group found out that cell adhesion occurred on all surfaces except a few very hydrophobic ones, and the higher rates of collagen synthesis were observed on the most hydrophobic surfaces. Collagen synthesis is an essential function of skin fibroblasts and tissue engineers are interested in promotion of such cell-specific functions. It should be noted that cell proliferation was not affected by the hydrophilicity of the substrate (Lanza et al., 2014). However, migration of the surface-attached fibroblasts was found to be dependent on surface chemistry (Saltzman et al., 1991).

Several methods have been used to increase the scaffold surface hydrophilicity. For example, modification of PS or PET by radiofrequency plasma deposition was shown to enhance attachment and spreading of fibroblasts (Chinn et al., 1989). In another study, the anhydrous ammonia plasma treatment of a PLA dermal scaffold increased the surface hydrophilicity and the attachment of human skin fibroblasts with a 99% cell seeding efficiency (Yang et al., 2002). A similar positive effect on the adhesion and growth of HaCaT (human adult low calcium high temperature) keratinocytes was observed when the nanofibrous polymeric scaffolds were treated with oxygen plasma (Bacakova et al., 2014). Addition of hydrophilic macromolecules was also used as a means to increase the surface hydrophilicity of the dermal scaffolds. These molecules can transform the current biodegradable polymers used in biomedical applications into hydrophilic ones, causing the suppression of non-specific protein adsorption on the polymer surface. The subsequent covalent attachment of cell-adhesion–mediating peptides to the hydrophilic surface promotes

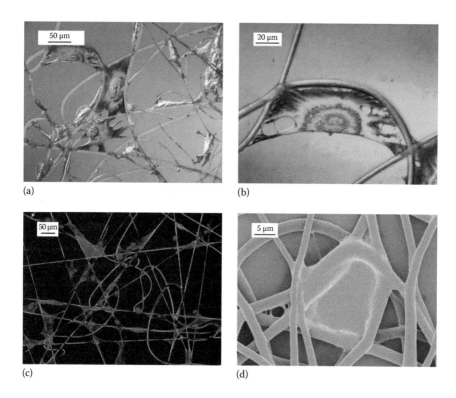

FIGURE 6.2 Fibroblast adhesion on **RGDS**-modified fibers. (a) Optical microscope image of human dermal fibroblasts after 24 h in cell culture on electrospun and **GRGDS**-functionalized PLGA/sP(EO-stat-PO) fibers. (b) A magnification of one cell. (c) Fluorescence microscope image (nuclei blue, actin filaments red) of human dermal fibroblasts after 24 h in cell culture on GRGDS-functionalized PLGA/sP(EO-stat-PO) fibers. (d) SEM image of a single cell on GRGDS-functionalized PLGA/sP(EO-stat-PO) fibers. (From Grafahrend, D. et al., *Nat. Mater.*, 10, 67, 2011. With permission.)

specific bioactivation and enables adhesion of cells through exclusive recognition of the immobilized binding motifs (Figure 6.2) (Grafahrend et al., 2011). For example, when chitosan was incorporated into a PCL scaffold, its hydrophilicity significantly increased, resulting in good fibroblast attachment and proliferation, making it suitable for skin tissue engineering (Shalumon et al., 2011). Likewise, addition of hyaluronan transformed the PCL fibers into hydrophilic ones, thus enhancing the infiltration and proliferation of human skin fibroblasts *in vitro* and enhancing tissue in-growth *in vivo* (Li et al., 2012).

Like surface hydrophobicity, surface charge was also shown to influence the behavior of cells. In fact, tissue culture polystyrene surfaces used for cell culture are obtained by the surface treatment of polystyrene by glow discharge or exposure to chemicals such as sulfuric acid so that the number of charged groups at the surface increases and so does the number of cells attached to that surface. Surface charge is important for skin tissue engineering. Human dermal fibroblasts were observed to adhere, spread, and proliferate better on cationic surfaces compared to neutral base surface (De Rosa et al., 2004). In addition, the presence of cationic charges on cell adhesion–resistant neutral surface increased the synthesis of collagen I and III, the release of their metabolites, and the expression of their mRNA by dermal fibroblasts. Interestingly, the scarce collagen deposits on neutral poly(hydroxyethyl methacrylate) (pHEMA) polymer consisted, for the most part, of collagen I while collagen III was present only in trace amounts probably due to the secretion of metalloproteinase-2 by nonadherent fibroblasts. These findings indicate that surface charge and hydrophilicity should be considered when designing the cell–scaffold interface for skin tissue engineering applications.

6.3 PHYSICAL PROPERTIES OF THE CELL–SCAFFOLD INTERFACE IN SKIN TISSUE ENGINEERING

6.3.1 POROSITY AND PORE SIZE AT THE CELL–SCAFFOLD INTERFACE

Porosity is an important property of a dermal scaffold that should possess an optimum pore size and distribution to allow fibroblast infiltration and proliferation, and also cell communication and medium perfusion. High-porosity scaffolds were reported to support active dermal fibroblast migration and infiltration into the scaffold. A dermal scaffold suitable for intrinsic vascularization must have a high porosity (>40%–60%) and an interconnected pore structure (Will et al., 2008). Cellular infiltration is particularly of importance and pore size, porosity, and pore interconnectivity dictate the extent of cellular infiltration and tissue in-growth into the scaffold. They influence a range of cellular processes and are crucial for diffusion of nutrients, metabolites, and waste products. Dermal substitute scaffolds are expected to promote dermal fibroblast adhesion, growth, and infiltration as the presence of fibroblasts in dermal substitutes accelerates and enhances dermal and epidermal regeneration (Rnjak-Kovacina and Weiss, 2011).

For different cell types, there are suitable pore sizes to accommodate their biological activity. It was reported that the scaffold pore size should be in the range of 20–150 μm for optimal skin regeneration (Maquet and Jerome, 1997). Likewise, O'Brien and colleagues have shown that the critical pore size range of collagen-based dermal scaffolds allowing optimal cellular activity and simultaneous blocking of wound contraction was between 20 and 120 μm (O'Brien et al., 2005). Furthermore, considering skin regeneration and wound healing, a porosity >90% was found to be ideal for dermal tissue engineering (Wang et al., 2013). Compared to nonporous scaffolds or sham wounds, dermal scaffolds with adequate porosity induced accelerated wound closure and stimulated regeneration of healthy dermal tissue, evidenced by a more normal-appearing matrix architecture, blood vessel in-growth, and hair follicle development (Bonvallet et al., 2014).

Scaffold porosity can be analyzed qualitatively by using low-voltage (1.3 kV) scanning electron microscopy (SEM) and quantitatively by using mercury intrusion porosimetry (MIP) to determine pore size distribution, specific pore area, median pore diameter, and porosity (Atala and Lanza, 2002; Kinikoglu et al., 2011b). Porosity can be introduced into polymeric scaffolds by phase separation, freeze-drying, salt leaching, and a variety of other methods. It is now possible to make porous, biodegradable scaffolds with controlled pore size and architectures and oriented pores (Saltzman, 2004). Freeze-drying of aqueous solutions of natural biopolymers such as collagen has been reported for the production of well-defined porous matrices, pore sizes and orientation, achieved by the controlled growth of ice crystals during the freeze-drying process (Chen et al., 2002). In this process, the solution to be frozen contains the polymer such as collagen and the solvent; freezing traps the polymer in the spaces between the growing ice crystals and forms a continuous interpenetrating network of ice and the polymer. A reduction in the chamber pressure causes the ice to sublimate, leaving behind the polymer as a highly porous foam (Figure 6.3). Freezing temperature, solute and polymer concentration were shown to strongly influence the porous structure of the scaffold obtained by freeze-drying. Freezing of a collagen solution in a −20°C freezer resulted in larger pore sizes than fast freezing using a mixture of dry ice and ethanol (−80°C), and the most rapid freezing procedure, using liquid nitrogen, lead to the smallest pores (−196°C) (Faraj et al., 2007). When the freezing temperature was kept constant, and the collagen was dissolved either in water or in acetic acid, it was observed that the morphology of a scaffold from a collagen suspension in water displayed more thin thread-like structures than a scaffold from a collagen suspension in diluted acetic acid. The walls of the pores and lamellae were more compact and smoother in the diluted acetic acid scaffold (Faraj et al., 2007). The same authors showed that the addition of ethanol (2.8%) in a collagen solution resulted in closed surfaced foams. Solute concentration was also shown to influence the pore size in scaffolds produced by freeze-drying. An inverse relationship was found between collagen concentration and pore size (Madaghiele et al., 2008).

Electrospinning is another technique to create relatively porous, fibrous scaffolds (Figure 6.3); however, it should be noted that cells cultured on electrospun scaffolds may not always penetrate into the scaffold and may accumulate at the surface due to short distances between the fibers of these scaffolds. But even this may be acceptable because the cells may receive nutrients and growth cues from the 3D structure, whereas the cells on 2D surfaces do not have this opportunity (Nisbet et al., 2009). Besides, it is possible to increase the porosity of these electrospun scaffolds, whereas it is not possible to do it on 2D scaffolds. Two techniques were used in skin tissue engineering to increase the pore size of the electrospun scaffolds for better infiltration of dermal fibroblasts: increasing the fiber diameter, which leads to greater pore size, up to 11.8 μm for gelatin fibers (Powell and Boyce, 2008) and 27.9 μm for tropoelastin fibers (Rnjak-Kovacina et al., 2011); and the use of a rotating mandrel collector, which increased the pore size of PLGA fibers up to 132 μm (Zhu et al., 2008). It was possible to increase the fiber diameter of synthetic human elastin scaffolds by increasing the flow rate (from 1 to 3 mL/h), which resulted in greater average pore size and more than two-fold increase of overall scaffold porosity (Rnjak-Kovacina et al., 2011).

6.3.2 Nanoscale Topography at the Cell–Scaffold Interface

Cells cultured in 3D environments behave differently than those cultured in a 2D environment, adopting more *in vivo* like morphologies. The architecture affects the cell–receptor ligation, intercellular signaling, cell migration, and also the diffusion

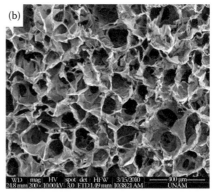

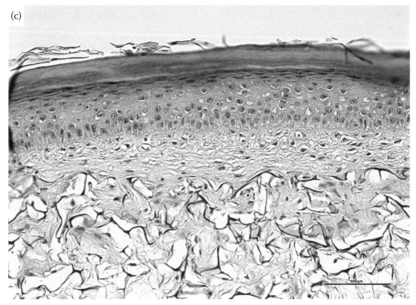

FIGURE 6.3 (a) Scanning electron microscopic (SEM) analysis of an electrospun, porous collagen-based dermal scaffold (a), and a freeze-dried, porous collagen-based dermal scaffold (b). (c) Histological analysis of a tissue-engineered skin based on a freeze-dried, porous collagen-based scaffold. Cell nuclei were stained in blue by hematoxylin, cytoplasm in pink by phloxine, and extracellular matrix of connective tissue in orange/yellow by saffron. In the tissue-engineered skin, fibroblasts seeded into the porous, collagen-based foam migrated into foam, proliferated, and finally populated the foam. The pores were filled with newly synthesized extracellular matrix. Keratinocytes formed a keratinized, pluristratified epithelium on the surface, completely differentiated with a stratum corneum.

and adhesion of proteins, growth factors, and enzymes needed for cell survival and function (Nisbet et al., 2009). The 3D fibrous scaffolds composed of nanoscale multifibrils prepared with the aim of mimicking the supramolecular architecture and the biological functions of the natural ECM as much as possible have attracted a great deal of attention in skin tissue engineering. They have shown great potential to mimic skin ECM (which has fibers in the range of 10–50 nm) in both morphology and composition, and many studies using fibrous, electrospun dermal scaffolds have yielded promising results. Interfiber distances between 5 and 10 µm appear to yield the most favorable skin substitute *in vitro*, demonstrating high cell viability, optimal cell organization, and good barrier formation (Powell and Boyce, 2008). Fibroblasts appear to respond more to surface topography compared to keratinocytes. Fibroblasts, unlike keratinocytes, were observed to orient on grooved surfaces, particularly when the texture dimensions were 1–8 µm (Brunette, 1986; Dunn and Brown, 1986).

The fibrous electrospun scaffolds used in the construction of skin equivalents were found to be superior to freeze-dried foams in terms of cellular organization and reduced wound contraction (Powell and Boyce, 2008). These advantages are expected to lead to reduced morbidity in patients treated with fibrous skin substitutes. Another study pointed out that collagen nanofibrous matrices were very effective as wound-healing accelerators in early-stage wound repair (Rho et al., 2006). The authors reported that cross-linked collagen nanofibers coated with ECM proteins, particularly type I collagen, might be a good candidate for skin tissue engineering applications, such as wound dressings and dermal scaffolds. Combination of collagen with PCL (70:30) yielded biocompatible dermal scaffolds with good tensile strength that degraded within 3–4 weeks post implantation, which is an optimal time frame for degradation and wound healing *in vivo* (Bonvallet et al., 2014). Other noncollagenous nanofibrous materials were also shown to be effective as skin substitutes. Indeed, high cell attachment and spreading of human oral keratinocytes and fibroblasts were observed on nanofibrous chitin scaffolds, and the cellular response was even higher when the scaffold was treated with collagen type I (Noh et al., 2006). PLGA–PLLA electrospun scaffolds were able to support keratinocyte, fibroblast, and endothelial cell growth, and ECM production (Blackwood et al., 2008). Nanofibrous scaffolds based on collagen/silk fibroin, carboxyethyl chitosan/poly(vinyl alcohol), gelatin, PLGA/chitosan, and poly(ε-caprolactone) were also found to promote keratinocyte and/or fibroblast attachment and proliferation, indicating the potential of nanofibrous mats as future wound dressings for skin regeneration.

6.4 CONCLUSION

Skin tissue engineering is a maturing field, and there is still room for improvement. It has been shown that the cell–scaffold interface plays an important role in regulating cell behavior and affecting the overall behavior of the tissue-engineered skin. The physicochemical properties of the cell–scaffold interface, such as surface biochemistry, nature of the biopolymer, protein immobilization, hydrophilicity/hydrophobicity, surface charge, surface topography, porosity, three-dimensionality, and mechanical properties have been shown to influence cellular response. Therefore, in the future, it will be important to design new biodegradable scaffolds by engineering and controlling this interface to enhance specific and desirable cell behavior and also rapid vascularization, which is a major challenge for the tissue engineering of skin. Tissue-engineered skin models have improved over the years by the incorporation of different cell types such as fibroblasts, keratinocytes, melanocytes, Langerhans cells, endothelial cells, and stem cells. For the future, advanced biomaterials and scaffolds interacting specifically with each cell type and a 3D environment that mimics the ECM will be a core challenge and a prerequisite for the organization of living cells to functional skin tissue that would be useful for the treatment of burns and chronic wounds.

REFERENCES

Atala A, Lanza RP. *Methods of Tissue Engineering.* San Diego, CA: Academic Press, Elsevier, 2002.

Auxenfans C, Fradette J, Lequeux C, Germain L, Kinikoglu B, Bechetoille N, Braye F, Auger FA, Damour O. Evolution of three dimensional skin equivalent models reconstructed in vitro by tissue engineering. *Eur J Dermatol* March–April 2009;19(2):107–113.

Bacakova M, Lopot F, Hadraba D, Varga M, Zaloudkova M, Stranska D, Suchy T, Bacakova L. Effects of fiber density and plasma modification of nanofibrous membranes on the adhesion and growth of HaCaT keratinocytes. *J Biomater Appl* 2015;29(6):837–853.

Berthod F, Hayek D, Damour O, Collombel C. Collagen synthesis by fibroblasts cultured within a collagen sponge. *Biomaterials* 1993;14:749–754.

Blackwood KA, McKean R, Canton I, Freeman CO, Franklin KL, Cole D, Brook I, Farthing P, Rimmer S, Haycock JW, Ryan AJ, MacNeil S. Development of biodegradable electrospun scaffolds for dermal replacement. *Biomaterials* 2008;29:3091–3104.

Bonvallet PP, Culpepper BK, Bain JL, Schultz MJ, Thomas SJ, Bellis SL. Microporous dermal-like electrospun scaffolds promote accelerated skin regeneration. *Tissue Eng A* September 2014;20(17–18):2434–2445.

Böttcher-Haberzeth S, Biedermann T, Reichmann E. Tissue engineering of skin. *Burns* 2010;36:450–460.

Braye FM, Stefani A, Venet E, Pieptu D, Tissot E, Damour O. Grafting of large pieces of human reconstructed skin in a porcine model. *Br J Plast Surg* 2001;54:532–538.

Brunette D. Fibroblasts on micromachined substrata orient hierarchically to grooves of different dimensions. *Exp Cell Res* 1986;164:11–26.

Bullock AJ, Pickavance P, Haddow DB, Rimmer S, MacNeil S. Development of a calcium-chelating hydrogel for treatment of superficial burns and scalds. *Regen Med* January 2010;5(1):55–64.

Chang HI, Wang Y. Cell responses to surface and architecture of tissue engineering scaffolds, in: D. Eberli (Ed.), *Regenerative Medicine and Tissue Engineering—Cells and Biomaterials*, InTech, Rijeka, Croatia, 2011, ISBN: 978-953-307-663-8.

Chen G, Ushida T, Tateishi T. Scaffold design for tissue engineering. *Macromol Biosci* 2002;2:67–77.

Chilkoti A, Christensen T, MacKay JA. Stimulus responsive elastin biopolymers: Applications in medicine and biotechnology. *Curr Opin Chem Biol* 2006;10:652–657.

Chinn J, Horbett T et al. Enhancement of serum fibronectin adsorption and the clonal plating efficiencies of Swiss mouse 3T3 fibroblast and MM14 mouse myoblast cells on polymer substrates modified by radiofrequency plasma deposition. *J Colloid Interf Sci* 1989;127:67–87.

Damodaran G, Tiong WH, Collighan R, Griffin M, Navsaria H, Pandit A. In vivo effects of tailored laminin-332 α3 conjugated scaffolds enhances wound healing: A histomorphometric analysis. *J Biomed Mater Res A* October 2013;101(10):2788–2795.

Damour O, Gueugniaud PY, Berthin-Maghit M, Rousselle P, Berthod F, Sahuc F et al. A dermal substrate made of collagen–GAG–chitosan for deep burn coverage: First clinical uses. *Clin Mater* 1994;15:273–276.

Dang JM, Leong KW. Natural polymers for gene delivery and tissue engineering. *Adv Drug Deliv Rev* 2006;58:487–499.

De Rosa M, Carteni M, Petillo O, Calarco A, Margarucci S, Rosso F et al. Cationic polyelectrolyte hydrogel fosters fibroblast spreading, proliferation, and extracellular matrix production: Implications for tissue engineering. *J Cell Physiol* January 2004;198(1):133–143.

Dunn GA, Brown AF. Alignment of fibroblasts on grooved surfaces described by a simple geometric transformation. *J Cell Sci* 1986;83:313–340.

Faraj KA, van Kuppevelt TH, Daamen WF. Construction of collagen scaffolds that mimic the three-dimensional architecture of specific tissues. *Tissue Eng* 2007;13:2387–2394.

Fu X, Xu M, Liu J, Qi Y, Li S, Wang H. Regulation of migratory activity of human keratinocytes by topography of multiscale collagen-containing nanofibrous matrices. *Biomaterials* February 2014;35(5):1496–1506.

Gautam S, Chou CF, Dinda AK, Potdar PD, Mishra NC. Surface modification of nanofibrous polycaprolactone/gelatin composite scaffold by collagen type I grafting for skin tissue engineering. *Mater Sci Eng C: Mater Biol Appl* January 1, 2014;34:402–409.

Glowacki J, Mizuno S. Collagen scaffolds for tissue engineering. *Biopolymers* 2008;89:338–344.

Grafahrend D, Heffels KH, Beer MV, Gasteier P, Möller M, Boehm G et al. Degradable polyester scaffolds with controlled surface chemistry combining minimal protein adsorption with specific bioactivation. *Nat Mater* 2011;10:67–73.

Groeber F, Holeiter M, Hampel M, Hinderer S, Schenke-Layland K. Skin tissue engineering—*In vivo* and *in vitro* applications. *Adv Drug Deliv Rev* 2011;128:352–366.

Hendrick E, Frey M. Increasing surface hydrophilicity in poly(lactic acid) electrospun fibers by addition of Pla-b-Peg Co-polymers. *J Eng Fibers Fabrics* 2014;9:153–164.

Ishihara M, Saito Y, Yura H, Ono K, Ishikawa K, Hattori H et al. Heparin-carrying polystyrene to mediate cellular attachment and growth via interaction with growth factors. *J Biomed Mater Res* May 2000;50(2):144–152.

Jin G, Prabhakaran MP, Kai D, Annamalai SK, Arunachalam KD, Ramakrishna S. Tissue engineered plant extracts as nanofibrous wound dressing. *Biomaterials* January 2013;34(3):724–734.

Johnson JK. Nanofiber scaffolds for biological structures. U.S. Patent US20140030315 A1, 2014.

Kamel RA, Ong JF, Eriksson E, Junker JP, Caterson EJ. Tissue engineering of skin. *J Am Coll Surg* September 2013;217(3):533–555.

Kinikoglu B, Rodríguez-Cabello JC, Damour O, Hasirci V. A smart bilayer scaffold of elastin-like recombinamer and collagen for soft tissue engineering. *J Mater Sci Mater Med* June 2011a;22(6):1541–1554.

Kinikoglu B, Rodríguez-Cabello JC, Damour O, Hasirci V. The influence of elastin-like recombinant polymer on the self-renewing potential of a 3D tissue equivalent derived from human lamina propria fibroblasts and oral epithelial cells. *Biomaterials* September 2011b;32(25):5756–5764.

Lanza R, Langer R, Vacanti JP. *Principles of Tissue Engineering.* San Diego, CA: Academic Press, Elsevier Inc., 2014.

LeBaron RG and Athanisou KA. Extracellular matrix cell adhesion peptides: Functional applications in orthopaedic materials. *Tissue Eng* 2000;6:85–103.

Lee YJ, Kim GM. Tunable electrospun scaffolds on cell behaviour for tissue regeneration and drug delivery system. *OA Tissue Eng* May 1, 2013;1(1):5.

Li L, Qian Y, Jiang C, Lv Y, Liu W, Zhong L et al. The use of hyaluronan to regulate protein adsorption and cell infiltration in nanofibrous scaffolds. *Biomaterials* April 2012;33(12):3428–3445.

Machado R, da Costa A, Sencadas V, Garcia-Arévalo C, Costa CM, Padrão J et al. Electrospun silk-elastin-like fibre mats for tissue engineering applications. *Biomed Mater* December 2013;8(6):065009.

MacNeil S. Biomaterials for tissue engineering of skin. *Mater Today* 2008;11:26–35.

MacNeil S. Progress and opportunities for tissue-engineered skin. *Nature* 2007;445:874–880.

Madaghiele M, Sannino A, Yannas IV, Spector M. Collagen-based matrices with axially oriented pores. *J Biomed Mater Res A* 2008;85A:757–767.

Maquet V, Jerome R. Design of macroporous biodegradable polymer scaffolds for cell transplantation. *Mater Sci Forum* 1997;250:15–42.

Mata A, Boehm C, Fleischman AJ, Muschler GF, Roy S. Connective tissue progenitor cell growth characteristics on textured substrates. *Int J Nanomed* 2007;2(3): 389–406.

Nagiah N, Madhavi L, Anitha R, Anandan C, Srinivasan NT, Sivagnanam UT. Development and characterization of coaxially electrospun gelatin coated poly (3-hydroxybutyric acid) thin films as potential scaffolds for skin regeneration. *Mater Sci Eng C: Mater Biol Appl* October 2013;33(7):4444–4452.

Nisbet DR, Forsythe JS, Shen W, Finkelstein DI, Horne MK. A review of the cellular response on electrospun nanofibers for tissue engineering. *J Biomater Appl* 2009;24:7–29.

Noh HK, Lee SW, Kim JM, Oh JE, Kim KH, Chung CP et al. Electrospinning of chitin nanofibers: Degradation behavior and cellular response to normal human keratinocytes and fibroblasts. *Biomaterials* 2006;27:3934–3944.

O'Brien FJ, Harley B, Yannas IV, Gibson LJ. The effect of pore size on cell adhesion in collagen–GAG scaffolds. *Biomaterials* 2005;26:433–441.

Ohga Y, Katagiri F, Takeyama K, Hozumi K, Kikkawa Y, Nishi N et al. Design and activity of multifunctional fibrils using receptor-specific small peptides. *Biomaterials* December 2009;30(35):6731–6738.

Parenteau-Bareil R, Gauvin R, Cliche S, Gariépy C, Germain L, Berthod F. Comparative study of bovine, porcine and avian collagens for the production of a tissue engineered dermis. *Acta Biomater* October 2011;7(10):3757–3765.

Powell HM, Boyce ST. Fiber density of electrospun gelatin scaffolds regulates morphogenesis of dermal-epidermal skin substitutes. *J Biomed Mater Res A* March 15, 2008;84(4):1078–1086.

Priya SG, Jungvid H, Kumar A. Skin tissue engineering for tissue repair and regeneration. *Tissue Eng B: Rev* March 2008;14(1):105–118.

Rheinwald JG, Green H. Serial cultivation of strains of human epidermal keratinocytes: The formation of keratinizing colonies from single cells. *Cell* 1975;6:331–343.

Rho KS, Jeong L, Lee G, Seo BM, Park YJ, Hong SD et al. Electrospinning of collagen nanofibers: Effects on the behavior of normal human keratinocytes and early-stage wound healing. *Biomaterials* 2006;27:1452–1461.

Rnjak J, Li Z, Maitz PK, Wise SG, Weiss AS. Primary human dermal fibroblast interactions with open weave three-dimensional scaffolds prepared from synthetic human elastin. *Biomaterials* November 2009;30(32):6469–6477.

Rnjak-Kovacina J, Weiss AS. Increasing the pore size of electrospun scaffolds. *Tissue Eng B: Rev* 2011;17(5):365–372.

Rnjak-Kovacina J, Wise SG, Li Z, Maitz PK, Young CJ, Wang Y et al. Tailoring the porosity and pore size of electrospun synthetic human elastin scaffolds for dermal substitute bioengineering. *Biomaterials* October 2011;32(28):6729–6736.

Rodriguez-Cabello JC, Prieto S, Reguera J, Arias FJ, Ribeiro A. Biofunctional design of elastin-like polymers for advanced applications in nanobiotechnology. *J Biomater Sci Polymer Edn* 2007;18:269–286.

Saltzman WM. *Tissue Engineering: Engineering Principles for the Design of Replacement Organs and Tissues.* New York: Oxford University Press, Inc., 2004.

Saltzman WM, Parsons-Wingerter P, Leong KW, Lin S. Fibroblast and hepatocyte behavior on synthetic polymer surfaces. *J Biomed Mater Res* 1991;25:741–759.

Shalumon KT, Anulekha KH, Chennazhi KP, Tamura H, Nair SV, Jayakumar R. Fabrication
of chitosan/poly(caprolactone) nanofibrous scaffold for bone and skin tissue engineering. *Int J Biol Macromol* May 1, 2011;48(4):571–576.

Shoulders MD, Raines RT. Collagen structure and stability. *Annu Rev Biochem*
2009;78:929–958.

Tamada Y, Ikada Q. Effect of preadsorbed proteins on cell adhesion to polymer surfaces.
J Colloid Interf Sci 1993;155:334–339.

Tamada Y, Ikada Y. Fibroblast growth on polymer surfaces and biosynthesis of collagen.
J Biomed Mater Res 1994;28:783–789.

Tomakidi P, Breitkreutz D, Fusenig NE, Zöller J, Kohl A, Komposch G. Establishment of
oral mucosa phenotype in vitro in correlation to epithelial anchorage. *Cell Tissue Res*
1998;292:355e66.

Tsuruta D, Kobayashi H, Imanishi H, Sugawara K, Ishii M, Jones JC. Laminin-332-integrin
interaction: A target for cancer therapy? *Curr Med Chem* 2008;15(20):1968–1975.

VandeVondele S, Vörös J, Hubbell JA. RGD-grafted poly-ʟ-lysine-graft-(polyethylene glycol) copolymers block non-specific protein adsorption while promoting cell adhesion.
Biotechnol Bioeng 2003;82(7):784–790.

Waldeck H, Chung AS, Kao WJ. Interpenetrating polymer networks containing gelatin modified with PEGylated RGD and soluble KGF: Synthesis, characterization, and application in in vivo critical dermal wound. *J Biomed Mater Res A* 2007;82(4):861–871.

Wang TW, Wu HC, Huang YC, Sun JS, Lin FH. The effect of self-designed bifunctional
RGD-containing fusion protein on the behavior of human keratinocytes and dermal
fibroblasts. *J Biomed Mater Res B: Appl Biomater* 2006;79(2):379–387.

Wang X, You C, Hu X, Zheng Y, Li Q, Feng Z et al. The roles of knitted mesh-reinforced
collagen-chitosan hybrid scaffold in the one-step repair of full-thickness skin defects
in rats. *Acta Biomater* 2013;9(8):7822–7832.

Wei J, Igarashi T, Okumori N, Maetani T, Liu BS, Yoshinari M. Influence of surface wettability on competitive protein adsorption and initial attachment of osteoblasts. *Biomed
Mater* 2009;4:045002.

Will J, Melcher R, Treul C, Travitzky N, Kneser U, Polykandriotis E et al. Porous ceramic
bone scaffolds for vascularized bone tissue regeneration. *J Mater Sci Mater Med*
2008;19:2781–2790.

Wong VW, Rustad KC, Galvez MG, Neofytou E, Glotzbach JP, Januszyk M et al. Engineered
pullulan-collagen composite dermal hydrogels improve early cutaneous wound healing. *Tissue Eng A* 2011;17(5–6):631–644.

Yang J, Shi G, Bei J, Wang S, Cao Y, Shang Q et al. Fabrication and surface modification of
macroporous poly(ʟ-lactic acid) and poly(ʟ-lactic-co-glycolic acid) (70/30) cell scaffolds for human skin fibroblast cell culture. *J Biomed Mater Res* 2002;62(3):438–446.

Yannas IV, Burke JF. Design of an artificial skin. I. Basic design principles. *J Biomed Mater
Res* 1980;14:65–81.

Zhu X, Cui W, Li X, Jin Y. Electrospun fibrous mats with high porosity as potential scaffolds
for skin tissue engineering. *Biomacromolecules* 2008;9:1795–1801.

7 Cell and Materials Interface in Cryobiology and Cryoprotection

Kazuaki Matsumura, Minkle Jain, and Robin Rajan

CONTENTS

7.1 INTRODUCTION

Nature governs whether biological material will decay or die. The structure and function of living organelles and cells can change and be lost with time, which is a matter of concern for the researchers studying these systems. Several attempts have been made to stop the biological clock since ancient times, which was successfully achieved by controlling temperature and water content.

Refrigeration is one of the everyday life processes that have been extensively used because it provides us the means for slowing the rate of deterioration of perishable goods. Removal of water from various biological materials paves another way for arresting biological degradation, which initiates again by the addition of water.

The pioneering work in this field was conducted in 1949 by Polge and coworkers, who stored fowl semen in a freezer by adding glycerol as cryoprotectant.[1] Afterward, many successful experiments were carried out, such as cryopreservation of bull spermatozoa,[2] plant cultures,[3] plant callus,[4] and human embryos for *in vitro* fertilization programs.[5]

The application of low-temperature preservation to living organisms has revolutionized several areas of biotechnology such as plant and animal breeding. The most interesting feature of cryopreservation is that both prokaryotic and eukaryotic cells can be cryopreserved at temperatures down to −200°C, which is a remarkable milestone for structural and molecular biologists. The most important ingredient required to achieve this goal is a cryoprotectant (CPA).

In the context of tissue cultures, simple preservation techniques like refrigeration cause limited shell life, high risk of contamination, and genetic drift. Therefore, cryopreservation has become indispensable in biological, medical, and agricultural research fields, and in the clinical practice of reproductive medicine. In the era of microbial contaminations, natural disasters, or alteration of genetic expressions in the latter generations, cryopreservation of sperm and embryos helps to maintain a backup of the microorganisms proliferating on animals, thus saving significant space and resources that could be used to better manage the microorganisms currently used for research. Moreover, it is an important tool to preserve strains that are not currently being used but could have potent applications in the future.

7.2 FATE OF CELLS AT ULTRA-LOW TEMPERATURE

How can a man who's warm understand one who's freezing?

Alexander Solzhenistyn

Life is a complex process that happens in water. When the temperature is below −0.6°C, biological water under isotonic conditions becomes thermodynamically unstable and tends to be in the crystalline state. Since biological systems are almost entirely made of water, the water–ice phase transition in these systems is a subject of great interest for cryopreservation. In particular, what are the effects of extremely low temperatures on cells?

7.3   BIOPHYSICAL ASPECTS OF ICE FORMATION

When biological systems are cooled to temperatures below the equilibrium melting point, ice begins to form in the extracellular medium. This extracellular ice plays an important role in the cryopreservation process because it alters the chemical environment of the cells, exerts mechanical constraints, and leads to the development of ice inside the cells.[6] The formation of extracellular ice has synergistic effect on the unfrozen fraction composition in the extracellular solution. Dropping temperature leads to an increase in the solute concentration in the extracellular solution, which is a driving force for the diffusion of solutes into and water flux out of the cell. At low temperature, the plasma membrane, which is more permeable to water than to the solute, behaves like a semipermeable membrane in time scale of cryopreservation.[7] Cells respond to this by releasing water via osmosis and undergo dehydration during freezing, the kinetic model of which was first given by Mazur.[8]

Solidification of the external medium can cause cell deformation because the ice matrix surrounding the cell acts as a mechanical constraint. During freezing, this mechanical force squeezes the cells into the channels of unfrozen liquid between ice crystals. Rapatz et al.[9] have directly measured the width of the unfrozen liquid channels between ice crystals and observed that channels' diameters decrease with temperature and the cells present in the channels get deformed as the channel width reaches the cell dimensions.

Besides these two processes, another effect of extracellular ice onto cells is the initiation of ice formation inside the cells. It has been experimentally proven that extracellular ice catalyzes the intracellular ice formation.[10,11] Toner[10] proposed that ice is formed inside the cells by nucleation on intracellular catalytic sites.

Since the cytoplasmic supercooling and diffusion constant of intracellular water depend on the instantaneous properties of the intracellular solution, the dynamics of ice formation inside the cells are highly affected by the corresponding dehydration process.[6] The fate of the cellular water during cryopreservation depends on the relative magnitudes of water transport and rate of nucleation. When cells are cooled slowly, the rate of water coming out of the cells is relatively fast, thus preventing intracellular ice formation and favoring cell dehydration. At rapid cooling rates, exosmosis of water is slow in comparison of intracellular water being supercooled, thus resulting in intracellular ice formation.

7.4   CORRELATION BETWEEN CRYOINJURY AND THE TWO PHENOMENA OCCURRING DURING FREEZING

Injury occurring because of intracellular ice formation during rapid cooling is believed to be due to mechanical forces.[12] Possible sites of injury are the plasma membrane[13] and the membrane of intracellular organelles.[14]

Cell dehydration during slow cooling is also a source of cell damage.[15] Lovelock[16] reported that hypertonic solutions cause denaturation of lipoproteins, which leads to hemolysis in red blood cells (RBC). Other theories proposed cell shrinkage as a response toward highly concentrated extracellular solution.

The two approaches used for cryopreservation are slow-rate freezing and vitrification. The core objective of the two methods is to minimize cryoinjury, intracellular ice formation, and dehydration.

Slow-rate freezing involves the pre-equilibration of cells in cryoprotectant solutions followed by slow cooling at the rate required for the particular type of cell being used. However, during the whole process, care must be taken to prevent intracellular ice formation. This complete process requires special equipment and takes 3–6 h to complete.

Vitrification is the conversion of liquid into glass. In this approach, an attempt is made to prevent ice formation throughout the entire sample. This process avoids the damaging effects of intra- and extracellular ice formation.

7.5 COOLING RATES

There are various factors affecting the efficiency of cryopreservation. One of the principal factors is the rate of freezing, which should be optimum. The relation between cell survival and cooling rate shows an inverted U-shaped curve (Figure 7.1). Each system has an optimum cooling rate, the efficiency of which is greatly affected by whether the rate of cooling is too fast or too slow.[17,18] When the rate of cooling is very slow, there is minimal intracellular ice formation, which implies a high degree of cell dehydration. On the other hand, at very high cooling rates, rapid water flow through the membrane can result in rough pressure distribution across the membrane[19] in sudden change in size and shape of the membrane.[20]

7.6 SLOW COOLING

When cells are frozen/cooled at slow rate (controlled rate), formation of extracellular ice takes place first, followed by a differential water gradient across the cell membrane, which results in the movement of intracellular water to the outside.

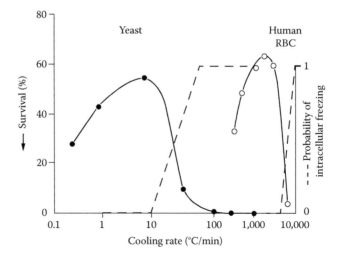

FIGURE 7.1 Plot of the survival percent versus the cooling rate for different cell types of cells. (From Mazur, P., *Cryobiology*, 14, 251, 1977.)

This has an important cryoprotective effect because it reduces the amount of water available to form ice. This process reduces the amount of water inside the cells, which could potentially form ice, thereby protecting the cells. Intracellular ice formation is lethal for cells and is the most important cause for cell death during cryopreservation. As the system is further cooled down, no further crystallization of ice is observed due to a tremendous increase in the viscosity of the unfrozen fraction (solutes), which turns into an amorphous solid lacking any ice crystals. On the other hand, slow cooling results in the increase of the solution effect, which can be damaging to the cells. The amount and rate at which water is lost from the inside of the cells depends on cell permeability; tolerance toward fast cooling is better for more permeable cells than for less permeable cells.[15,21] Interestingly, there is interplay between ice crystal formation and solution effects on cell damage.

Generally, a cooling rate of 1°C/min is preferred. However, there are exceptions to this requirement such as for yeast,[22,23] liver,[24] and higher plant cells,[25] which shrink or become plasmolyzed when the rate of cooling is 1°C/min, but when the rate is increased to about 200°C/min or more, these cells remain in their normal state. In the case of yeast, shrinkage is the result of water loss and not of solutes loss[26,27]; therefore, in these cells, water content is an estimate of the volume of the cell. However, faster cooling rates render the cells unable to maintain equilibrium with the extracellular solution due to the inability of water to leave the cells, which causes intracellular ice formation to preserve the equilibrium (Figure 7.2).

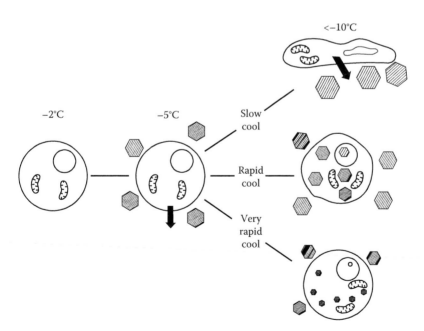

FIGURE 7.2   Schematics of the physical events occurring in cells during freezing. (From Mazur, P., *Cryobiology*, 14, 251, 1977.)

7.7 CRYOPROTECTIVE ADDITIVES

CPAs are additives provided to cells before freezing to enhance post-thaw survival.[6,28] CPA can be divided into two different groups[29,30]:

1. Low-molecular-weight CPAs such as glycerol, ethylene (propylene) glycol, and dimethyl sulfoxide (DMSO), which can penetrate the cell membrane.
2. Non-cell-membrane-penetrating CPAs that usually do not enter inside the cells, like polymers such as polyvinyl propyliodone, hydroxyl ethyl starch (HES), and various sugars.

7.8 HISTORY OF CPAs

Ever since the discovery of the role of glycerol in cryopreservation by Polge et al. in 1949, the use of CPAs has become common practice. Lovelock and Bishop[31] later found that the protective property shown by glycerol is due to its nontoxicity, high solubility in aqueous electrolyte solutions, and its ability to permeate living cells. However, he found that glycerol is impermeable to bovine red blood cells and proposed DMSO as an alternative solute with greater permeability to living cells and exerting protective action against the freezing damage to human and bovine red blood cells. This CPA gained considerable attention globally and was mentioned as miracle compound. In the same period, Garzon et al.[32] and Knorpp et al.[33] independently proposed HES as a cryoprotectant for erythrocytes.[34] Thereafter, HES continued to be used as the CPA for RBCs,[31–35] granulocytes,[36] cultured hamster cells,[37] and pancreatic islets.[38,39]

7.9 PROBLEMS ASSOCIATED WITH CURRENT CPAs

The cryoprotective properties of glycerol are relatively weak and DMSO, which is considered the most effective CPA, shows high cytotoxicity[40] and disturbs the differentiation of neuron-like cells,[41] cardiac myocytes,[42] and granulocytes.[43] When DMSO is used at low concentrations, it can decrease the membrane thickness and induce temporary water pores, while, when it is used at higher concentrations, it causes the disintegration of the bilayer structure of the lipid membrane.[44] Therefore, after thawing, it is necessary to remove it. In the current scenario, the most efficient cryopreservation technique, which is being used worldwide in cell banks, is 10% DMSO in the presence of fetal bovine serum (FBS). Importantly, it has been reported that post-thaw survival of cells decreases without the addition of FBS to the cryopreserving solution.[45,46] Indeed, it is the interplay between FBS and DMSO at the appropriate ratio that makes DMSO a potent cryoprotectant. However, the pit hole of this technique is that animal-derived proteins should be avoided for clinical usage due to high risk of infection. Therefore, these issues stimulated the development of new CPAs.

7.10 POLYAMPHOLYTES AS NEWER LOW-TOXIC CPAs

In 2009, Matsumura and Hyon[47] developed polyampholytes based on poly-L-lysine with an appropriate ratio of amino and carboxyl groups as a new CPA (COOH-PLL) (Figure 7.3a), which is a nonpenetrating CPA and shows very low cytotoxicity.

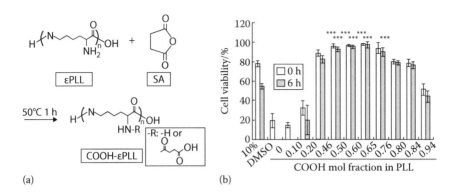

(a)   (b)

FIGURE 7.3 Cryoprotective property of carboxylated poly-L-lysine (PLL). (a) Schematic representation of the reaction of PLL succinylation. (b) L929 cells were cryopreserved using 10% DMSO and 7.5% (w/w) PLL at different ratios of COOH. Cell viability immediately (white bars), and 6 h (gray bars) after thawing at 37°C. Data are expressed as mean ± standard deviation (SD) for three independent experiments (n = 5). ***p < 0.001 vs. 10% DMSO for the corresponding time period (0 or 6 h). (From Matsumura, K. and Hyon, S.H., *Biomaterials*, 30, 4842, 2009.)

COOH-PLL shows high cryopreservation efficiency without the addition of any other CPAs. The authors reported that L929 cells were cryopreserved using 7.5% COOH-PLL as the sole CPA and in the absence of animal-derived proteins. Moreover, COOH-PLL shows high post-thaw survival efficiency when used at 7% or more compared with 10% DMSO (Figure 7.3b). In addition, compared with DMSO, cell attachment efficiency is higher with COOH-PLL than with 10% DMSO.

The development of this CPA opened the gateways toward new strategies of cryopreservation. Recently, Matsumura et al.[48] showed that the period of cryopreservation using COOH-PLL was not restricted to 1 week by successfully cryopreserving human bone marrow cells (hBMSCs) for up to 24 months. This result showed that this novel CPA did not alter the phenotypic characteristics and proliferative ability of the cells following thawing after cryopreservation for 24 months.

7.11 SYNTHETIC POLYAMPHOLYTES: AN APPROACH TOWARD SOLVING THE MYSTERY BEHIND THE MECHANISM OF CRYOPROTECTION BY POLYAMPHOLYTES

Using reversible addition-fragmentation chain transfer (RAFT), Rajan et al.[49] synthesized polyampholytes, which exhibit cryoprotective properties and show relationship between their cryoprotective properties and cell membrane protection. The authors reported on a copolymer synthesized at 1:1 ratio of methacrylic acid (MAA) to 2-(dimethylamino)ethyl methacrylate (DMAEMA), which showed the highest cell viability compared with copolymers synthesized at 2:3 and 3:2 ratios. In order to elucidate the cryoprotection mechanism, the authors introduced different hydrophobic and hydrophilic moieties in the polymer. They found that the cell viability after cryopreservation with hydrophilic and hydrophobic polyampholyte solutions at the same polymer concentration (10%) was higher in the case of hydrophobic polymer. They also varied the molecular weight of the polymers through RAFT polymerization and found that the protective properties of these CPAs were related to water absorption, which depends on the molecular weight of the compound. Thus, careful control of the molecular weight using RAFT polymerization may be an effective tool for the development of polyampholyte CPAs. The authors also found that synthetic CPAs protected the cell membrane during cryopreservation, whose effect was enhanced by the introduction of hydrophobic moieties.[49] However, the cryoprotection mechanism of theses CPAs is still not clear and needs to be further investigated in the future.

7.12   CELL ENCAPSULATION AND CRYOPROTECTIVE HYDROGELS

Successful application of hydrogels in tissue engineering is due in part to the possibility of encapsulating cells without adversely affecting their viability. Most of the connective tissues in our body are composed of tissue-specific cells encapsulated within a specific extracellular matrix (ECM), which has a complex function-orientated structure.[50]

Vrana et al. reported that a poly(vinyl alcohol) cryogel containing many crosslinking points encapsulated vascular endothelial cells and smooth muscle cells through physical hydrogel formation via crystallization during freezing.[51] In that report, the authors found that cells could be stored for the desired period using the cryogelation process, during which gelation proceeds after thawing occurs. When this cryogel is used to encapsulate cells in a freezing process, CPAs must be added to avoid freezing damage to the cells.

Since CPAs are generally cytotoxic,[52,53] they need to be removed after thawing. However, their complete removal from hydrogels is challenging. However, since hydrogels are the material of choice for various applications in regenerative medicine because of their unique properties like biocompatibility, flexible methods of synthesis, range of constituents, and desirable physical characteristics, if hydrogels having cryoprotective properties could be developed, then the problem of CPA removal would be solved. Moreover, tissue engineering applications in regenerative medicines could be successful if further advances could be made in low-temperature preservation. To produce tissue-engineered products off-the-shelf, cryopreservation of cells containing constructs is in high demand. Therefore, hydrogels with cryoprotective properties could be a good alternative for the storage of cell-based systems. Jain et al.[54] reported for the first time on the successful cryopreservation of cells encapsulated in a hydrogel having its own cryoprotective properties. The authors prepared dextran-based polyampholytes with cryoprotective properties, and *in situ* hydrogelation was performed using Cu-free click chemistry. In their report, they prepared a polyampholyte based on azide-amino-dextran, which shows excellent cryoprotective properties and performs better than DMSO (Figure 7.4). The gelation was achieved by mixing polyampholyte and dibenzylcyclooctyne-substituted dextran (DBCO-Dex) via click chemistry. This biocompatible hydrogel showed >90% viability after thawing. This system could be used for biomedical therapeutic applications, for example, stem cells cryopreserved with azide-Dex-polyampholyte can be mixed with DBCO-Dex immediately after thawing for injection into defect sites

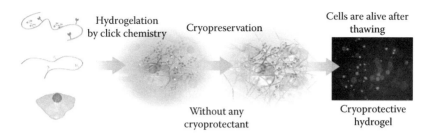

FIGURE 7.4 Schematic image of *in situ* cell-encapsulated hydrogel having cryoprotective properties. (From Jain, M., Rajan, R., Hyon, S.H., and Matsumura, K., Hydrogelation of dextran-based polyampholytes with cryoprotective properties via click chemistry, *Biomater. Sci.*, 2, 308–317, 2014. Reproduced with permission from The Royal Society of Chemistry.)

to form a scaffold for cell growth and tissue repair. This system does not require any pretreatment of the stem cells before injection, making cell maintenance, harvesting, and mixing with hydrogel-forming media unnecessary. Using this system, the stem cells could be preserved until just before usage and could be injected just after thawing without washing out the cryoprotectant.[54]

7.13 VITRIFICATION

7.13.1 Principle of Vitrification

The word vitrification originated from the Latin word *vitreum*, which means glass. Vitrification refers to the process of transformation of any substance into a glassy state, involves cooling at very fast rate, and causes enormous increase in viscosity.

Rapid cooling leads to the formation of amorphous ice, that is, water molecules lacking long-range order unlike in crystalline ice. As such, during the process of vitrification, water transforms directly to a glassy state, thereby preventing crystallization because the rate of freezing is so high that water molecules do not have time to form ice crystals. In vitrification, the viscosity of the intracellular fluid is elevated to such an extent that molecules get arrested and cease to behave as a liquid. By averting mechanical damage caused by ice crystals and resisting the change in salt concentration, vitrification prevents major damages to cells. Vitrification and freezing are not entirely different because the crystalline and vitreous phases often coexist. Indeed, during standard cryopreservation using controlled freezing of cells, a fragment of the system also undergoes vitrification.

Although vitrification studies had been performed since the nineteenth century when Tammann vitrified the molecules of carbon compounds in 1898,[55] one of the earliest studies on the use of vitrification in biological applications was carried out by Luyet in 1937, who is sometimes referred to as the founder of cryobiology. Luyet tried to vitrify living matter without the addition of any CPA and identified the prospective of attaining structurally immobile states for cryopreservation by rapid cooling.[56]

Vitrification has been shown to provide effective preservation for a number of cells, including monocytes, ova, early embryos, and pancreatic islets.[57–59]

Matsumura et al.[60] successfully vitrified human induced pluripotent (iPS) cells using COOH-PLL. The development of an efficient method to cryopreserve stem cells is very important because iPS cells enable the production of disease- and patient-specific pluripotent stem cells for cell therapy applications without using human oocytes or embryos.[61,62] The authors developed a vitrification solution comprising ethylene glycol (EG) and sucrose as well as COOH-PLL, and found that this solution inhibited devitrification. Cells retained their pluripotency when frozen and warmed on a relatively large scale in cryovials. In addition, this vitrification solution does not require DMSO or serum proteins.

The process of vitrification involves exposure to very high concentration of CPA and subsequent cooling in liquid nitrogen. At the initial stage of conventional freezing, low concentration (15%–20% w/v) of CPA is used because it will reach very high values once ice starts to grow at very low temperatures and most of the water is frozen. However, in the process of vitrification, the initial concentration of CPA is very high because there is no change in its value as the temperature is decreased; in fact, at very low temperatures, the concentration remains the same because the glassy state is obtained without any ice formation. While thawing, the whole sequence of vitrification is reversed.

7.13.2 VITRIFICATION OF OOCYTES

Oocytes are immature egg cells such as female gametocytes or germ cells, which are involved in the reproduction process. Oocyte matures during the menstrual cycle and develops into an ovum by the process of oogenesis.

The preservation of oocytes is of utmost importance for humans[63] because it is one of the most useful methods among assisted reproductive techniques (ART). A woman's capacity to reproduce is limited and is related to the number of oocytes she had at birth.[64,65] Women undergoing chemotherapy for cancer treatment[66] or suffering from other malignant diseases risk accelerated atresia, which could cause infertility due to ovarian insufficiency. Therefore, it is always advisable for those women before undergoing such treatments to cryopreserve their oocytes, which can be fertilized after recovery. Moreover, it is equally important for working women and women who have difficulty in finding a suitable partner to postpone child bearing by cryopreserving their oocytes at early age, when they have a sufficient amount and can later opt to fertilize them according to their needs.

While the first reported sperm (semen) cryopreservation was performed in 1776 by the Italian physiologist Spallanzani,[67,68] it took a long time for the cryopreservation of female eggs (oocytes) because they are more difficult to cryopreserve. Slow cooling of oocytes leads to both intracellular and extracellular crystallization of ice, which is fatal to cells. Moreover, slow cooling could result in chilling injury, to which oocytes are very sensitive.[69–71]

The first successful study on the cryopreservation of human oocytes, which lead to pregnancy was done by Chen,[72] who used DMSO as a cryoprotectant. They cryopreserved 40 oocytes, and 80% of these survived freezing and thawing and retained their morphology. Out of the 40 oocytes, 30 were inseminated and 83% maintained their ability to be fertilized. Since then, numerous research centers worldwide have published reports on successful pregnancies after cryopreservation.[73–77] However, the clinical efficiency remained very low, that is, 2%–4%,[78] probably due to poor survival, fertilization, and development of the cryopreserved oocytes.[79] Many reports on animals showed that vitrification was used at both the germinal vesicle stage and metaphase II stage.[80,81]

Hunter et al.[82] attempted vitrification of human oocytes using mixed CPAs and achieved survival rate of 65% and fertilization rate of 45%. EG has been frequently used for the vitrification of oocytes because of its low toxicity and ability to permeate cells.[80,81,83]

Recently, Zhang et al. developed a novel ejector-based droplet vitrification system to continuously cryopreserve oocytes in nanoliter droplets.[84] The droplet-based vitrification technique helps in achieving higher cooling and warming rates than other techniques due to the enhanced heat transfer of a droplet in the absence of a carrier, which reduces the likelihood of ice formation and the need for high CPA concentrations.[85,86] The employment of this technology overcame one of the primary problems with conventional vitrification methods, that is, inability to control the freezing volume. This system can limit and control the droplet volumes encapsulating the vitrified oocytes.

Watanabe et al.[87] used COOH-PLL for the vitrification of mouse oocytes to produce a live offspring. Their study revealed that the employment of COOH-PLL alone for the vitrification of oocytes resulted in very low survival rate, probably because of the low permeability of COOH-PLL, which they showed by confocal laser scanning microscopy using fluorescein isothiocyanate (FITC)-labeled COOH-PLL. The authors observed that the developmental ability of the oocytes *in vitro* was very high when a solution of EG and COOH-PLL was used. Vitrification of oocytes in presence of the E20P10 group (10% EG + 5% COOH-PLL as the equilibration medium and 20% EG + 10% COOH-PLL as the vitrification medium) showed a significant improvement in the *in vivo* development, which was due to the decrease in the invisible damage by vitrification as a result of using COOH-PLL and decreasing the amount of EG. This may result in the increase of the *in vivo* developmental ability beyond the blastocyst stage. A potential justification is that COOH-PLL controls the rapid increase of osmolality. Hence, the use of the combined solution of COOH-PLL and other CPA such as EG is very effective for oocyte vitrification. Moreover, there are many advantages of using COOH-PLL such as its low cytotoxicity even up to a concentration of 20%.[88]

7.13.3 VITRIFICATION OF TISSUE-ENGINEERED CONSTRUCTS

The success of tissue engineering applications in regenerative medicine requires further advances in low-temperature preservation. Preservation of tissues and tissue engineering products is one of the most important techniques for the clinical and industrial application of tissue engineering. However, cryopreservation of regenerated tissues including cell sheets and cell constructs is not easy compared with the cryopreservation of cell suspensions. Moreover, conventional freezing methods destroy the membranous structures of cultured sheets during the freezing and thawing process.

Cryopreservation of cell-containing constructs is in high demand for tissue engineering applications such as producing off-the-shelf tissue-engineered products.

Kuleshova and Hutmacher reviewed the studies on the vitrification of tissue-engineered constructs for preservation.[89] In the case of cartilage tissue, isolated chondrocytes could be cryopreserved using the slow-cooling method with cell survival rates as high as 93%.[90] In contrast, chondrocytes embedded in their ECM were quite difficult to preserve using the slow-cooling method, with cell death of 80%–100%.[90–92] The findings of Pegg and coworkers pointed out the need to establish vitrification protocols for cartilage tissue to avoid ice formation.[92–94] The vitrification approach has been successfully applied to the cryopreservation of cartilage grafts using a mixture of DMSO, formamide, and 1,2-propanediol.[95,96] Song et al. have demonstrated that rabbit cartilage could be preserved by vitrification with >80% cell viability.[95] *In vivo* studies of the cartilage preserved with different approaches demonstrated that the vitrified grafts performed considerably better than the frozen grafts, but did not differ from the fresh control sample.[96]

7.13.4   VITRIFICATION OF TISSUE-ENGINEERED BONE

Studies of preservation of tissue-engineered bone using the slow-cooling method in alcohol bath showed limited success.[97] Low cell viability (about 50%) after freezing was obtained for the mineralized constructs in poly(lactide-co-glycolyde) films. Fahy et al. reported that osteoblast-seeded hydroxyapatite scaffolds were vitrified in the vitrification solution called V_{EG} (i.e., 1,2-propandial and ethylene glycol) with ice brockers. The ice brocker inhibited ice recrystallization and maintained the solution amorphous state at lower cooling and warming rates, leading to the success of vitrification of osteoblast-hydroxyapatite complexes.[98]

Cao's group reported on the vitrification of tissue-engineered bone composed of osteoinduced canine bone marrow mesenchymal stem cells (cBMSCs) and partially demineralized bone matrix scaffold in DMSO containing vitrification solution.[99] In that report, cell viability, proliferative ability, alkaline phosphatase expression, and osteocalcin deposition after vitrification in VS442 (40% DMSO, 40% EuroCollins solution, 20% culture medium) was higher than that in a commercially available vitrification solution (i.e., VS55) (Figure 7.5).

7.13.5   VITRIFICATION OF TISSUE-ENGINEERED INTESTINE

Tissue-engineered intestine has been studied to recover the short bowel syndrome, which is highly morbid and mortal and results from the loss of intestinal tissue. In the past decades, many attempts have been performed to freeze intestinal tissues, but the results were unsatisfactory. Intestinal tissue engineering requires healthy epithelial and mesenchymal cellular contributions, but whole-organ preservation is neither feasible nor clinically relevant. Instead, isolating organoid units (OU), which are multicellular clusters including intestinal epithelial and mesenchymal progenitor cells, from healthy margin of resected bowel for tissue-engineered small intestine generation may be the key technique.[100] Grikscheit et al. reported for the first time on the successful cryopreservation of murine and human OU using vitrification,

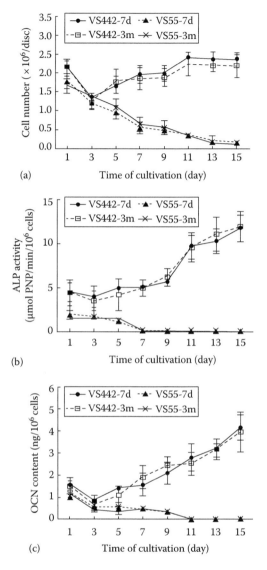

(a)

(b)

(c)

FIGURE 7.5 *In vitro* cellular activity and osteogenic potential of tissue-engineered bone (TEB) after 7 days or 3 months of vitrification. (a) Cellular number measurements were obtained at days 1, 3, 5, 7, 9, 11, 13, and 15 post thawing. Numerous viable cells were found in the VS442 groups compared to those found in the VS55 groups ($p < 0.05$). No significant difference in cell number was found between the 7-day groups and 3-month groups ($p > 0.05$; the error bars indicate the standard deviation [SD] for $n = 4$) both treated with VS442. TEB osteogenic differentiation was evaluated by measuring the expression of alkaline phosphatase (ALP) (b) and the release of osteocalcin (OCN) (c) at days 1, 3, 5, 7, 9, 11, 13, and 15 post thawing. Significantly increased osteogenesis was found in the VS442 groups compared to that found in the VS55 groups. No significant difference in osteogenesis was detected between the 7-day groups and 3-month groups ($p > 0.05$; the error bars indicate the standard deviation [SD] for $n = 4$). (Reproduced from Yin, H. et al., *Cryobiology*, 59, 180, 2009. With permission.)

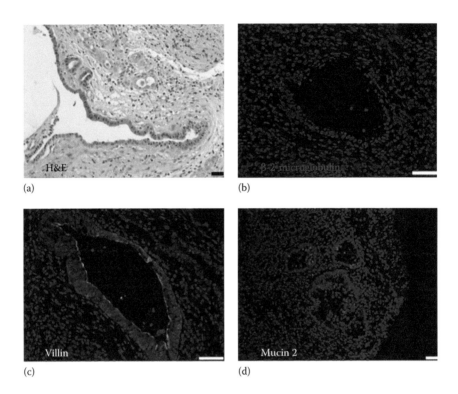

FIGURE 7.6 Human tissue-engineered small intestine (TESI) generated from vitrified organoid units (OU). (a) Hematoxylin and eosin staining showing a basic epithelial layer. (b) Immunofluorescence staining for β-2-microglobulin confirming the human origin of the TESI construct. Immunofluorescence staining for villin (c) and mucin 2 (d) showing the presence of differentiated enterocytes and goblet cells. 4′,6-Diamidino-2-phenylindole (DAPI) stains all the nuclei in blue. Scale bars = 50 μm. (From Spurrier, R. et al., *J. Surg. Res.*, 190, 399, 2014.)

which resulted in >90% viability. The OU were grown *in vitro* and generated tissue-engineered intestine from cryopreserved human cells (Figure 7.6). Vitrification may play a key role in the preservation of progenitor cells, which are necessary for tissue engineering-based therapies.

7.13.6 VITRIFICATION OF TISSUE-ENGINEERED BLOOD VESSELS

Cryopreservation of blood vessel and valves using the slow-cooling method showed unsatisfactory results after implantation.[101] Considering that the structure of the vessels is quite important for successful implantation, injury in the vascular endothelial cell layer, which causes severe problems leading to thrombus formation, should be avoided. Fractures in the blood vessels during cryopreservation and vitrification have been associated with the cooling rate and thermal nonuniformity in the samples. Pegg et al. performed the slow-warming method at less than 50°C/min down to −100°C, and subsequent thawing was rapidly carried out to prevent devitrification.[102] The method could prevent the fracturing in the vascular tissues.

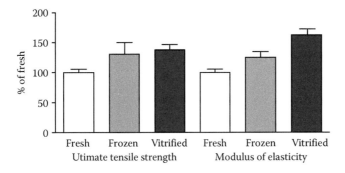

FIGURE 7.7 Ultimate tensile strength (n = 31 fresh, n = 31 frozen, and n = 42 vitrified) and modulus of elasticity (n = 60 fresh, n = 60 frozen, and n = 84 vitrified) of fresh, frozen, and vitrified engineered blood vessel constructs. (From Elder, E. et al., *Transpl. Proc.*, 37, 4625, 2005.)

Collagen-based tissue-engineered blood vessels were used to study vitrification for the first time. The vessel vitrified with DMSO, formamide, and propylene glycol showed good ice formation inhibition but low (39%) viability.[103]

When tissue-engineered blood vessels were prepared using a polyglycolic acid mesh sewn into a cylindrical construct, positive results were obtained through vitrification. Metabolic and apoptosis assays revealed that the vitrified blood vessels, which showed no ice formation, had similar viability to that of the fresh control.[104]

Maintenance of the biomechanical properties after preservation is also important for the application of tissue-engineered blood vessels. The ultimate tensile strength of collagen-based blood vessels increase after both slow freezing and vitrification.[104] Freezing has been shown to affect the matrix structure of collagen gels, specifically by increasing the void area and porosity of the matrix (Figure 7.7). Another study showed that human native blood vessels after vitrification showed no significant difference in their viscoelastic properties compared with fresh specimens, whereas slow freezing increased the tensile strength and elasticity of the vessels, probably because of the disruption of the ECM of the vessel by extracellular ice formation, which caused poor function of the cryopreserved blood vessels.

7.13.7 VITRIFICATION OF ENCAPSULATED TISSUE-ENGINEERED CELL CONSTRUCTS

Vitrification of microencapsulated hepatocytes was studied as a model for cell–matrix complex preservation by the Kuleshova group.[105] They used rat hepatocytes encapsulated with collagen hydrogels and vitrified with EG and sucrose–based vitrification solution. It was found that microencapsulated hepatocytes could be completely preserved without damage to their integrity, and migration was also observed. Pig hepatocytes, which are more similar to those of human, were also vitrified well even though they are smaller and more sensitive than rat hepatocytes.

Zhang et al. reported that water in small (100 μm) alginate microcapsules is preferentially vitrified over water in the bulk solution using 10% or higher DMSO with a cooling rate of 100°C/min.[106] The preferential vitrification of water in small microcapsules was found to significantly increase the viability of cells in a small quartz micro capillary, suggesting that the small alginate microcapsule is a good system for preserving cells.

In a study by the Sambanis group,[107] recombinant insulin-secreting C2C12 myoblasts were encapsulated in oxidized RGD-alginate and cryopreserved with slow cooling or vitrification. As a quality control for the preservation procedures, the metabolic activity and insulin secretion levels were assessed 3 days after thawing of the myoblasts containing samples. The results showed that vitrified cells have lower metabolic activity than freshly prepared myoblasts. On the other hand, no differences in the metabolic activity were observed between the freshly prepared myoblasts and cells cryopreserved using the slow-cooling technique. In addition, the insulin secretion rate was fully maintained in the slowly cooled samples. Therefore, the authors concluded that the conventional slow freezing is more appropriate for cryopreservation of C2C12 cells encapsulated in a partially oxidized RGD-alginate matrix. Interestingly, in stark contrast with these results, Murua et al.[108] showed that freezing of transfected C2C12 cells encapsulated in alginate-poly-L-lysine-alginate beads decreased by 42% in erythropoietin secretion compared to a fresh control.

7.13.8 VITRIFICATION OF CHONDROCYTE CELL SHEETS

Various cell sheets have been investigated in the field of regenerative medicine as a potential treatment for various lesions.[109,110] For example, Okano et al. developed a method for preparing various types of cell sheets using temperature-responsive polymer-immobilized culture dishes.[111] Cryopreservation of cell sheets would simplify the coordination of transplantation timing and would also allow to stock cell sheets for tissue banking or repeating treatments. An indispensable prerequisite for cryopreservation methods is to maintain the integrity of the membranous structure of the cell sheet. However, this is challenging and, although the viability of the cells comprising the sheets can be maintained, damage to the integrity of the sheet often occurs.[112] Maehara et al. reported on the development of an effective vitrification method that does not impair either the macro- or microstructures of cell sheets, and thereby possesses significant potential for applications related to clinical cell sheet therapy.[113] In this study, cell sheets were vitrified in liquid nitrogen vapor, rather than by direct immersion in liquid nitrogen, using COOH-PLL in the vitrification solution.

Recently, Vorontsov et al.[114] studied COOH-PLL for its antifreeze effect on the growth of ice crystals. They investigated the crystallization of ice in the presence of COOH-PLL by carrying out free-growth experiments of ice crystals in solutions at various COOH-PLL concentrations. They used a growth cell (Figure 7.8a) and different degrees of supercooling to reveal the characteristics and mechanism of its antifreeze effect. The results revealed that ice crystals adopt a dendritic morphology in the presence of COOH-PLL and that the presence of COOH-PLL results in hysteresis of growth rates and depression of the freezing point. In addition, the analysis of the inhibitory effect of COOH-PLL on crystal growth using the Gibbs–Thomson law and the Langmuir's adsorption isotherm suggested that the adsorption of large biological molecules has a non-steady-state character and occurs at a slower rate than does the process of embedding of crystal growth units (Figure 7.8b).

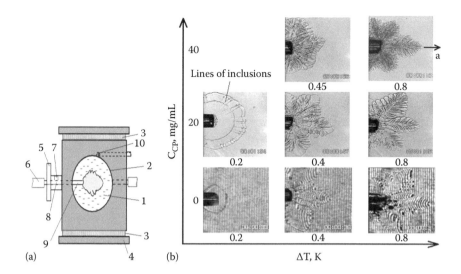

FIGURE 7.8 (a) Schematic of the cell growth: 1, growth chamber containing water or carboxyl poly-L-lysine (COOH-PLL) solution; 2, triple glass window; 3, Peltier elements; 4, heat reservoir with circulating water; 5, capillary holder; 6, tube with water or COOH-PLL solution; 7, inlet for cold spray; 8, seed crystal; 9, capillary; 10, thermistor. (b) Morphology of ice crystals at different supercooling temperatures (ΔT) and concentrations of COOH-PLL (C_{CP}). (From Vorontsov, D. et al., *J. Phys. Chem. B*, 118, 10240, 2014.)

The usage of COOH-PLL for the stabilization of the glassy state during vitrification has been also described by Maehara et al., who vitrified a cell sheet in a solution containing 20% DMSO, 20% EG, 0.5 M sucrose, and 10% COOH-PLL.[113] In particular, they coated the cell sheet with a viscous vitrification solution containing permeable and nonpermeable CPAs before vitrification in liquid nitrogen vapor, thereby preventing fracturing of the fragile cell sheet after vitrification and rewarming. Both the macro- and microstructures of the vitrified cell sheets were maintained without damage or loss of major components. Cell survival in the vitrified sheets was comparable to that in nonvitrified samples.

In the conventional slow-freezing method, cultured cell sheets are frozen in the presence of a relatively low concentration of a CPA.[112] Thus, extra- and intracellular ice crystal formation is inevitable during freezing, which destroys the cell sheet structure and decreases cell viability.[112] In contrast, with the vitrification method, a solution containing a high concentration of a CPA is rapidly cooled to achieve the transition from the liquid phase to the solid phase (amorphous) without ice crystal formation.[115] Therefore, cell sheets could be sealed in a glassy state maintaining their macro- and microstructures and retaining high cell viability (Figure 7.9).

This also has critical influence on the maintenance of the membranous structure of the cell sheet during vitrification. In preliminary experiments, in which cell sheets were vitrified by direct immersion in liquid nitrogen, all the sheets cracked. Very likely, direct immersion in liquid nitrogen has a more drastic impact on the membrane integrity than does the immersion in nitrogen gas, therefore resulting in the disruption of the cell sheet structure.

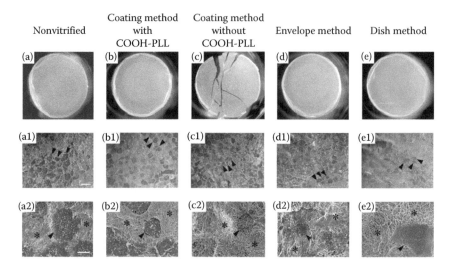

FIGURE 7.9 The macro- and microstructures of triple-layered rabbit chondrocyte sheets after vitrification. (a–e) Morphological appearance of chondrocyte sheets after vitrification and rewarming. (a1–e2) Scanning electron microscopic images of the surfaces of cell sheets recovered after vitrification and rewarming. (a1–e1) Magnification, ×300; scale bar = 50 μm. (a2–e2) Magnification, ×2000, scale bar = 10 μm. The surfaces (a1–e2) are irregular, featuring pavement-like cell populations (arrowheads in a1–e1 indicate three representative cells) and well-developed extracellular matrices with dense fibrous structure (*). The microstructures of the cell sheets vitrified by any of the methods described in this review were similar to those of the nonvitrified control samples. (From Maehara, M. et al., *BMC Biotechnol.*, 13, 58, 2013.)

The immersion in liquid nitrogen vapor technique enables vitrification of cell sheets using minimum amounts of vitrification solution. Interestingly, the technique was employed to produce vitrified chondrocyte cell sheets, which retained their normal characteristics upon thawing.

This cryopreservation technology may open new avenues to the industrialization of tissue engineering applications by facilitating long-term storage of tissue-engineered constructs.

7.14 CONCLUSION

The field of cryopreservation has advanced greatly in the last two centuries, especially in the past 50–60 years. A plethora of new technologies and methods have emerged, and cryopreservation no longer remains a science fiction fantasy. It has evolved since when it was performed using only cooling to now that it involves various intricacies and new advanced technologies have surfaced to make it more efficient. Cryopreservation of cells has minimized the need to culture them for a long time, which could make the cells prone to genetic drift. Culturing the cells for a long time may also lead to cell contamination. Therefore, with the advancement of cryopreservation technology, cells can be preserved for a very long time and restored to their original state whenever required. This technology has really helped in various cellular studies of different dreadful diseases and for the testing of new drugs or drug delivery vehicles because it is now easy to store cell banks.

Especially, cryopreservation (vitrification) of tissues and tissue-engineered constructs must become the fundamental technology for the commercialization and industrialization of tissue engineering. It will change the lives of thousands of people who need to regenerate some tissues due to injury and disease. With the advancement in technology, post-thaw survival efficiency of these cells, tissues, and tissue-engineered constructs is very high and, thus, a great success rate has been achieved.

Although this field has seen tremendous development in the last six decades, many studies need to be further explored. Cryopreservation (vitrification) requires storage at very low temperature for a long time and transportation at ultra-low temperatures. Therefore, it leads to damages caused by ice crystal formation if the samples are to be stored above their glass transition temperatures. Another major concern is the high concentration of CPAs involved in the vitrification methods, which could be potentially toxic during the process of thawing because the cells may be exposed to these high concentrations at higher temperatures than in those used with the freezing methods of cryopreservation. Therefore, there is a pressing need to develop new cryoprotective agents for cells and to develop new techniques and methods for cryopreservation.

REFERENCES

1. Polge, C., Smith, A. U., and Parkes, A. S. Revival of spermatozoa after vitrification and dehydration at low temperatures. *Nature* 164 (1949): 666.
2. Smith, A. U. and Polge, C. Storage of bull spermatozoa at low temperatures. *Vet. Record* 62 (1950): 115–117.
3. Latta, R. Preservation of suspension cultures of plant cells by freezing. *Canad. J. Bot.* 49 (1971): 1253–1254.
4. Bannier, L. J. and Steponkus, P. L. Freeze preservation of callus cultures of *Chrysanthemum morifolium Ramat. Hortscience* 7 (1972): 411–414.
5. Cohen, J., Simons, R., Fehily, C. B. et al. Birth after replacement of hatching blastocyst cryopreserved at expanded blastocyst stage. *Lancet* 325 (1985): 647.
6. Karlsson, J. O. M. and Toner, M. Long-term storage of tissues by cryopreservation: Critical issues. *Biomaterials* 17 (1996): 243–256.
7. Lynch, M. E. and Diller, K. R. Analysis of the kinetics of cell freezing with cytoplasmic additives. *Trans. ASME* 81 (1981): 229–232.
8. Mazur, P. Kinetics of water loss from cells at subzero temperatures and the likelihood of intracellular freezing. *J. Gen. Physiol.* 47 (1963): 347–369.
9. Rapatz, G., Menz, L. J., and Luyet, B. J. Anatomy of freezing process in biological materials. *Cryobiology* 3 (1966): 139–162.
10. Toner, M. Nucleation of ice crystals inside biological cells. In *Advances in Low-Temperature Biology*, Steponkus, P.L. ed., London, U.K.: JAI Press, Vol. 2, 1993, pp. 1–51.
11. Mazur, P. The role of cell membranes in the freezing of yeast and other single cells. *Ann. NY Acad. Sci.* 125 (1965): 658–676.
12. Mazur, P. The role of intracellular freezing in the death of cells cooled at supraoptimal rates. *Cryobiology* 14 (1977): 251–272.
13. Fujikawa, S. Freeze fracture and etching studies on membrane damage on human erythrocytes caused by formation of intracellular ice. *Cryobiology* 17 (1980): 351–362.
14. Mazur, P. Physical and chemical basis of injury of single-celled micro-organisms subjected to freezing and thawing. *Cryobiology* 3 (1966): 214–315.

15. Mazur, P., Leibo, S., and Chu, EHY. A two-factor hypothesis of freezing injury. *Exp. Cell Res.* 71 (1972): 345–355.
16. Lovelock, E. The haemolysis of red blood cells by freezing and thawing. *Biochim. Biophys. Acta* 10 (1953): 414–426.
17. Mazur, P. Freezing and low temperature storage of living cells. In *Proceedings of the Workshop on Preservation of Mouse Strains.* Miihlbock, O., ed. Stuttgart, Germany: Gustov Fischer Verlag, 1976, pp. 1–12.
18. Jacobs, M., Glassman, H., and Parpart, A. Osmotic properties of the erythrocyte. VII. The temperature coefficients of certain hemolytic processes. *J. Cell. Comp. Physiol.* 7 (1935): 197–225.
19. Woelders, H., Matthijs, A. and Engel, B. Effects of trehalose and sucrose, osmolality of the freezing medium and cooling rate on viability and intactness of bull sperm after freezing and thawing. *Cryobiology* 35 (1997): 93–105.
20. Muldrew, K., McGann, L. The osmotic rupture hypothesis of intracellular freezing injury. *Biophysics* 166 (1994): 532–541.
21. Nei, T., Araki, T., Matsusaka, T. Freezing injury to aerated and non-aerated cultures of *Escherichia coli.* In *Freezing and Drying of Microorganisms.* Nei, T., ed. Tokyo, Japan: University of Tokyo Press, 1969, p. 3.
22. Nei, T. Some factors which influence the stability of freeze-dried cultures. In *Recent Research in Freezing and Drying.* Parkes, A. and Smith, A., eds. Oxford, U.K.: Blackwell Scientific Publications, 1960, p. 78.
23. Mazur, P., Manifestations of injury in yeast cells exposed to subzero temperatures. I. Morphological changes in freeze-substituted and in frozen-thawed cells. *J. Bacteriol.* 82 (1961): 662–672.
24. Meryman, H.T. and Platt, W.T. The distribution and growth of ice crystals in frozen mammalian tissue. Naval Medical Research Institute Report Project NM 000 018.01.08, 13, 1, 1955.
25. Asahina, E. The freezing process of plant cell. *Contrib. Inst. Low Temp. Sci., Hokkaido Univ.* 10 (1956): 83.
26. Mazur, P. Studies on rapidly frozen suspensions of yeast cells by differential thermal analysis and conductometry. *Biophys. J.* 3 (1963): 323–353.
27. Mazur, P., Leibo, S., Farrant, J., Chu, E., Hanna, M., and Smith, L. Interactions of cooling rate, warming rate and protective additive on the survival of frozen mammalian cells. In *The Frozen Cell.* Wolstenholme, G. E. W. and O'Connor, M., eds. London, U.K.: J & A Churchill, 1970, pp. 69–88.
28. Lovelock, E. The denaturation of lipid-protein complexes as a cause of damage by freezing. *Proc. Roy. Soc. Ser. B* 147 (1957): 427–433.
29. Karow, A. Cryoprotectants—A new class of drugs. *J. Pharm. Pharmacol.* 21 (1969): 209–223.
30. Stolzing, A., Naaldijk, Y., Fedorova, V., and Seethe, S. Hydroxyethylstarch in cryopreservation—Mechanisms, benefits and problems. *Transf. Apheresis Sci.* 46 (2012): 137–147.
31. Lovelock, E. and Bishop, M. W. Prevention of freezing damage to living organisms by dimethyl sulfoxide. *Nature* 183 (1959): 1349–1355.
32. Garzon, A. A., Cheng, C., Lerner, B., Lichtenstein, S., and Karlson, K. E. Hydroxyethyl starch (HES) and bleeding. An experimental investigation of its effect on hemostasis. *J. Trauma* 7 (1967): 757–766.
33. Knorpp, C. T., Merchant, W. R., Gikas, P. W., Spencer, H. H., and Thompson, N. W. Hydroxyethyl starch: Extracellular cryophylactic agent for erythrocytes. *Science* 157 (1967): 1312–1313.
34. Lionetti, F. J. and Hunt, S. M. Cryopreservation of human red cells in liquid nitrogen with hydroxyethyl starch. *Cryobiology* 12 (1975): 110–118.

35. Kim, H., Tanaka, S., Une, S., Nakaichi, M., Sumida, S., and Taura, Y. A. Comparative study of the effects of glycerol and hydroxyethyl starch in canine red blood cell cryopreservation. *J. Vet. Med. Sci.* 66 (2004): 1543–1547.

36. Lionetti, F. J., Hunt, S. M., Gore, J. M., and Curby, W. A. Cryopreservation of human granulocytes. *Cryobiology* 12 (1975): 181–191.

37. Ashwood, M. J., Warby, C., Connor, K. W., and Becker, G. Low-temperature preservation of mammalian cells in tissue culture with polyvinylpyrrolidone (PVP), dextrans, and hydroxyethyl starch (HES). *Cryobiology* 9 (1972): 441–449.

38. Kenmochi, T., Asano, T., Maruyama, M. et al. Cryopreservation of human pancreatic islets from non-heart-beating donors using hydroxyethyl starch and dimethyl sulfoxide as cryoprotectants. *Cell Transpl.* 17 (2008): 61–67.

39. Maruyama, M., Kenmochi, T., Sakamoto. K., Arita. S., Iwashita, C., and Kashiwabara, H. Simplified method for cryopreservation of islets using hydroxyethyl starch and dimethyl sulfoxide as cryoprotectants. *Transpl. Proc.* 36 (2004): 1133–1134.

40. Fahy, G. M. The relevance of cryoprotectant "toxicity" to cryobiology. *Cryobiology* 123 (1986): 1–13.

41. Oh, J. E., Karlmark, R. K., Shin, J. H., Pollak, A., Hengstschlager, M., and Lubec, G. Cyto-skeleton changes following differentiation of N1E-115 neuroblastoma cell line. *Amino Acids* 31 (2006): 289–298.

42. Young, D. A., Gavrilov, S., Pennington, C. J. et al. Expression of metalloproteinases and inhibitors in the differentiation of P19CL6 cells into cardiac myocytes. *Biochem. Biophys. Res. Commun.* 322 (2004): 759–765.

43. Jiang, G., Bi, K., Tang, T. et al. Down-regulation of TRRAP-dependent hTERT and TRRAP-independent CAD activation by Myc/Max contributes to the differentiation of HL60 cells after exposure to DMSO. *Int. Immunopharmacol.* 6 (2006): 1204–1213.

44. Gurtovenko, A. A. and Anwar, J. Modulating the structure and properties of cell membranes: The molecular mechanism of action of dimethyl sulfoxide. *J. Phys. Chem. B* 111 (2011): 10453–10460.

45. Men, H., Agca, Y., Critser, E. S., and Critser, J. K. Beneficial effects of serum supplementation during in vitro production of porcine embryos on their ability to survive cryopreservation by open pulled straw vitrification. *Theriogenology* 64 (2005): 1340–1349.

46. Son, J. H., Kim, K. H., Nam, Y. K., Park, J. K., and Kim, S. K. Optimization of cryoprotectants for cryopreservation of rat hepatocyte. *Biotechnol. Lett.* 26 (2004): 829–833

47. Matsumura, K. and Hyon, S. H. Polyampholytes as low toxic efficient cryoprotective agents with antifreeze protein properties. *Biomaterials* 30 (2009): 4842–4849.

48. Matsumura, K., Hayashi, F., Nagashima, T., and Hyon, S. H. Long term cryopreservation of human mesenchymal stem cells using carboxylated poly-l-l-lysine without the addition of proteins or dimethyl sulfoxide. *J. Biomater. Sci. Polym. Edn.* 24 (2013): 1484–1497.

49. Rajan, R., Jain, M., and Matsumura, K. Cryoprotective properties of completely synthetic polyampholytes via reversible addition-fragmentation chain transfer (RAFT) polymerization and the effects of hydrophobicity. *J. Biomater. Sci. Polym. Edn.* 24 (2013): 1767–1780.

50. Goh, K. L., Meakin, J. R., Aspden, R. M. et al. Stress transfer in collagen fibrils reinforcing connective tissues: Effects of collagen fibril slenderness and relative stiffness. *J. Theor. Biol.* 245 (2007): 305–311.

51. Vrana, N., Matsumura, K., Hyon, S. H. et al., Cell encapsulation and cryostorage in PVA–gelatin cryogels: Incorporation of carboxylated ε-poly-L-lysine as cryoprotectant. *J. Tissue Eng. Regen. Med.* 6 (2012): 280–290.

52. Lowenthal, R. M., Park, D. S., Goldman, J. M., Hill, R. S., Whyte, G., and Th'ng, K. H. The cryopreservation of leukaemia cells: Morphological and functional changes. *Brit. J. Haematol.* 34 (1976): 105–117.

53. Douay, L., Gorin, N. C., David, R. et al. Study of granulocyte-macrophage progenitor (CFUc) preservation after slow freezing of bone marrow in the gas phase of liquid nitrogen. *Exp. Haematol.* 10 (1982): 360–366.

54. Jain, M., Rajan, R., Hyon, S. H., and Matsumura, K. Hydrogelation of dextran-based polyampholytes with cryoprotective properties via click chemistry. *Biomater. Sci.* 2 (2014): 308–317.

55. Tammann, G. Ueber die abhangkeit der Kernr, welche sich in verschiedenen flussigkeiten bilden, von der temperature. *Zeitschr. Physik. Chem.* 25 (1898): 441–479.

56. Luyet, B. The vitrification of organic colloids and protoplasm. *Biodynamica* 1 (1937): 1–14.

57. Takahashi, T., Hirsh, A., Erbe, E., Bross, J., Steere, R., and Williams, R. Vitrification of human monocytes. *Cryobiology* 23 (1986): 103–115.

58. Jutte, N., Heyse, P., Jansen, H., Bruining, G., and Zeilmaker, G. Vitrification of mouse islets of Langerhans: Comparison with a more conventional freezing method. *Cryobiology* 24 (1987): 292–302.

59. Leeuw, A., Daas, J., Kruip, T., and Rall, W. Comparison of the efficacy of conventional slow freezing and rapid cryopreservation methods for bovine embryos. *Cryobiology* 32 (1995): 157–167.

60. Matsumura, K., Bae, J., Kim, H. and Hyon, S. H. Effective vitrification of human induced pluripotent stem cells using carboxylated ε-poly-L-lysine *Cryobiology* 63 (2011): 76–83.

61. Meng, X., Shen, J., Kawagoe, S., Ohashi, T., Brady, R., and Eto, Y. Induced pluripotent stem cells derived from mouse models of lysosomal storage disorders. *Proc. Natl Acad. Sci. USA* 107 (2010): 7886–7891.

62. Park, I., Arora, N., Huo, H. et al. Disease-specific induced pluripotent stem cells. *Cell* 134 (2008): 877–886.

63. Oktay, K., Newton, H., Aubard, Y., Salha, O., and Gosden, R. Cryopreservation of immature human oocytes and ovarian tissue: An emerging technology? *Fertil. Steril.* 69 (1998): 1–7

64. Pfeifer, S., Goldberg, J., McClure, R. et al. Mature oocyte cryopreservation: A guideline. *Fertil. Steril.* 99 (2013): 37–43.

65. Welt, C. Primary ovarian insufficiency: A more accurate term for premature ovarian failure. *Clin. Endocrinol.* 68 (2008): 499–509.

66. Meirow, D. Reproduction post-chemotherapy in young cancer patients. *Mol. Cell. Endocrinol.* 169 (2000): 121–131.

67. Spallanzani, L. Osservazioni e spezienze interno ai vermicelli spermatici dell' uomo e degli animali. In *Opusculi di Fisica Animale e Vegetabile*, Modena, Italy, 1776.

68. Spallanzani, L. Esperimenti che servono nella storio della generazione di animali e piante. In *Barthelmi Ciro*, Genova, Italy, 1785.

69. Vincent, C. and Johnson, J. Cooling cryoprotectants and the cytoskeleton of the mammalian oocyte. *Oxford Rev. Reprod. Biol.* 14 (1992): 73–100.

70. Bernard, A. and Fuller, B. Cryopreservation of human oocytes: A review of current problems and perspectives. *Hum. Reprod. Update* 2 (1996): 193–207.

71. Leibo, S., Martino, A., Kobayashi, S., and Pollard, J. Stage-dependent sensitivity of oocytes and embryos to low temperatures. *Anim. Reprod. Sci.* 42 (1996): 45–53.

72. Chen, C. Pregnancy after human oocyte cryopreservation. *Lancet* 327 (1986): 884–886.

73. Siebzehnruebl, E., Todorow, S., Uem, J., Roch, R., Wildt, L., and Lang, N. Cryopreservation of human and rabbit oocytes and one-cell embryos: A comparison of DMSO and propanediol. *Hum. Reprod.* 4 (1989): 312–317.

74. Tucker, M., Wright, G., Morton, P., Shanguo, L., Massey, J., and Kort, H. Preliminary experience with human oocyte cryopreservation using 1,2-propanediol and sucrose. *Hum. Reprod.* 11 (1996): 1513–1515.

75. Young, E., Kenny, A., Puigdomenech, E., Thillo, G., Tiveron, M., and Piazza, A. Triplet pregnancy after intracytoplasmic sperm injection of cryopreserved oocyte: Case report. *Fertil. Steril.* 70 (1998): 360–361.

76. Porcu, E., Fabbri, R., Seracchioli, R., Ciotti, P., Magrini, O., and Flamigni, C. Birth of healthy female after intracytoplasmic sperm injection of cryopreserved human oocytes. *Fertil. Steril.* 68 (1997): 724–726.

77. Porcu, E., Fabbri, R., and Ciotti, P. Ongoing pregnancy intracytoplasmic injection of testicular spermatozoa into cryopreserved human oocytes. *Am. J. Obstet. Gynecol.* 180 (1999): 1044–1045.

78. Oktay, K., Cil, A., and Bang, H. Efficiency of oocyte cryopreservation: A meta-analysis. *Fertil. Steril.* 86 (2006): 70–80.

79. Chen, S., Lien, Y., Chao, K., Lu, H., Ho, H., and Yang, Y. Cryopreservation of mature human oocytes by vitrification with ethylene glycol in straws. *Fertil. Steril.* 74 (2000): 804–808.

80. Martino, A., Songsasen, N., and Leibo, S. Development into blastocysts of bovine oocytes cryopreserved by ultrarapid cooling. *Biol. Reprod.* 54 (1996): 1059–1069.

81. Otoi, T., Yamamoto, K., Koyama, N., Yachikawa, S, and Suzuki T. Cryopreservation of mature bovine oocytes by vitrification in straws. *Cryobiology* 37 (1998): 77–85.

82. Hunter, J., Fuller, B., Bernard, A., Jackson, A., and Shaw, R. Vitrification of human oocytes following minimal exposure to cryoprotectants; initial studies on fertilization and embryonic development. *Hum. Reprod.* 10 (1995): 1184–1188.

83. Vajta, G., Holm, P., Kuwayama, M. et al. Open pulled straw (OPS) vitrification: A new way to reduce cryoinjuries of bovine ova and embryos. *Mol. Reprod. Dev.* 51 (1998): 53–58.

84. Zhang, X., Khimji, I., and Shao, L. Nanoliter droplet vitrification for oocyte cryopreservation. *Nanomedicine (London)* 7 (2012): 553–564.

85. Xu, F., Moon, S., Zhang, X., Shao, L., Song, Y., and Demirci, U. Multi-scale heat and mass transfer modelling of cell and tissue cryopreservation. *Philos. Trans. A: Math. Phys. Eng. Sci.* 368 (2010): 561–583.

86. Song, Y., Adler, D., Xu, F. et al. Vitrification and levitation of a liquid droplet on liquid nitrogen. *Proc. Natl. Acad. Sci. USA* 107 (2010): 4596–4600.

87. Watanabe, H., Kohaya, N., Kamoshita, M. et al. Efficient production of live offspring from mouse oocytes vitrified with a novel cryoprotective agent, carboxylated e-poly-L-lysine. *PLoS ONE* 8 (2013): 1–5.

88. Matsumura, K., Bae, J., and Hyon, S. H. Polyampholytes as cryoprotective agents for mammalian cell cryopreservation. *Cell Transpl.* 19 (2010): 691–699.

89. Kuleshova, L. L., Gouk, S., and Hutmacher, D. Vitrification as a prospect for cryopreservation of tissue-engineered constructs. *Biomaterials* 28 (2007): 1585–1596.

90. Smith, A. U. Cartilage. In *Organ Preservation for Transplantation.* Karow, A. M., Abouna, G. J. M., and Humphries, L, eds. Boston, MA: Little, Brown, 1976, p. 214.

91. Ohlendorf, C., Tomford, W. W., and Mankin, H. J. Chondrocyte survival in cryopreserved osteochondral articular cartilage. *J. Orthop. Res.* 14 (1996): 413–416.

92. Pegg, D. E., Wusteman, M. C., and Wang, L. Cryopreservation of articular cartilage 1: Conventional cryopreservation methods. *Cryobiology* 52 (2006): 335–346.

93. Pegg, D. E., Wang, L. H., Vaughan, D., and Hunt, C. J. Cryopreservation of articular cartilage. Part 2: Mechanisms of cryoinjury. *Cryobiology* 52 (2006): 347–359.

94. Pegg, D. E, Wang, L. H, and Vaughan, D. Cryopreservation of articular cartilage. Part 3: The liquidus-tracking method. *Cryobiology* 52 (2006): 360–368.

95. Song, Y. C., An, Y. H., Kang, Q. K. et al. Vitreous preservation of articular cartilage grafts. *J. Invest. Surg.* 17 (2004): 65–70.

96. Song, Y. C., Lightfoot, F. G., Chen, Z., Taylor, M. J., and Brockbank, K. G. M. Vitreous preservation of rabbit articular cartilage. *Cell Preserv. Technol.* 2 (2004): 67–74.
97. Konfron, M. D., Opsitnick, N. C., Attawia, M. A, and Laurencin, C. T. Cryopreservation of tissue engineered constructs for bone. *J. Orthop. Res.* 21 (2003): 1005–1010.
98. Wowk, B., Leitl, E., Rasch, C. M., Mesbah-Karimi, N., Harris, S. B., and Fahy, G. M. Vitrification enhancement by synthetic ice blocking agents. *Cryobiology* 40 (2000): 228–236.
99. Yin, H., Cui, L., Liu, G., Cen, L., and Cao, Y. Vitreous cryopreservation of tissue engineered bone composed of bone marrow mesenchymal stem cells and partially demineralized bone matrix. *Cryobiology* 59 (2009): 180–187.
100. Spurrier, R., Speer, A., Grant, C., Levin, D., and Grikscheit, T. Vitrification preserves murine and human donor cells for generation of tissue-engineered intestine. *J. Surg. Res.* 190 (2014): 399–406.
101. Gall, K. L., Smith, S. E., Willmette, C. A., and O'Brien, M. F. Allograft heart valve viability and valve-processing variables. *Ann. Thorac. Surg.* 65 (1998): 1032–1038.
102. Pegg, D. E., Wusteman, M. C., and Boylan, S. Fractures in cryopreservedelastic arteries. *Cryobiology* 34 (1997): 183–192.
103. Elder, E., Chen, Z., Ensley, A., Nerem, R., Brockbank, K., and Song, Y. Enhanced tissue strength in cryopreserved, collagen-based blood vessel constructs. *Transpl. Proc.* 37 (2005): 4625–4629.
104. Dahl, S. L., Chen, Z., Solan, A. K., Brockbank, K. G., Niklason, L. E, and Song, Y. C. Feasibility of vitrification as a storage method for tissue-engineered blood vessels. *Tissue Eng.* 12 (2006): 291–300.
105. Magalhaes, R., Nugraha, B., Pervaiz, S., Yu, H., and Kuleshova, L. Influence of cell culture configuration on he post-cryopreservation viability of primary rat hepatocytes. *Biomaterials* 33 (2012): 829–836.
106. Zhang, W., Yang, G., Zhang, A., Xu, L., and He, X. Preferential vitrification of water in small alginate microcapsules significantly augments cell cryopreservation by vitrification. *Biomed. Microdev.* 12 (2010): 89–96.
107. Ahmad, H., and Sambanis, A. Cryopreservation effects on recombinant myoblasts encapsulated in adhesive alginate hydrogels. *Acta Biomater.* 9 (2013): 6814–6822.
108. Murua, A., Orive, G., Hernandez, R. M., and Pedraz, J. L. Cryopreservation based on freezing protocols for the long-term storage of microencapsulated myoblasts. *Biomaterials* 30 (2009): 3495–4501.
109. Yamato, M. and Okano, T. Cell sheet engineering. *Mater. Today* 7 (2004): 42–47.
110. Elloumi-Hannachi, I., Yamato, M., and Okano, T. Cell sheet engineering: A unique nanotechnology for scaffold-free tissue reconstruction with clinical applications in regenerative medicine. *J. Intern. Med.* 267 (2010): 54–70.
111. Yamada, N., Okano, T., Sakai, H., Karikusa, F., Sawasaki, Y., and Sakurai, Y. Thermoresponsive polymeric surfaces; control of attachment and detachment of cultured cells. *Macromol. Rapid Commun.* 11 (1990): 571–576.
112. Kito, K., Kagami, H., Kobayashi, C., Ueda, M., and Terasaki, H. Effects of cryopreservation on histology and viability of cultured corneal epithelial cell sheets in rabbit. *Cornea* 24 (2005): 735–741.
113. Maehara, M., Sato, M., Watanabe, M. et al. Development of a novel vitrification method for chondrocyte sheets. *BMC Biotechnol.* 13 (2013): 58.
114. Vorontsov, D., Sazaki, G., Hyon, S., Matsumura, K., and Furukawa, Y. Antifreeze effect of carboxylated ε-poly-L-lysine on the growth kinetics of ice crystals. *J. Phys. Chem. B* 118 (2014): 10240–10249.
115. Rall, W. F, and Fahy, G. M. Ice-free cryopreservation of mouse embryos at −196°C by vitrification. *Nature* 313 (1985): 573–575.

8 Micropatterning Techniques to Control Cell–Biomaterial Interface for Cardiac Tissue Engineering

Harpinder Saini, Feba S. Sam,
Mahshid Kharaziha, and Mehdi Nikkhah

CONTENTS

Abstract: Recent advancements in microengineering technologies have enabled creating biomimetic *in vitro* tissue substitutes for the repair and regeneration of injured myocardium. In this chapter, we will broadly overview the applications of micropatterning techniques in cardiac tissue engineering. We will first describe the cellular organization, microarchitecture, and the fundamentals of cell–matrix interactions within the native myocardium. Our discussion will be followed by a brief presentation on the use of natural and synthetic biomaterials in cardiac regeneration. Then, we will highlight the recent reports on the development of two- (2D) and three- (3D) dimensional micropatterned cardiac tissue constructs. This chapter will be concluded with a brief summary on the effects of electrical and mechanical stimulation on structural and functional properties of engineered cardiac tissues.

Keywords: Micropatterning, topographical cues, biomaterials, electrical and mechanical stimulation, cardiac tissue engineering

8.1 INTRODUCTION

Across different tissues in human body, cells are surrounded by an intricate meshwork known as "extracellular matrix" (ECM) (Hay 1991; Alberts et al. 2007; Frantz et al. 2010). The ECM is composed of complex macromolecules such as proteins and polysaccharides organized in a highly ordered fashion along with bioactive molecules such as growth factors and cytokines (Alberts et al. 2007; Frantz et al. 2010). Various compositions of the macromolecules within the ECM structure impart a broad range of mechanical properties in different tissues (Alberts et al. 2007; Frantz et al. 2010). For instance, the ECM in bone tissue is composed of collagenous proteins providing a microenvironment with high mechanical strength (Clarke 2008), while the ECM in the myocardial tissue mainly embodies collagen to impart sufficient elasticity to support contractile properties of the cardiac cells (Gupta and Grande-Allen 2006; Shamhart and Meszaros 2010). In addition to providing mechanical support to its respective tissue, the ECM induces a set of biophysical and biochemical cues to regulate cell functions (Dalby et al. 2007; Park et al. 2007b; Huang and Li 2011; Nikkhah et al. 2012a). In particular, the presence of various molecules with well-defined organization, along with the topography of the ECM, directs cell growth, differentiation, migration, and intracellular signal transduction pathways (Hay 1991; Kane et al. 1999; Frantz et al. 2010; Song et al. 2011; Nikkhah et al. 2012a). Due to the unique role of the ECM in regulating cellular functions, replicating its topography and composition is paramount for the development of native like engineered tissue constructs (Nikkhah et al. 2012a).

The emergence of micro- and nanofabrication technologies in the past few years has enabled researchers to create platforms with precise geometrical features resembling the native *in vivo* cellular microenvironment (Khademhosseini et al. 2006; Chung et al. 2007; Murtuza et al. 2009; Dvir et al. 2011b; Nikkhah et al. 2012a; Zorlutuna et al. 2012). These technologies have been successfully adopted by biologists and bioengineers for a wide range of applications from tissue engineering to biosensors design and fundamental biological studies (Park and Shuler 2003; Khademhosseini et al. 2006; Park et al. 2007a; Gillette et al. 2008; Atala et al. 2012). In particular, photolithography techniques adapted from the semiconductor industry have gained significant attention to fabricate scalable topographical features using advanced biomaterials such as hydrogels (Khademhosseini et al. 2006; Nikkhah et al. 2012a). Using these techniques, it is possible to develop well-ordered structures (e.g., grooves, pillars, ridges) with precisely defined geometrical dimensions to control cell–substrate interactions (Bettinger et al. 2009; Nikkhah et al. 2012a). Alternatively, soft lithography techniques such as microcontact printing and microfluidic patterning have been successfully employed to create 2D patterned features of ECM proteins to generate geometrically defined arrangement of cells (Xia and Whitesides 1998; Whitesides et al. 2001).

In the past decade, microfabrication techniques have found significant applications in numerous aspects of tissue engineering in general (Khademhosseini et al. 2006) and cardiovascular tissue engineering in particular (Khademhosseini et al. 2007; Zhang et al. 2011; Iyer et al. 2012; Nikkhah et al. 2012b; Annabi et al. 2013; Camci-Unal et al. 2014). These techniques have enabled the development of *in vitro* bioengineered cardiac tissue substitutes to mimic the anisotropic architecture of native myocardium for regenerative medicine applications and fundamental biological studies (Bursac et al. 2002; Kim et al. 2010; Zhang et al. 2011). Furthermore, using these technologies, it is possible to generate physiologically relevant cardiac-related disease models for high throughput drug screening (Natarajan et al. 2011; Agarwal et al. 2013b). This chapter covers the applications of micropatterning techniques to control cell–biomaterial interactions for cardiac tissue engineering. We will briefly discuss cellular organization and the architecture of native myocardium, followed by an overview on the application of various types of biomaterials in cardiac tissue engineering. Lastly, we will highlight the recent use of microengineering technology, micropatterning techniques in particular, to develop highly organized cardiac tissue constructs.

8.2 NATIVE CARDIAC TISSUE

8.2.1 MAJOR CELL TYPES WITHIN THE NATIVE CARDIAC TISSUE

There are three main cell types embedded within the native myocardium: cardiomyocytes, cardiac fibroblasts, and endothelial cells (Severs 2000; Iyer et al. 2009; Fleischer and Dvir 2013). Cardiomyocytes are well distributed throughout the heart, but those that beat fastest and determine the natural frequency of cardiac muscle are known as pacemaker cells (Mark and Strasser 1966). In the human heart, these cells are located in the sinoatrial node (Malmivuo and Plonsey 1995). Cardiomyocytes have active machinery of myofibrils that under the effect of propagating electrical impulses lead to their contraction and relaxation (Nag 1980). These cells act coherently with each other through intracellular junctions (e.g., gap junctions), thus forming a 3D syncytium (Radisic et al. 2007). Among noncardiomyocytes, cardiac fibroblast and endothelial cells are the other most abundant cells within the myocardial tissue (Radisic et al. 2007). Cardiac fibroblast cells are crucial for ECM synthesis and degradation within the myocardium (Souders et al. 2009; Castaldo et al. 2013). Furthermore, they are involved in cardiac tissue remodeling in response to numerous biophysical/biochemical signals and to various diseased states (Souders et al. 2009; Castaldo et al. 2013). On the other hand, endothelial cells are mainly responsible for the formation of blood vessels and capillaries for oxygen supply, and also for waste removal throughout the tissue similar to other organs (Radisic et al. 2007). The paracrine signaling between the endothelial cells and cardiomyocytes has been shown to significantly influence the functional properties such as contractility and rhythmicity of the myocardial tissue (Ramaciotti et al. 1992; Brutsaert 2003; Narmoneva et al. 2004).

8.2.2 MYOCARDIAL ECM

The myocardial ECM has a well-defined anisotropic structure, composed of major proteins such as collagen, elastin, vitronectin, fibronectin, laminin etc. (Castaldo et al. 2013). Various composition of these proteins notably influence the characteristics of the matrix within healthy and diseased states and contribute to the contractile capacity of the heart (Engler et al. 2008; Marsano et al. 2010; Castaldo et al. 2013). For instance, collagen is the main load-bearing protein that transmits the force generated by the cardiomyocytes in systole phase while imparting the passive stiffness within the diastole phase. Collagen also prevents the dilation and edema of the muscle over a long period of time (Chen et al. 2008; Godier-Furnemont and Vunjak-Novakovic 2013). Different types of collagen that have been identified in myocardial ECM consist of collagen type I, III, IV, and VI. Collagen type I comprise around 85% of the fibrillar collagen, affecting the overall rigidity of the heart muscle. Alternatively, collagen type III modulates matrix elasticity (Chen et al. 2008; Engler et al. 2008; Marsano et al. 2010). Topography of the cardiac muscle can also be attributed to the folded and highly ordered structure of its components maintained by disulphide and hydrophobic bonds (Wang and Carrier 2011). Overall, the stiffness and the architecture of the myocardial tissue provide the necessary signaling cues to support cardiac cells' phenotype and functions (Tandon et al. 2013).

8.2.3 IMPACT OF CELLULAR ORGANIZATION ON CARDIAC TISSUE FUNCTIONS

The interaction of cardiac cells with the anisotropic structure of myocardium is paramount for regulation of the tissue properties such as synchronous contractility (Au et al. 2007, 2009; Feinberg et al. 2007; Zhang et al. 2012). In particular, cellular organization and the orientation of the actin fibers, through a process known as contact guidance, significantly influence the contractile force generated throughout the tissue (Au et al. 2007, 2009; Zhang et al. 2012). For instance, in a study by Kim et al., it was shown that nonaligned cardiomyocytes generate 65%–85% less contractile force in comparison to highly organized cells that exhibit rod-like morphologies (Kim et al. 2008). In this regard, cardiac tissue remodeling due to a diseased state (e.g., arrhythmia) can affect the ECM composition (e.g., excessive collagen deposition) and consequently lead to poor cellular organization and tissue contractility (Baig et al. 1998; Biernacka and Frangogiannis 2011). Intracellular calcium dynamics, which is a crucial factor during systole and diastole phases, get also altered due to the changes in cellular alignment (Yin et al. 2004). Equivalently, it has been shown that junctional markers such as N-cadherin and connexin 43, which are responsible for mechanical and electrical signal propagations, are significantly influenced by cellular organization (Patel et al. 2011). These markers, in particular, which regulate synchronous beating and contraction of the cells, will be highly expressed when cells are properly aligned within the tissue matrix (Patel et al. 2011). Therefore, well-ordered arrangement of cardiac cells is essential for viable structural integrity and proper functioning of the myocardium in a healthy state.

8.3 BIOMATERIALS IN CARDIAC REGENERATION

Ventricular-specific cardiomyocytes within the adult mammalian heart exhibit significantly limited self-regenerative capacity (Bergmann et al. 2009). Myocardial infarction often results in a significant cell loss within the infarcted region and ultimately leads to catastrophic heart failure (Segers and Lee 2008). In the past few years, significant amount of efforts have been devoted to exploit efficient therapeutic strategies to induce long-term myocardial regeneration. These approaches span from stem cell–based transplantation to tissue engineering strategies (Soonpaa et al. 1994; Orlic et al. 2001; Zimmermann et al. 2002a; Nugent and Edelman 2003; Eschenhagen and Zimmermann 2005; Nelson et al. 2009; Wang et al. 2010b; Tous et al. 2011; Yamada et al. 2013). The common goal of these approaches is to maintain structural integrity of the injured myocardium and to restore native-like tissue functionalities (Zimmermann et al. 2002a; Nugent and Edelman 2003; Nelson et al. 2009; Wang et al. 2010b; Tous et al. 2011; Yamada et al. 2013). Despite marginal clinical success, there are still numerous challenges to induce efficient cardiac regeneration. Stem cell–based therapies are particularly appealing as incorporation of stem/progenitor cells within the infarcted region of the myocardium initiates the regeneration process with minimum risk of unfavorable tissue remodeling (Segers and Lee 2008). However, the success of stem cell–based transplantation has been notably hindered due to significant cell loss, poor integration with the host tissue, and the lack of control over cellular distribution upon transplantation (Zhang et al. 2001; Chen et al. 2008). As a potential remedy to these shortcomings, a number of recent strategies have proposed the use of natural and synthetic biomaterials to improve cellular engraftment, survival, and differentiation (Segers and Lee 2011). Alternatively, purely injectable biomaterials have also shown great promises, as bulking agents, to inhibit global remodeling of injured myocardium and to alleviate the stress distribution within the scar tissue upon infarction (Tous et al. 2011). Injected biomaterials could potentially align along the native ECM of the infarcted zone and integrate with the defected regions (Tous et al. 2011). On the other hand, the ultimate goal of cardiac tissue engineering is to recapitulate the architectural complexities of native myocardium and develop biomimetic *in vitro* constructs for the repair and the regeneration of the injured myocardium (Vunjak-Novakovic et al. 2010; Zhang et al. 2011). An ideal cardiac tissue substitute should exhibit proper contractility, mechanical robustness, and contain appropriate signaling cues for cellular assembly and vascular formation (Vunjak-Novakovic et al. 2010; Camci-Unal et al. 2014).

The success of stem cell–based transplantation and tissue engineering strategies for cardiac repair and regeneration is highly dependent on the selection of appropriate scaffolding biomaterials with enhanced biomechanical/biological properties (Vunjak-Novakovic et al. 2010; Camci-Unal et al. 2014). In particular, the selected biomaterial should be biocompatible and exhibit suitable mechanical strength while supporting cyclic contraction of the cardiac cells (Walsh 2005; Camci-Unal et al. 2014; Lanza et al. 2014). Biomaterials can be obtained naturally by either using cardiac tissue ECM (decellularized matrix) (Wainwright et al. 2010; Duan et al. 2011) or can be chemically synthesized (Chen et al. 2008). Natural biomaterials such as collagen are considered to be suitable candidates for cardiac regeneration since they are biocompatible in nature and induce different signals to cells through surface receptor interactions (Vunjak-Novakovic et al. 2010). Furthermore, these materials can be patterned using microcontact printing or micromolding techniques to provide sufficient anisotropy for generation of cardiac organoids (Kofidis et al. 2002; Zimmermann et al. 2002b; Black et al. 2009; Chiu et al. 2012). To date, numerous natural biomaterials such as collagen and matrigel (Zimmermann et al. 2002b; Chiu et al. 2012; Simpson and Dudley 2013), hyaluronic acid (Yoon et al. 2009; Ifkovits et al. 2010), gelatin (Li et al. 1999, 2000), chitosan (Fujita et al. 2005; Karp et al. 2006), alginate (Amir et al. 2009; Zieber et al. 2014), laminin (McDevitt et al. 2002), elastin (Annabi et al. 2013), fibrin (Birla et al. 2005; Black et al. 2009), cellulose-based scaffolds (Entcheva et al. 2004), plant origin polysaccharide (Venugopal et al. 2013), silk fibroin (Yang et al. 2009; Patra et al. 2012), as well as self-assembling peptides (Davis et al. 2006; Hsieh et al. 2006; Soler-Botija et al. 2014) have been used for cardiac regeneration. For instance, in a study by Li et al., fetal rat ventricular cells were seeded on commercial gelatin-based foams (Gelfoam®) and implanted onto rats at scarred area (Li et al. 1999). Seven days after implantation, it was found that the cells on the implanted graft were beating, and the density of cardiomyocytes in the area was higher than initial seeding density (Li et al. 1999). In another study, artificial myocardial tissue was created by seeding cardiomyocytes on collagen scaffold. The cells started beating by 36 h of culture and maintained contractility for 12 weeks *in vitro* (Kofidis et al. 2002). Similarly, Zimmermann et al. created an engineered heart tissue using collagen type I seeded with neonatal rat cardiomyocytes. Their findings demonstrated that the engineered tissue exhibited synchronous contractility (Zimmermann et al. 2002b). Self-assembling peptides integrated with heparin-binding domain sequences have also been shown to be suitable candidates to generate nanofibrous scaffolds under physiological conditions and induce enhanced cardiac functions along with sustained delivery of angiogenic growth factors (Guo et al. 2012).

In a recent work, Patra et al. used silk protein fibroin obtained from *Antheraea mylitta* as a potentially suitable biomaterial to develop cardiac patches. Their findings demonstrated enhanced cardiomyocytes attachment, spreading, and synchronous contraction on the silk fibroin similar to fibronectin-coated substrates (Patra et al. 2012). In a similar study, Yang et al. fabricated hybrid cardiac patches using chitosan or hyaluronic acid microparticles, incorporated with silk fibroin, to study the cardiomyogenic differentiation of rat mesenchymal stem cell (MSCs) (Yang et al. 2009). Higher expressions of cardiac-specific genes (Gata4, Nkx2.5) and proteins (cardiotin, connexin 43) were observed within the hybrid patches as compared to the scaffolds made of only silk fibroin (Yang et al. 2009). Although natural biomaterials have numerous advantages, they suffer from poor mechanical properties

(Kofidis et al. 2007; Chen et al. 2008). Furthermore, their degradation rate may not be optimal to develop engineered cardiac tissue substitutes to allow for sufficient ECM deposition (Chen et al. 2008). Therefore, researchers have tried to overcome these limitations by developing synthetic biomaterials.

Biodegradable synthetic polymers have had extensive use in medical field for numerous applications including the development of patches and scaffolds for cellular transplantation and myocardial repair (Chen et al. 2008). In particular, various synthetic polymers such as poly(ethylene glycol) (Jongpaiboonkit et al. 2008; Kraehenbuehl et al. 2008), polyglycolic acid (PGA) (Solan et al. 2003), poly(lactic-co-glycolic acid) (PLGA) (Ayaz et al. 2014), poly-glycolide-co-caprolactone (PGCL) (Piao et al. 2007), poly(L-lactic acid) (PLLA) (Zong et al. 2005; Caspi et al. 2007), and its copolymers with PLGA (Zong et al. 2005; Caspi et al. 2007), poly(glycerol-sebacate) (PGS) (Engelmayr et al. 2008; Radisic et al. 2008b), poly-N-isopropylacrylamide (PNIPAAm), and their copolymers (Naito et al. 2004; Miyagawa et al. 2005; Wang et al. 2010a), polyurethane (Rockwood et al. 2008), and polycaprolactone (PCL) (Shin et al. 2004; Soler-Botija et al. 2014) have been widely utilized in cardiac regeneration. For instance, in a study by Piao et al., PGCL scaffolds were seeded with bone marrow–derived mononuclear cells (BMMNC) for the treatment of myocardial infarction in animal model (Piao et al. 2007). The developed scaffolds were found to be biocompatible while promoting vascular formation as well as the migration of BMMNCs into the epicardial region (Piao et al. 2007). Another study by Fujimoto et al. utilized NIPAAm-based polymer as a temperature-responsive biomaterial for cardiac repair (Fujimoto et al. 2009). This polymer was used in conjunction with hydroxyl methacrylate-poly(trimethylene carbonate) and acrylic acid to be injected into the myocardial infarcted areas in a rat model. The injected hydrogel resulted in a thicker left ventricular wall and enhanced capillary density and contractility compared to the control condition (injection of phosphate buffered saline [PBS]) (Fujimoto et al. 2009). Similarly in another study, PNIPAAm-based dishes were used to generate monolayers of adipose tissue-derived mesenchymal stem cells. The engineered cell sheets were subsequently transplanted on the infarcted region of the rat heart for efficient cardiac regeneration (Miyahara et al. 2006).

With the advancements in cardiac tissue engineering, there has been an increasing demand toward the development of innovative biomaterials, which can respond to the microenvironmental cues and provide sufficient signaling to the surrounding cells for rapid regeneration of the injured tissue (Sakiyama-Elbert and Hubbell 2001). Particularly, composite biomaterials have been proposed to address these needs (Ozawa et al. 2004; Kharaziha et al. 2013, 2014; McGann et al. 2013; Qazi et al. 2014). For instance, the blends of natural and synthetic polymers such as copolymers made of gelatin with PCL, PLA, and PGS are among the examples of composite biomaterials that have been widely used for cardiac regeneration (Ozawa et al. 2004; Ifkovits et al. 2009; Kharaziha et al. 2013). Hybrid biomaterials incorporated with nanoparticles, such as carbon nanotubes (CNTs) (Shin et al. 2013; Kharaziha et al. 2014), gold nanoparticles (GNPs) (Dvir et al. 2011a), and graphene oxide (GO) (Shin et al. 2014), also belong to the emerging class of innovative biomaterials for cardiac tissue engineering.

For instance, Shin et al. embedded CNTs within photocrosslinkable gelatin methacrylate (GelMA) hydrogel to develop functional cardiac patches. Incorporation of CNTs significantly enhanced the electrical conductivity and mechanical robustness of the hybrid hydrogel and ultimately improved electrophysiological functionalities of cardiomyocytes. In a similar study, an electrically conductive cardiac patch was developed via integration of gold nanowires within alginate hydrogel. The developed constructs significantly enhanced protein expression, alignment, and synchronous contraction of cardiac cells (Dvir et al. 2011a). Oxygen-releasing materials are also considered to be suitable candidates that provide the cardiac cells with sufficient oxygen to maintain their viability and functionality (Oh et al. 2009).

In the following section, we will focus our attention toward the integration of microengineering technology and biomaterials, as a potentially powerful approach, to develop 2D and 3D micropatterned cardiac tissue constructs.

8.4 MICROPATTERNING TECHNIQUES IN CARDIAC TISSUE ENGINEERING

8.4.1 MICROFABRICATED 2D *IN VITRO* MODELS

To date, numerous studies have utilized 2D *in vitro* cardiac tissue models for various applications ranging from fundamental biological studies (e.g., cell–substrate interactions) to regenerative medicine, disease modeling, and drug screening (Bursac et al. 2002; McDevitt et al. 2002, 2003; Khademhosseini et al. 2007; Cimetta et al. 2009; Alford et al. 2010; Bray et al. 2010; Thery 2010; Natarajan et al. 2011; Feinberg et al. 2012, 2013; Grosberg et al. 2012; Kuo et al. 2012; Serena et al. 2012; Shim et al. 2012; Agarwal et al. 2013a; Yasukawa et al. 2013; Salick et al. 2014). The 2D models are mainly engineered through micropatterning of ECM proteins using soft lithography techniques such as microcontact printing and microfluidic patterning (Zhang et al. 2011; Folch 2012). Microcontact printing is a well-respected technique where the proteins of interest can be easily transferred from a microfabricated rubber stamp to the desired areas of the substrate, which comes in contact with the stamp (Xia and Whitesides 1998; Whitesides et al. 2001). Stamps with various geometrical features are usually fabricated in PDMS using soft lithography techniques. The transfer of the desired proteins on the substrate is mainly governed due to differences in the hydrophilicity of the surfaces (Xia and Whitesides 1998; Whitesides et al. 2001). In a recent study by Salick et al., PDMS stamps with different aspect ratios (width/length ratio) were used to pattern fibronectin and matrigel on glass slides (Salick et al. 2014). Human embryonic stem cell-differentiated cardiomyocytes (hESC-CMs) were seeded on the patterned features, and the effect of constructs' aspect ratio on sarcomere alignment was investigated. Based on the findings of this study, the width of the constructs had a pronounced effect on sarcomere alignment as compared to the aspect ratio (Salick et al. 2014). Furthermore, it was shown that that constructs with a width in the range of 30–80 μm notably enhanced sarcomere alignment (Figure 8.1A) (Salick et al. 2014). Similarly, this technique was

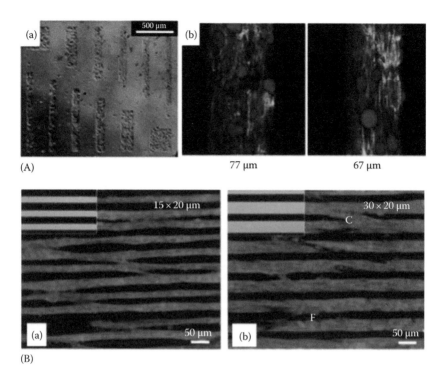

FIGURE 8.1 Application of 2D micropatterning techniques in cardiac tissue engineering: (A) micropatterned matrigel-fibronectin features with different aspect ratios seeded with hESC-CMs; (a) phase contrast images of the patterned cardiac cells, (b) confocal images demonstrating sarcomere organization (green stain) and nuclei (blue) orientation across different widths. (Adapted from *Biomaterials*, 35(15), Salick, M.R., Napiwocki, B.N., Sha, J., Knight, G.T., Chindhy, S.A., Kamp, T.J., Ashton, R.S., and Crone, W.C., Micropattern width dependent sarcomere development in human ESC-derived cardiomyocytes, 4454–4464, Copyright 2014, with permission from Elsevier.) (B) Microcontact printed lanes of laminin demonstrating the actin cytoskeleton (red) and nuclei (blue) organization of aligned cardiomyocytes across patterns with variable width. Inset images show laminin-printed lanes. (McDevitt, T.C., Angello, J.C., Whitney, M.L., Reinecke, H., Hauschka, S.D., Murry, C.E., and Stayton, P.S.: In vitro generation of differentiated cardiac myofibers on micropatterned laminin surfaces. *Journal of Biomedical Materials Research*. 2002. 60(3). 472–479. Copyright Wiley-VCH Verlag GmbH & Co. KGaA. Adapted with permission.)

successfully employed by McDevitt et al., where laminin was micropatterned on nonadhesive polystyrene surfaces to study the effect of geometrical constraint of laminin lanes (width) on synchronous beating of neonatal cardiomyocytes (McDevitt et al. 2002). According to this study, narrower laminin lanes with 15–20 μm width, resulted in aligned and bipolar cell–cell junctions similar to native myocardium (Figure 8.1B) (McDevitt et al. 2002). Although microcontact printing on 2D surfaces has been widely accepted as an efficient technique for the patterning of cardiac cells (Bursac et al. 2002; Cimetta et al. 2009; Bray et al. 2010; Thery 2010; Feinberg et al. 2012; Grosberg et al. 2012; Kuo et al. 2012;

Yasukawa et al. 2013; Salick et al. 2014), there are several disadvantages associated with this approach. Some limitations include denaturation of patterned proteins, necessity of using multiple stamps to pattern several proteins onto the same substrate, stamp deformation etc. (Perl et al. 2009; Zhang et al. 2011; Folch 2012). Another popular technique for 2D micropatterning is microfluidic patterning that addresses some of the limitations of microcontact printing (Folch 2012; Wang et al. 2014). The advantages of microfluidic patterning are delivery of proteins on the selective areas of the substrate in their natural form and thus preserving them from denaturation (Folch 2012). This technique utilizes the microchannels formed when a fabricated PDMS stamp comes into contact with the substrate. The proteins to be patterned are then delivered, with a fluidic carrier, to the areas of substrate that do not come into contact with the PDMS stamp (Folch 2012). Khademhosseini et al. successfully employed microfluidic patterning technique to pattern hyaluronic acid onto glass substrates and generate contractile cardiac constructs (Khademhosseini et al. 2007). They showed successful alignment of neonatal rat cardiomyocytes along with the patterns layout. Interestingly, the patterned cells were detached from the substrate and formed contractile organoids after 3 days of culture (Khademhosseini et al. 2007). The use of electrical field in microfluidic systems has been also suggested to develop highly organized cellular constructs mimicking the anisotropy of the native cardiac tissue (Yang and Zhang 2007). For instance, in a study by Yang et al., dielectrophoresis in a microfluidic device was used to induce cellular alignment along the direction of electrical field. The results of this study demonstrated successful cellular orientation in between the electrodes (Yang and Zhang 2007). Similar to microcontact printing, microfluidic patterning also suffers from many limitations such as buckling of the PDMS stamp and leakage of ECM proteins from microfluidic channels (Folch 2012).

Overall, microcontact printing and microfluidic patterning techniques have both shown great potential as simple and efficient approaches to form highly organized cardiac constructs to study fundamental biological questions regarding cardiac cells and ECM interactions (Bursac et al. 2002; Cimetta et al. 2009; Bray et al. 2010; Thery 2010; Feinberg et al. 2012; Grosberg et al. 2012; Kuo et al. 2012; Yasukawa et al. 2013; Salick et al. 2014). However, many of the geometrical cues, which modulate native-like cellular functionalities, are missing on 2D patterned surfaces. For instance, the lack of third dimension in these approaches significantly influences cellular phenotype and function (Thery 2010). Therefore, in the past few years, there has been tremendous efforts to engineer physiologically relevant cardiac tissue models using 3D topographical surfaces or micropatterned hydrogels. These approaches are summarized in the following section.

8.4.2 Microfabricated 3D Tissue Constructs

3D tissue constructs, fabricated using polymeric biomaterials, provide a more realistic microenvironment compared to 2D models for various tissue engineering applications (Khademhosseini et al. 2006; Thery 2010; Zorlutuna et al. 2012; Camci-Unal et al. 2014). To date, 3D microfabricated models such as surface topographies, micropatterned hydrogel constructs, and microengineered polymeric biomaterials have been utilized to impart necessary biophysical cues to control cardiac cells phenotype, cytoskeletal organization, and contractility (Deutsch et al. 2000; Bursac et al. 2002; Entcheva and Bien 2003; Motlagh et al. 2003; Yin et al. 2004; Au et al. 2007; Arai et al. 2008; Engelmayr et al. 2008; Kajzar et al. 2008; Guillemette et al. 2010; Luna et al. 2011; Patel et al. 2011; Zhang et al. 2012; Annabi et al. 2013; Kolewe et al. 2013; Rao et al. 2013; Wang et al. 2013; Yu et al. 2013; Bian et al. 2014a,b; Chen et al. 2014; Rodriguez et al. 2014). Different topographical features can be fashioned in polymeric or rigid biomaterials (e.g., silicon) using photolithography and micromolding techniques (Bettinger et al. 2009; Folch 2012; Nikkhah et al. 2012a). Patterned substrates can be etched to generate 3D surface topographies with the desired geometrical features to proceed with biological studies (Bettinger et al. 2009; Nikkhah et al. 2012a). On the other hand, patterned or etched substrates can be used as master molds to replicate the features in polymeric biomaterials (e.g., PDMS) using micromolding techniques (Xia and Whitesides 1998). For instance, in an early study by Desai group, microtextured surfaces in silicone including, two distinct designs of microposts (micropegged features) and microgrooves coated with laminin, were fabricated for cardiac tissue engineering (Deutsch et al. 2000). The goal of this study was to develop a physiologically relevant cell culture substrate that can enhance cardiac cell attachment and alignment. Cardiac cells oriented, bridged the gaps, and attached to the end points of the pegged features. Alternatively, the cells on microgrooved features aligned along the direction of the grooves. Overall, cellular attachment was higher on the micropegs as compared to flats surfaces (control) (Deutsch et al. 2000). In another study by Luna et al., a nonphotolithographic technique was used to generate parallel wrinkles in the range of nano- and microscale on a PDMS substrate to study cardiac cell organization and protein marker expression (Luna et al. 2011). In this approach, a layer of metal (gold palladium) was first deposited on a pre-stressed polystyrene sheet to generate wrinkles. Thereafter, the wrinkled substrates were used as molds to cast PDMS. Seeding of neonatal mouse cardiomyocytes and hESC-CMs resulted in the formation of highly organized cellular constructs on the wrinkled surfaces. Further analysis confirmed the orientation of N-cadherin as well as connexin 43 along the major axis of the wrinkles (Luna et al. 2011). Rodriguez et al. fabricated an array of patterned microposts to quantify passive tension, twitch force and the frequency of spontaneous beating of human induced pluripotent stem cells derived cardiomyocytes (hiPSC-CMs) (Rodriguez et al. 2014). The elastomeric microposts were fabricated in PDMS using soft lithography technique. The microposts were then stamped with different ECM proteins such as collagen type IV, fibronectin, and laminin to enhance cellular attachment. The findings of this study demonstrated that microposts coated with laminin resulted in enhanced cellular attachment with higher spreading as compared to collagen and fibronectin-coated substrates. This platform was reported to be versatile to measure the contractility of the cardiac cells in healthy, diseased, differentiated, and undifferentiated states. However, a major limitation of this study was the lack of cellular alignment on the micropost array (Rodriguez et al. 2014). In a similar context, Rao et al. studied the effects of fibronectin coated PDMS microgrooves on calcium cycling of hiPSC-CMs. Comparing to flat substrates (control condition), the cells exhibited enhanced alignment and sarcomere organization

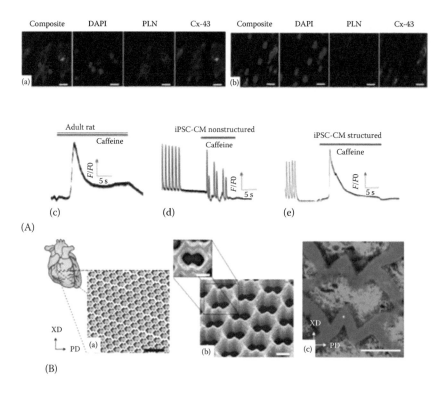

FIGURE 8.2 Illustrative examples for 3D scaffolds for cardiac tissue engineering. (A) Microgrooved PDMS structures seeded with hiPSC-CM; (a) immunostained hiPSC-CM on (a) flat and (b) structured PDMS substrates demonstrating the expression of cardiac specific and nuclei markers, (c–e) representative traces of Ca²⁺ release from sarcoplasmic reticulum in response to caffeine for (c) adult rat heart, cells seeded on (d) unstructured, and (e) structured PDMS substrates. (Adapted from *Biomaterials*, 34(10), Rao, C., Prodromakis, T., Kolker, L., Chaudhry, U.A.R., Trantidou, T., Sridhar, A., Weekes, C. et al., The effect of microgrooved culture substrates on calcium cycling of cardiac myocytes derived from human induced pluripotent stem cells, 2399–2411, Copyright 2013, with permission from Elsevier.) (B) Honeycomb structure fabricated in PGS; (a) representative images showing the honeycomb structure of collagen fibers in native cardiac tissue, (b) microfabricated accordion-like honeycomb structure, and (c) confocal images demonstrating actin cytoskeleton (green) organization and cellular alignment. (Adapted with permission from Macmillan Publishers Ltd. *Nature Materials*, Engelmayr, G.C., Jr., Cheng, M., Bettinger, C.J., Borenstein, J.T., Langer, R., and Freed, L.E., Accordion-like honeycombs for tissue engineering of cardiac anisotropy, 7(12), 1003–1010, Copyright 2008.)

on the microgrooved features. Furthermore, improved Ca²⁺ cycling, in response to caffeine, was observed on structured surfaces (Figure 8.2A) (Rao et al. 2013). In another interesting study by Engelmayr et al., accordion-like honeycomb structures were fabricated to provide a biomimetic microenvironment for cardiac tissue engineering (Engelmayr et al. 2008). The rationale behind the development of such structures was the honeycomb architecture of collagen fibers surrounding cardiomyocytes within the native myocardial tissue. These structures were fabricated using microablation technique through orienting two square-shaped pores at 45° on a PGS scaffold.

The microfabricated 3D scaffolds exhibited excellent anisotropic mechanical properties matching the native cardiac tissue. Seeding neonatal rat cardiomyocytes on the developed scaffolds, resulted in enhanced cellular orientation and directional contractile properties (Figure 8.2B) (Engelmayr et al. 2008).

Hydrogels are favorable biomaterials with attractive properties for cardiac tissue engineering applications (Dvir et al. 2011a; Zorlutuna et al. 2012; Shin et al. 2013; Camci-Unal et al. 2014). Hydrogels exhibit high water content, tunable mechanical properties (e.g., stiffness), and structural architecture (e.g., porosity) while providing a 3D native microenvironment to support cellular growth and assembly (Peppas et al. 2006). To date, several studies have used micropatterned hydrogel constructs to provide biomimetic topographical anisotropy for cardiac tissue engineering (Karp et al. 2006; Iyer et al. 2009; Aubin et al. 2010; Al-Haque et al. 2012; Chiu et al. 2012; Agarwal et al. 2013a; Annabi et al. 2013; Zhang et al. 2013). For instance, in a recent study by Annabi et al., methacrylated tropoelastin (MeTro) hydrogel, with suitable resilience, was used to develop micropatterned cardiac patches (Annabi et al. 2013). Patterns of 20×20 μm (width × spacing) and 50×50 μm channels were formed, using replica molding technique, and subsequently, photocrosslinked through UV exposure. Unpatterned MeTro and micropatterned GelMA hydrogel were used as controls. The findings of this study demonstrated that neonatal rat cardiomyocytes exhibited higher cellular attachment, elongation as well as cardiac marker expression (troponin I, connexin 43, sarcomeric α actinin) on micropatterned MeTro features as compared to control conditions. Furthermore, micropatterned MeTro substrates significantly promoted the spontaneous contractility of the cardiac cells for a long period of culture time (2 weeks) (Figure 8.3A) (Annabi et al. 2013). In another study, Zhang et al. studied functional and structural maturation of hESC-CMs on micropatterned fibrinogen and matrigel hydrogels (Zhang et al. 2013). The differentiated cardiac cells were encapsulated in the hydrogel solution and poured over the PDMS micromold, to polymerize at 37°C. Within 2 weeks of culture, cardiac cells reorganized, with aligned actin fibers, along the patterned layouts. In addition, the cells expressed high levels of sarcomeric α actinin, N-cadherin, troponin-T, and connexin 43 along with enhanced conduction velocity (action potential propagation) within the 3D micropatterned hydrogel patches as compared to 2D monolayer substrates (Figure 8.3B) (Zhang et al. 2013).

Electrospun and filamentous scaffolds have also been widely used in cardiac tissue engineering (Soliman et al. 2010; Orlova et al. 2011; Hsiao et al. 2013; Kharaziha et al. 2013; Venugopal et al. 2013; Ayaz et al. 2014; Kharaziha et al. 2014; Lin et al. 2014; Ma et al. 2014). However, the discussion of these studies is out of the scope of this book chapter.

8.4.3 EFFECT OF ELECTRICAL AND MECHANICAL STIMULATION ON MICROFABRICATED CARDIAC TISSUES

In addition to structural and topographical cues, electrical and mechanical stimulation enhance the maturity and functionality of engineered cardiac tissues (Zimmermann et al. 2006; Hsiao et al. 2013; Hirt et al. 2014; Miklas et al. 2014). Cells grown on scaffolds have to be stimulated via either electrical signals (Radisic et al. 2004) or

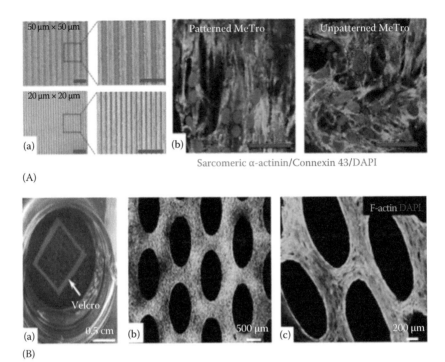

FIGURE 8.3 Representative examples of 3D micropatterned hydrogels for cardiac tissue engineering. (A) micropatterned MeTro hydrogels: (a) phase contrast images showing the patterned layouts with variable dimensions and (b) confocal images of cardiac-specific markers on patterned and unpatterned substrate. (Annabi, N., Tsang, K., Mithieux, S.M., Nikkhah, M., Ameri, A., Khademhosseini, A., and Weiss, A.S.: Highly elastic micropatterned hydrogel for engineering functional cardiac tissue. *Advanced Functional Materials*. 2013. 23(39). 4950–4959. Copyright Wiley-VCH Verlag GmbH & Co. KGaA. Adapted with permission.) (B) Fibrin-based cardiac tissue patch; (a) representative image of the patch, (b) elliptical pores within the patch, and (c) densely aligned cells with highly organized actin cytoskeleton fibers embedded within the patch. (Adapted from *Biomaterials*, 34(23), Zhang, D., Shadrin, I.Y., Lam, J., Xian, H.-Q., Snodgrass, H.R., and Bursac, N., Tissue-engineered cardiac patch for advanced functional maturation of human ESC-derived cardiomyocytes, 5813–5820, Copyright 2013, with permission from Elsevier.)

mechanical stimulation (Zimmermann et al. 2002b) to achieve optimal conditions similar to those in the native heart. Particularly, electrical stimulation results in the alignment of cardiac fibers, promotes cellular differentiation, and enhances contractile properties of the tissue (Radisic et al. 2004, Zimmermann et al. 2004; Radisic et al. 2007, 2008a; Kreutziger and Murry 2011; Annabi et al. 2013; Park et al. 2014). In this regard, numerous studies have incorporated the effects of electrical stimulation on micropatterned scaffolds (Tandon et al. 2009; Alford et al. 2010; Boudou et al. 2012; Chiu et al. 2012; Annabi et al. 2013; Thavandiran et al. 2013; Park et al. 2014). For instance, Park et al. utilized microfabricated PGS scaffold to study the individual and combined effects of insulin like growth factor (IGF-1) and electrical stimulation on maturation of engineered cardiac tissues (Park et al. 2014). PGS scaffolds, with excellent mechanical and biodegradability properties, were fabricated with rectangular shape pores using photolithography and micromolding techniques. Neonatal cardiac cells were then seeded on the scaffolds under four different conditions including IGF-1 only, electrical stimulation only, with IGF-1 and electrical stimulation, and without electrical stimulation and IGF-1. Monophasic electrical stimulations, with 5 V/cm amplitude and 1 Hz frequency for duration of 2 ms were used to induce contractility within the engineered tissue constructs. Electrical stimulation enhanced the orientation of tissue like bundles, parallel to the electrical field, and significantly improved the expression of matrix metalloprotease-2 (MMP-2). The presence of IGF-1 reduced the excitation threshold, while the integration of IGF-1 and electrical stimulation further promoted the expression of cardiac gap junction markers (connexin 43) and sarcomere organization (Park et al. 2014). In another study, Chiu et al. studied the combined effects of topographical cues and electrical stimulation on the engineered cardiac tissues fabricated in collagen-chitosan hydrogels (Chiu et al. 2012). The topographical features were composed of microgrooves with the width in the range of 10 µm, 20 µm, and 100 µm. A custom-made bioreactor system was used to induce 2.5 V/cm biphasic electrical pulses with 1 Hz frequency on the engineered tissue constructs seeded with neonatal rat cardiomyocytes. Cardiac cells reorganized along the major axis of the microgrooved features upon 6 days of culture, while cellular alignment significantly reduced the electrical stimulation threshold. Specifically, 10 µm width microgrooves resulted in the formation of complete contractile tissues comprised of mature gap junctions while the presence of electrical stimulation promoted cellular density (Chiu et al. 2012). Au et al. also investigated the combinatorial effects of electrical field stimulation and surface topography on cardiomyocyte organization on polyvinyl substrates consisting of V-shaped grooves of 13 µm width and 700 nm high. Their findings demonstrated the topographical cues and electrical field stimulation resulted in enhanced cellular elongation and alignment along the direction of microgrooves (Au et al. 2007).

Similar to electrical stimulation, a number of other studies have utilized mechanical stimulation to enhance the maturity and functionalities of engineered cardiac tissues (Zimmermann et al. 2002b; Shachar et al. 2012; Zhang et al. 2012; Miklas et al. 2014). Zimmermann and Eschenhagen performed numerous studies on this subject using cardiac tissues fabricated in collagen and matrigel (Eschenhagen et al. 1997; Fink et al. 2000; Zimmermann et al. 2002a,b, 2006). Their findings demonstrated that mechanical stimulation could lead to enhanced cardiomyocyte organization with increased mitochondrial density and improved length of myofilaments. They further concluded that under the effect of mechanical stimulation, highly differentiated cardiac muscle syncytium will be developed with contractile and electrophysiological characteristics similar to the native myocardium (Fink et al. 2000; Zimmermann et al. 2002b, 2006). In another study by Miklas et al., a custom-made bioreactor setup was used to simultaneously induce electrical and mechanical stimulations on patterned cardiac tissues (Miklas et al. 2014). The bioreactor design consisted of eight individual microwells fabricated in PDMS. Each microwell had two end posts acting as fixation points to the tissue along with two electrodes for electrical stimulation. Mechanical stimulation (5% cyclic stretch) was induced using a pneumatically actuated stretching setup while electrical stimulation was generated through paired carbon electrodes within each chamber. Neonatal rat cardiomyocytes were encapsulated in collagen type I hydrogel and then injected within each microwell for subsequent experimental analysis. Following cell culture for 3 days, electrical and mechanical stimulation in individual and combinatorial settings were applied to the micro-tissues for duration of 3 days. Cyclic mechanical stretch in combination with electrical stimulation significantly enhanced sarcomere and troponin-T expression throughout the tissues as compared to individualized stimulation conditions. Furthermore, the contractility of the microtissues was promoted in the presence of coupled electrical and mechanical stimulations (Miklas et al. 2014).

8.5 CONCLUSION AND PERSPECTIVES

Cardiac tissue engineering is a growing field due to its significant promises for the regeneration and repair of injured heart. This field has recently gained a unique focus for specific applications in organ-on-chip and disease modeling aspects. Despite significant progresses from both biological and engineering standpoints, there are still numerous difficulties in creating engineered tissue substitutes that can be safely transplanted in patients. A major challenge is the access to the optimal cell sources to successfully perform the required *in vitro* studies prior to clinical assessments. In this regard, a large number of previously reported studies have relied on the use of animal derived cells. Although the use of stem cells such as human iPSCs (Nelson et al. 2009; Yamada et al. 2013) has gained significant attention for cardiac regeneration, the homogeneous differentiation of these cells toward ventricular-specific cardiac lineage is still a major challenge. Furthermore, there is an unmet need for the development of a new generation of biomaterials with optimal degradability, electrical conductivity, robust elasticity, and angiogenic properties. Lastly, the design of improved bioreactor systems are crucially required to induce tunable regimes of electrical and mechanical stimulations to the engineered cardiac tissues in presence of a perfusion system.

ABBREVIATIONS

3D	Three-dimensional
2D	Two-dimensional
BMMNC	Bone marrow–derived mononuclear cells
CMs	Cardiomyocytes
CNT	Carbon nanotube
ECM	Extracellular matrix
GelMA	Gelatin methacrylate
GNPs	Gold nanoparticles
GO	Graphene oxide
HA	Hyaluronic acid
hESC	Human embryonic stem cell
IGF	Insulin like growth factor
iPSCs	Induced pluripotent stem cells
MeTro	Methacrylated tropoelastin

MMP-2	Matrix metalloprotease-2
MSC	Mesenchymal stem cell
PANI	Polyaniline
PCL	Polycaprolactone
PDMS	Polydimethylsiloxane
PEG	Polyethylene glycol
PGCL	Poly-glycolide-co-caprolactone
PGS	Poly(glycerol-sebacate)
PLA	Poly(lactic acid)
PLGA	Poly(lactic-co-glycolic acid)
PLLA	Poly(L-lactide acid)
PLN	Phospholamban
PNIPAAm	Poly-N-isopropylacrylamide
RLP	Resilin-like polypeptide
UV	Ultraviolet

REFERENCES

Agarwal, A., Y. Farouz, A. P. Nesmith, L. F. Deravi, M. L. McCain, and K. K. Parker. 2013a. Micropatterning alginate substrates for in vitro cardiovascular muscle on a chip. *Advanced Functional Materials* 23(30):3738–3746.

Agarwal, A., J. A. Goss, A. Cho, M. L. McCain, and K. K. Parker. 2013b. Microfluidic heart on a chip for higher throughput pharmacological studies. *Lab on a Chip* 13(18):3599–3608.

Al-Haque, S., J. W. Miklas, N. Feric, L. L. Y. Chiu, W. L. K. Chen, C. A. Simmons, and M. Radisic. 2012. Hydrogel substrate stiffness and topography interact to induce contact guidance in cardiac fibroblasts. *Macromolecular Bioscience* 12(10):1342–1353.

Alberts, B., A. Johnson, J. Lewis, M. Raff, K. Roberts, and P. Walter. 2007. *Molecular Biology of the Cell*, 5th edn. New York: Garland Science—Taylor & Francis Group.

Alford, P. W., A. W. Feinberg, S. P. Sheehy, and K. K. Parker. 2010. Biohybrid thin films for measuring contractility in engineered cardiovascular muscle. *Biomaterials* 31(13):3613–3621.

Amir, G., L. Miller, M. Shachar, M. S. Feinberg, R. Holbova, S. Cohen, and J. Leor. 2009. Evaluation of a peritoneal-generated cardiac patch in a rat model of heterotopic heart transplantation. *Cell Transplantation* 18(3):275–282.

Annabi, N., K. Tsang, S. M. Mithieux, M. Nikkhah, A. Ameri, A. Khademhosseini, and A. S. Weiss. 2013. Highly elastic micropatterned hydrogel for engineering functional cardiac tissue. *Advanced Functional Materials* 23(39):4950–4959.

Arai, K., M. Tanaka, S. Yamamoto, and M. Shimomura. 2008. Effect of pore size of honeycomb films on the morphology, adhesion and cytoskeletal organization of cardiac myocytes. *Colloids and Surfaces A—Physicochemical and Engineering Aspects* 313–314:530–535.

Atala, A., F. Kurtis Kasper, and A. G. Mikos. 2012. Engineering complex tissues. *Science Translational Medicine* 4(160):160rv12.

Au, H. T. H, I. Cheng, M. F. Chowdhury, and M. Radisic. 2007. Interactive effects of surface topography and pulsatile electrical field stimulation on orientation and elongation of fibroblasts and cardiomyocytes. *Biomaterials* 28(29):4277–4293.

Au, H. T. H., B. Cui, Z. E. Chu, T. Veres, and M. Radisic. 2009. Cell culture chips for simultaneous application of topographical and electrical cues enhance phenotype of cardiomyocytes. *Lab on a Chip* 9(4):564–575.

Aubin, H., J. W. Nichol, C. B. Hutson, H. Bae, A. L. Sieminski, D. M. Cropek, P. Akhyari, and A. Khademhosseini. 2010. Directed 3D cell alignment and elongation in microengineered hydrogels. *Biomaterials* 31(27):6941–6951.

Ayaz, H. G. S., A. Perets, H. Ayaz, K. D. Gilroy, M. Govindaraj, D. Brookstein, and P. I. Lelkes. 2014. Textile-templated electrospun anisotropic scaffolds for regenerative cardiac tissue engineering. *Biomaterials* 35(30):8540–8552.

Baig, M. K., N. Mahon, W. J. McKenna, A. L. P. Caforio, R. O. Bonow, G. S. Francis, and M. Gheorghiade. 1998. The pathophysiology of advanced heart failure. *American Heart Journal* 135(6):S216–S230.

Bergmann, O., R. D. Bhardwaj, S. Bernard, S. Zdunek, F. Barnabe-Heider, S. Walsh, J. Zupicich et al. 2009. Evidence for cardiomyocyte renewal in humans. *Science* 324(5923):98–102.

Bettinger, C. J., R. Langer, and J. T. Borenstein. 2009. Engineering substrate topography at the micro- and nanoscale to control cell function. *Angewandte Chemie-International Edition* 48(30):5406–5415.

Bian, W., N. Badie, H. D. Himel, and N. Bursac. 2014a. Robust T-tubulation and maturation of cardiomyocytes using tissue engineered epicardial mimetics. *Biomaterials* 35(12):3819–3828.

Bian, W., C. P. Jackman, and N. Bursac. 2014b. Controlling the structural and functional anisotropy of engineered cardiac tissues. *Biofabrication* 6(2):024109.

Biernacka, A. and N. G. Frangogiannis. 2011. Aging and cardiac fibrosis. *Aging and Disease* 2(2):158–173.

Birla, R. K., G. H. Borschel, R. G. Dennis, and D. L. Brown. 2005. Myocardial engineering in vivo: Formation and characterization of contractile, vascularized three-dimensional cardiac tissue. *Tissue Engineering* 11(5–6):803–813.

Black, L. D., III, J. D. Meyers, J. S. Weinbaum, Y. A. Shvelidze, and R. T. Tranquillo. 2009. Cell-induced alignment augments twitch force in fibrin gel-based engineered myocardium via gap junction modification. *Tissue Engineering Part A* 15(10):3099–3108.

Boudou, T., W. R. Legant, A. Mu, M. A. Borochin, N. Thavandiran, M. Radisic, P. W. Zandstra, J. A. Epstein, K. B. Margulies, and C. S. Chen. 2012. A microfabricated platform to measure and manipulate the mechanics of engineered cardiac microtissues. *Tissue Engineering Part A* 18(9–10):910–919.

Bray, M. A. P., W. J. Adams, N. A. Geisse, A. W. Feinberg, S. P. Sheehy, and K. K. Parker. 2010. Nuclear morphology and deformation in engineered cardiac myocytes and tissues. *Biomaterials* 31(19):5143–5150.

Brutsaert, D. L. 2003. Cardiac endothelial-myocardial signaling: Its role in cardiac growth, contractile performance, and rhythmicity. *Physiological Reviews* 83(1):59–115.

Bursac, N., K. K. Parker, S. Iravanian, and L. Tung. 2002. Cardiomyocyte cultures with controlled macroscopic anisotropy—A model for functional electrophysiological studies of cardiac muscle. *Circulation Research* 91(12):E45–E54.

Camci-Unal, G., N. Annabi, M. R. Dokmeci, R. Liao, and A. Khademhosseini. 2014. Hydrogels for cardiac tissue engineering. *NPG Asia Materials* 6:e99.

Caspi, O., A. Lesman, Y. Basevitch, A. Gepstein, G. Arbel, I. Huber, M. Habib, L. Gepstein, and S. Levenberg. 2007. Tissue engineering of vascularized cardiac muscle from human embryonic stem cells. *Circulation Research* 100(2):263–272.

Castaldo, C., F. Di Meglio, R. Miraglia, A. M. Sacco, V. Romano, C. Bancone, A. D. Corte, S. Montagnani, and D. Nurzynska. 2013. Cardiac fibroblast-derived extracellular matrix (biomatrix) as a model for the studies of cardiac primitive cell biological properties in normal and pathological adult human heart. *Biomed Research International*. Article ID: 352370, 1–7.

Chen, A., E. Lee, R. Tu, K. Santiago, A. Grosberg, C. Fowlkes, and M. Khine. 2014. Integrated platform for functional monitoring of biomimetic heart sheets derived from human pluripotent stem cells. *Biomaterials* 35(2):675–683.

Chen, Q.-Z., S. E. Harding, N. N. Ali, A. R. Lyon, and A. R. Boccaccini. 2008. Biomaterials in cardiac tissue engineering: Ten years of research survey. *Materials Science & Engineering R—Reports* 59(1–6):1–37.

Chiu, L. L. Y., K. Janic, and M. Radisic. 2012. Engineering of oriented myocardium on three-dimensional micropatterned collagen-chitosan hydrogel. *International Journal of Artificial Organs* 35(4):237–250.

Chung, B. G., L. Kang, and A. Khademhosseini. 2007. Micro- and nanoscale technologies for tissue engineering and drug discovery applications. *Expert Opinion on Drug Discovery* 2(12):1653–1668.

Cimetta, E., S. Pizzato, S. Bollini, E. Serena, P. De Coppi, and N. Elvassore. 2009. Production of arrays of cardiac and skeletal muscle myofibers by micropatterning techniques on a soft substrate. *Biomedical Microdevices* 11(2):389–400.

Clarke, B. 2008. Normal bone anatomy and physiology. *Clinical Journal of the American Society of Nephrology* 3:S131–S139.

Dalby, M. J., N. Gadegaard, R. Tare, A. Andar, M. O. Riehle, P. Herzyk, C. D. W. Wilkinson, and R. O. C. Oreffo. 2007. The control of human mesenchymal cell differentiation using nanoscale symmetry and disorder. *Nature Materials* 6(12):997–1003.

Davis, M. E., P. C. H. Hsieh, T. Takahashi, Q. Song, S. G. Zhang, R. D. Kamm, A. J. Grodzinsky, P. Anversa, and R. T. Lee. 2006. Local myocardial insulin-like growth factor 1 (IGF-1) delivery with biotinylated peptide nanofibers improves cell therapy for myocardial infarction. *Proceedings of the National Academy of Sciences of the United States of America* 103(21):8155–8160.

Deutsch, J., D. Motiagh, B. Russell, and T. A. Desai. 2000. Fabrication of microtextured membranes for cardiac myocyte attachment and orientation. *Journal of Biomedical Materials Research* 53(3):267–275.

Duan, Y., Z. Liu, J. O'Neill, L. Q. Wan, D. O. Freytes, and G. Vunjak-Novakovic. 2011. Hybrid gel composed of native heart matrix and collagen Induces cardiac differentiation of human embryonic stem cells without supplemental growth factors. *Journal of Cardiovascular Translational Research* 4(5):605–615.

Dvir, T., B. P. Timko, M. D. Brigham, S. R. Naik, S. S. Karajanagi, O. Levy, H. Jin, K. K. Parker, R. Langer, and D. S. Kohane. 2011a. Nanowired three-dimensional cardiac patches. *Nature Nanotechnology* 6(11):720–725.

Dvir, T., B. P. Timko, D. S. Kohane, and R. Langer. 2011b. Nanotechnological strategies for engineering complex tissues. *Nature Nanotechnology* 6(1):13–22.

Engelmayr, G. C., Jr., M. Cheng, C. J. Bettinger, J. T. Borenstein, R. Langer, and L. E. Freed. 2008. Accordion-like honeycombs for tissue engineering of cardiac anisotropy. *Nature Materials* 7(12):1003–1010.

Engler, A. J., C. Carag-Krieger, C. P. Johnson, M. Raab, H.-Y. Tang, D. W. Speicher, J. W. Sanger, J. M. Sanger, and D. E. Discher. 2008. Embryonic cardiomyocytes beat best on a matrix with heart-like elasticity: Scar-like rigidity inhibits beating. *Journal of Cell Science* 121(22):3794–3802.

Entcheva, E. and H. Bien. 2003. Tension development and nuclear eccentricity in topographically controlled cardiac syncytium. *Biomedical Microdevices* 5(2):163–168.

Entcheva, E., H. Bien, L. H. Yin, C. Y. Chung, M. Farrell, and Y. Kostov. 2004. Functional cardiac cell constructs on cellulose-based scaffolding. *Biomaterials* 25(26):5753–5762.

Eschenhagen, T., C. Fink, U. Remmers, H. Scholz, J. Wattchow, J. Weil, W. Zimmerman et al. 1997. Three-dimensional reconstitution of embryonic cardiomyocytes in a collagen matrix: A new heart muscle model system. *FASEB Journal* 11(8):683–694.

Eschenhagen, T. and W. H. Zimmermann. 2005. Engineering myocardial tissue. *Circulation Research* 97(12):1220–1231.

Feinberg, A. W., P. W. Alford, H. Jin, C. M. Ripplinger, A. A. Werdich, S. P. Sheehy, A. Grosberg, and K. K. Parker. 2012. Controlling the contractile strength of engineered cardiac muscle by hierarchal tissue architecture. *Biomaterials* 33(23):5732–5741.

Feinberg, A. W., A. Feigel, S. S. Shevkoplyas, S. Sheehy, G. M. Whitesides, and K. Kit Parker. 2007. Muscular thin films for building actuators and powering devices. *Science* 317(5843):1366–1370.

Feinberg, A. W., C. M. Ripplinger, P. van der Meer, S. P. Sheehy, I. Domian, K. R. Chien, and K. K. Parker. 2013. Functional differences in engineered myocardium from embryonic stem cell- derived versus neonatal cardiomyocytes. *Stem Cell Reports* 1(5):387–396.

Fink, C., S. Ergun, D. Kralisch, U. Remmers, J. Weil, and T. Eschenhagen. 2000. Chronic stretch of engineered heart tissue induces hypertrophy and functional improvement. *FASEB Journal* 14(5):669–679.

Fleischer, S. and T. Dvir. 2013. Tissue engineering on the nanoscale: Lessons from the heart. *Current Opinion in Biotechnology* 24(4):664–671.

Folch, A. 2012. *Introduction to BioMEMS*. Boca Raton, FL: CRC Press.

Frantz, C., K. M. Stewart, and V. M. Weaver. 2010. The extracellular matrix at a glance. *Journal of Cell Science* 123(24):4195–4200.

Fujimoto, K. L., Z. W. Ma, D. M. Nelson, R. Hashizume, J. J. Guan, K. Tobita, and W. R. Wagner. 2009. Synthesis, characterization and therapeutic efficacy of a biodegradable, thermoresponsive hydrogel designed for application in chronic infarcted myocardium. *Biomaterials* 30(26):4357–4368.

Fujita, M., M. Ishihara, Y. Morimoto, M. Simizu, Y. Saito, H. Yura, T. Matsui et al. 2005. Efficacy of photocrosslinkable chitosan hydrogel containing fibroblast growth factor-2 in a rabbit model of chronic myocardial infarction. *Journal of Surgical Research* 126(1):27–33.

Gillette, B. M., J. A. Jensen, B. Tang, G. J. Yang, A. Bazargan-Lari, M. Zhong, and S. K. Sia. 2008. In situ collagen assembly for integrating microfabricated three-dimensional cell-seeded matrices. *Nature Materials* 7(8):636–640.

Godier-Furnemont, A. and G. Vunjak-Novakovic. 2013. Cardiac muscle engineering. In *Biomaterials Science: An Introduction to Materials in Medicine*, eds. B. D. Ratner, A. S. Hoffman, F. J. Schoen and J. E. Lemons. Waltham, MA: Academic Press.

Grosberg, A., A. P. Nesmith, J. A. Goss, M. D. Brigham, M. L. McCain, and K. K. Parker. 2012. Muscle on a chip: In vitro contractility assays for smooth and striated muscle. *Journal of Pharmacological and Toxicological Methods* 65(3):126–135.

Guillemette, M. D., H. Park, J. C. Hsiao, S. R. Jain, B. L. Larson, R. Langer, and L. E. Freed. 2010. Combined technologies for microfabricating elastomeric cardiac tissue engineering scaffolds. *Macromolecular Bioscience* 10(11):1330–1337.

Guo, H.-D., G.-H. Cui, J.-J. Yang, C. Wang, J. Zhu, L.-S. Zhang, J. Jiang, and S.-J. Shao. 2012. Sustained delivery of VEGF from designer self-assembling peptides improves cardiac function after myocardial infarction. *Biochemical and Biophysical Research Communications* 424(1):105–111.

Gupta, V. and K. J. Grande-Allen. 2006. Effects of static and cyclic loading in regulating extracellular matrix synthesis by cardiovascular cells. *Cardiovascular Research* 72(3):375–383.

Hay, E. D. 1991. *Cell Biology of Extracellular Matrix*, 2nd edn. New York: Springer.

Hirt, M. N., J. Boeddinghaus, A. Mitchell, S. Schaaf, C. Börnchend, C. Müller, H. Schulz et al. 2014. Functional improvement and maturation of rat and human engineered heart tissue by chronic electrical stimulation. *Journal of Molecular and Cellular Cardiology* 74:151–161.

Hsiao, C.-W., M.-Y. Bai, Y. Chang, M.-F. Chung, T.-Y. Lee, C.-T. Wu, B. Maiti, Z.-X. Liao, R.-K. Li, and H.-W. Sung. 2013. Electrical coupling of isolated cardiomyocyte clusters grown on aligned conductive nanofibrous meshes for their synchronized beating. *Biomaterials* 34(4):1063–1072.

Hsieh, P. C. H., M. E. Davis, J. Gannon, C. MacGillivray, and R. T. Lee. 2006. Controlled delivery of PDGF-BB for myocardial protection using injectable self-assembling peptide nanofibers. *Journal of Clinical Investigation* 116(1):237–248.

Huang, N. F. and S. Li. 2011. Regulation of the matrix microenvironment for stem cell engineering and regenerative medicine. *Annals of Biomedical Engineering* 39(4):1201–1214.

Ifkovits, J. L., J. J. Devlin, G. Eng, T. P. Martens, G. Vunjak-Novakovic, and J. A. Burdick. 2009. Biodegradable fibrous scaffolds with tunable properties formed from photo-cross-linkable poly(glycerol sebacate). *ACS Applied Materials & Interfaces* 1(9):1878–1886.

Ifkovits, J. L., E. Tous, M. Minakawa, M. Morita, J. D. Robb, K. J. Koomalsingh, J. H. Gorman, III, R. C. Gorman, and J. A. Burdick. 2010. Injectable hydrogel properties influence infarct expansion and extent of postinfarction left ventricular remodeling in an ovine model. *Proceedings of the National Academy of Sciences of the United States of America* 107(25):11507–11512.

Iyer, R. K., L. L. Y. Chiu, and M. Radisic. 2009. Microfabricated poly(ethylene glycol) templates enable rapid screening of triculture conditions for cardiac tissue engineering. *Journal of Biomedical Materials Research Part A* 89A(3):616–631.

Iyer, R. K., L. L. Y. Chiu, G. Vunjak-Novakovic, and M. Radisic. 2012. Biofabrication enables efficient interrogation and optimization of sequential culture of endothelial cells, fibroblasts and cardiomyocytes for formation of vascular cords in cardiac tissue engineering. *Biofabrication* 4(3):035002.

Iyer, R. K., J. Chui, and M. Radisic. 2009. Spatiotemporal tracking of cells in tissue-engineered cardiac organoids. *Journal of Tissue Engineering and Regenerative Medicine* 3(3):196–207.

Jongpaiboonkit, L., W. J. King, G. E. Lyons, A. L. Paguirigan, J. W. Warrick, D. J. Beebe, and W. L. Murphy. 2008. An adaptable hydrogel array format for 3-dimensional cell culture and analysis. *Biomaterials* 29(23):3346–3356.

Kajzar, A., C. M. Cesa, N. Kirchgessner, B. Hoffmann, and R. Merkel. 2008. Toward physiological conditions for cell analyses: Forces of heart muscle cells suspended between elastic micropillars. *Biophysical Journal* 94(5):1854–1866.

Kane, R. S., S. Takayama, E. Ostuni, D. E. Ingber, and G. M. Whitesides. 1999. Patterning proteins and cells using soft lithography. *Biomaterials* 20(23–24):2363–2376.

Karp, J. M., Y. Yeo, W. Geng, C. Cannizzaro, K. Yan, D. S. Kohane, G. Vunjak-Novakovic, R. S. Langer, and M. Radisic. 2006. A photolithographic method to create cellular micropatterns. *Biomaterials* 27(27):4755–4764.

Khademhosseini, A., G. Eng, J. Yeh, P. A. Kucharczyk, R. Langer, G. Vunjak-Novakovic, and M. Radisic. 2007. Microfluidic patterning for fabrication of contractile cardiac organoids. *Biomedical Microdevices* 9(2):149–157.

Khademhosseini, A., R. Langer, J. Borenstein, and J. P. Vacanti. 2006. Microscale technologies for tissue engineering and biology. *Proceedings of the National Academy of Sciences of the United States of America* 103(8):2480–2487.

Kharaziha, M., M. Nikkhah, S.-R. Shin, N. Annabi, N. Masoumi, A. K. Gaharwar, G. Camci-Unal, and A. Khademhosseini. 2013. PGS: Gelatin nanofibrous scaffolds with tunable mechanical and structural properties for engineering cardiac tissues. *Biomaterials* 34(27):6355–6366.

Kharaziha, M., S.-R. Shin, M. Nikkhah, S. N. Topkaya, N. Masoumi, N. Annabi, M. R. Dokmeci, and A. Khademhosseini. 2014. Tough and flexible CNT-polymeric hybrid scaffolds for engineering cardiac constructs. *Biomaterials* 35(26):7346–7354.

Kim, D.-H., E. A. Lipke, P. Kim, R. Cheong, S. Thompson, M. Delannoy, K.-Y. Suh, L. Tung, and A. Levchenko. 2010. Nanoscale cues regulate the structure and function of macroscopic cardiac tissue constructs. *Proceedings of the National Academy of Sciences of the United States of America* 107(2):565–570.

Kim, J., J. Park, K. Na, S. Yang, J. Baek, E. Yoon, S. Choi, S. Lee, K. Chun, J. Park, and S. Park. 2008. Quantitative evaluation of cardiomyocyte contractility in a 3D microenvironment. *Journal of Biomechanics* 41(11):2396–2401.

Kofidis, T., P. Akhyari, J. Boublik, P. Theodorou, U. Martin, A. Ruhparwar, S. Fischer et al. 2002. In vitro engineering of heart muscle: Artificial myocardial tissue. *Journal of Thoracic and Cardiovascular Surgery* 124(1):63–69.

Kofidis, T., K. Mueller-Stahl, and A. Haverich. 2007. Myocardial restoration and tissue engineering of heart structures. In *Methods in Molecular Medicine*, Hansjörg, H. and Martin, F. (eds.), Vol. 140, pp. 273–290, Humana Press.

Kolewe, M. E., H. Park, C. Gray, X. Ye, R. Langer, and L. E. Freed. 2013. 3D structural patterns in scalable, elastomeric scaffolds guide engineered tissue architecture. *Advanced Materials* 25(32):4459–4465.

Kraehenbuehl, T. P., P. Zammaretti, A. J. Van der Vlies, R. G. Schoenmakers, M. P. Lutolf, M. E. Jaconi, and J. A. Hubbell. 2008. Three-dimensional extracellular matrix-directed cardioprogenitor differentiation: Systematic modulation of a synthetic cell-responsive PEG-hydrogel. *Biomaterials* 29(18):2757–2766.

Kreutziger, K. L. and C. E. Murry. 2011. Engineered human cardiac tissue. *Pediatric Cardiology* 32(3):334–341.

Kuo, P.-L., H. Lee, M.-A. Bray, N. A. Geisse, Y.-T. Huang, W. J. Adams, S. P. Sheehy, and K. K. Parker. 2012. Myocyte shape regulates lateral registry of sarcomeres and contractility. *American Journal of Pathology* 181(6):2030–2037.

Lanza, R., R. Langer, and J. P. Vacanti. 2014. *Principles of Tissue Engineering*, 4th edn. Waltham, MA: Academic Press.

Li, R. K., Z. Q. Jia, R. D. Weisel, D. A. G. Mickle, A. Choi, and T. M. Yau. 1999. Survival and function of bioengineered cardiac grafts. *Circulation* 100(19):63–69.

Li, R. K., T. M. Yau, R. D. Weisel, D. A. G. Mickle, T. Sakai, A. Choi, and Z. Q. Jia. 2000. Construction of a bioengineered cardiac graft. *Journal of Thoracic and Cardiovascular Surgery* 119(2):368–375.

Lin, Y.-D., M.-C. Ko, S.-T. Wu, S.-F. Li, J.-F. Hu, Y.-J. Lai, H. I. C. Harn et al. 2014. A nanopatterned cell-seeded cardiac patch prevents electro-uncoupling and improves the therapeutic efficacy of cardiac repair. *Biomaterials Science* 2(4):567–580.

Luna, J. I., J. Ciriza, M. E. Garcia-Ojeda, M. Kong, A. Herren, D. K. Lieu, R. A. Li, C. C. Fowlkes, M. Khine, and K. E. McCloskey. 2011. Multiscale biomimetic topography for the alignment of neonatal and embryonic stem cell-derived heart cells. *Tissue Engineering Part C—Methods* 17(5):579–588.

Ma, Z., S. Koo, M. A. Finnegan, P. Loskill, N. Huebsch, N. C. Marks, B. R. Conklin, C. P. Grigoropoulos, and K. E. Healy. 2014. Three-dimensional filamentous human diseased cardiac tissue model. *Biomaterials* 35(5):1367–1377.

Malmivuo, J. and R. Plonsey. 1995. *Bioelectromagnetism: Principles and Applications of Bioelectric and Biomagnetic Fields*. New York: Oxford University Press.

Mark, G. E. and F. F. Strasser. 1966. Pacemaker activity and mitosis in cultures of newborn rat heart ventricle cells. *Experimental Cell Research* 44(2–3):217.

Marsano, A., R. Maidhof, L. Q. Wan, Y. Wang, J. Gao, N. Tandon, and G. Vunjak-Novakovic. 2010. Scaffold stiffness affects the contractile function of three-dimensional engineered cardiac constructs. *Biotechnology Progress* 26(5):1382–1390.

McDevitt, T. C., J. C. Angello, M. L. Whitney, H. Reinecke, S. D. Hauschka, C. E. Murry, and P. S. Stayton. 2002. In vitro generation of differentiated cardiac myofibers on micropatterned laminin surfaces. *Journal of Biomedical Materials Research* 60(3):472–479.

McDevitt, T. C., K. A. Woodhouse, S. D. Hauschka, C. E. Murry, and P. S. Stayton. 2003. Spatially organized layers of cardiomyocytes on biodegradable polyurethane films for myocardial repair. *Journal of Biomedical Materials Research Part A* 66A(3):586–595.

McGann, C. L., E. A. Levenson, and K. L. Kiick. 2013. Resilin-based hybrid hydrogels for cardiovascular tissue engineering. *Macromolecular Chemistry and Physics* 214(2):203–213.

Miklas, J. W., S. S. Nunes, A. Sofla, L. A. Reis, A. Pahnke, Y. Xiao, C. Laschinger, and M. Radisic. 2014. Bioreactor for modulation of cardiac microtissue phenotype by combined static stretch and electrical stimulation. *Biofabrication* 6:1–14.

Miyagawa, S., Y. Sawa, S. Sakakida, S. Taketani, H. Kondoh, I. A. Memon, Y. Imanishi, T. Shimizu, T. Okano, and H. Matsuda. 2005. Tissue cardiomyoplasty using bioengineered contractile cardiomyocyte sheets to repair damaged myocardium: Their integration with recipient myocardium. *Transplantation* 80(11):1586–1595.

Miyahara, Y., N. Nagaya, M. Kataoka, B. Yanagawa, K. Tanaka, H. Hao, K. Ishino et al. 2006. Monolayered mesenchymal stem cells repair scarred myocardium after myocardial infarction. *Nature Medicine* 12(4):459–465.

Motlagh, D., T. J. Hartman, T. A. Desai, and B. Russell. 2003. Microfabricated grooves recapitulate neonatal myocyte connexin43 and N-cadherin expression and localization. *Journal of Biomedical Materials Research Part A* 67A(1):148–157.

Murtuza, B., J. W. Nichol, and A. Khademhosseini. 2009. Micro- and nanoscale control of the cardiac stem cell niche for tissue fabrication. *Tissue Engineering Part B—Reviews* 15(4):443–454.

Nag, A. C. 1980. Study of non-muscle cells of the adult mammalian heart: A fine structural analysis and distribution. *Cytobios* 28(109):41–61.

Naito, H., Y. Takewa, T. Mizuno, S. Ohya, Y. Nakayama, E. Tatsumi, S. Kitamura, H. Takano, S. Taniguchi, and Y. Taenaka. 2004. Three-dimensional cardiac tissue engineering using a thermoresponsive artificial extracellular matrix. *ASAIO Journal* 50(4):344–348.

Narmoneva, D. A., R. Vukmirovic, M. E. Davis, R. D. Kamm, and R. T. Lee. 2004. Endothelial cells promote cardiac myocyte survival and spatial reorganization—Implications for cardiac regeneration. *Circulation* 110(8):962–968.

Natarajan, A., M. Stancescu, V. Dhir, C. Armstrong, F. Sommerhage, J. J. Hickman, and P. Molnar. 2011. Patterned cardiomyocytes on microelectrode arrays as a functional, high information content drug screening platform. *Biomaterials* 32(18):4267–4274.

Nelson, T. J., A. Martinez-Fernandez, S. Yamada, C. Perez-Terzic, Y. Ikeda, and A. Terzic. 2009. Repair of acute myocardial infarction by human stemness factors induced pluripotent stem cells. *Circulation* 120(5):408–416.

Nikkhah, M., F. Edalat, S. Manoucheri, and A. Khademhosseini. 2012a. Engineering microscale topographies to control the cell-substrate interface. *Biomaterials* 33(21):5230–5246.

Nikkhah, M., N. Eshak, P. Zorlutuna, N. Annabi, M. Castello, K. Kim, A. Dolatshahi-Pirouz et al. 2012b. Directed endothelial cell morphogenesis in micropatterned gelatin methacrylate hydrogels. *Biomaterials* 33(35):9009–9018.

Nugent, H. M. and E. R. Edelman. 2003. Tissue engineering therapy for cardiovascular disease. *Circulation Research* 92(10):1068–1078.

Oh, S. H., C. L. Ward, A. Atala, J. J. Yoo, and B. S. Harrison. 2009. Oxygen generating scaffolds for enhancing engineered tissue survival. *Biomaterials* 30(5):757–762.

Orlic, D., J. Kajstura, S. Chimenti, F. Limana, I. Jakoniuk, F. Quaini, B. Nadal-Ginard, D. M. Bodine, A. Leri, and P. Anversa. 2001. Mobilized bone marrow cells repair the infarcted heart, improving function and survival. *Proceedings of the National Academy of Sciences of the United States of America* 98(18):10344–10349.

Orlova, Y., N. Magome, L. Liu, Y. Chen, and K. Agladze. 2011. Electrospun nanofibers as a tool for architecture control in engineered cardiac tissue. *Biomaterials* 32(24):5615–5624.

Ozawa, T., D. A. G. Mickle, R. D. Weisel, K. Matsubayashi, T. Fujii, P. W. M. Fedak, N. Koyama, Y. Ikada, and R. K. Li. 2004. Tissue-engineered grafts matured in the right ventricular outflow tract. *Cell Transplantation* 13(2):169–177.

Park, H., C. Cannizzaro, G. Vunjak-Novakovic, R. Langer, C. A. Vacanti, and O. C. Farokhzad. 2007a. Nanofabrication and microfabrication of functional materials for tissue engineering. *Tissue Engineering* 13(8):1867–1877.

Park, H., B. L. Larson, M. E. Kolewe, G. Vunjak-Novakovic, and L. E. Freed. 2014. Biomimetic scaffold combined with electrical stimulation and growth factor promotes tissue engineered cardiac development. *Experimental Cell Research* 321:297–306.

Park, J., S. Bauer, K. von der Mark, and P. Schmuki. 2007b. Nanosize and vitality: TiO_2 nanotube diameter directs cell fate. *Nano Letters* 7(6):1686–1691.

Park, T. H. and M. L. Shuler. 2003. Integration of cell culture and microfabrication technology. *Biotechnology Progress* 19(2):243–253.

Patel, A. A., T. A. Desai, and S. Kumar. 2011. Microtopographical assembly of cardiomyocytes. *Integrative Biology* 3(10):1011–1019.

Patra, C., S. Talukdar, T. Novoyatleva, S. R. Velagala, C. Muehlfeld, B. Kundu, S. C. Kundu, and F. B. Engel. 2012. Silk protein fibroin from *Antheraea mylitta* for cardiac tissue engineering. *Biomaterials* 33(9):2673–2680.

Peppas, N. A., J. Z. Hilt, A. Khademhosseini, and R. Langer. 2006. Hydrogels in biology and medicine: From molecular principles to bionanotechnology. *Advanced Materials* 18(11):1345–1360.

Perl, A., D. N. Reinhoudt, and J. Huskens. 2009. Microcontact printing: Limitations and achievements. *Advanced Materials* 21(22):2257–2268.

Piao, H., J. S. Kwon, S. Piao, J. H. Sohn, Y. S. Lee, J. W. Bae, K. K. Hwang et al. 2007. Effects of cardiac patches engineered with bone marrow-derived mononuclear cells and PGCL scaffolds in a rat myocardial infarction model. *Biomaterials* 28(4):641–649.

Qazi, T. H., R. Rai, D. Dippold, J. E. Roether, D. W. Schubert, E. Rosellini, N. Barbani, and A. R. Boccaccini. 2014. Development and characterization of novel electrically conductive PANI-PGS composites for cardiac tissue engineering applications. *Acta Biomaterialia* 10(6):2434–2445.

Radisic, M., A. Marsano, R. Maidhof, Y. Wang, and G. Vunjak-Novakovic. 2008a. Cardiac tissue engineering using perfusion bioreactor systems. *Nature Protocols* 3(4):719–738.

Radisic, M., H. Park, S. Gerecht, C. Cannizzaro, R. Langer, and G. Vunjak-Novakovic. 2007. Biomimetic approach to cardiac tissue engineering. *Philosophical Transactions of the Royal Society B-Biological Sciences* 362(1484):1357–1368.

Radisic, M., H. Park, T. P. Martens, J. E. Salazar-Lazaro, W. Geng, Y. Wang, R. Langer, L. E. Freed, and G. Vunjak-Novakovic. 2008b. Pre-treatment of synthetic elastomeric scaffolds by cardiac fibroblasts improves engineered heart tissue. *Journal of Biomedical Materials Research Part A* 86A(3):713–724.

Radisic, M., H. Park, H. Shing, T. Consi, F. J. Schoen, R. Langer, L. E. Freed, and G. Vunjak-Novakovic. 2004. Functional assembly of engineered myocardium by electrical stimulation of cardiac myocytes cultured on scaffolds. *Proceedings of the National Academy of Sciences of the United States of America* 101(52):18129–18134.

Ramaciotti, C., A. Sharkey, G. McClellan, and S. Winegrad. 1992. Endothelial cells regulate cardiac contractility. *Proceedings of the National Academy of Sciences of the United States of America* 89(9):4033–4036.

Rao, C., T. Prodromakis, L. Kolker, U. A. R. Chaudhry, T. Trantidou, A. Sridhar, C. Weekes et al. 2013. The effect of microgrooved culture substrates on calcium cycling of cardiac myocytes derived from human induced pluripotent stem cells. *Biomaterials* 34(10):2399–2411.

Rockwood, D. N., R. E. Akins, Jr., I. C. Parrag, K. A. Woodhouse, and J. F. Rabolt. 2008. Culture on electrospun polyurethane scaffolds decreases atrial natriuretic peptide expression by cardiomyocytes in vitro. *Biomaterials* 29(36):4783–4791.

Rodriguez, M. L., B. T. Graham, L. M. Pabon, S. J. Han, C. E. Murry, and N. J. Sniadecki. 2014. Measuring the contractile forces of human induced pluripotent stem cell-derived cardiomyocytes with arrays of microposts. *Journal of Biomechanical Engineering—Transactions of the ASME* 136(5):051005.

Sakiyama-Elbert, S. E. and J. A. Hubbell. 2001. Functional biomaterials: Design of novel biomaterials. *Annual Review of Materials Research* 31:183–201.

Salick, M. R., B. N. Napiwocki, J. Sha, G. T. Knight, S. A. Chindhy, T. J. Kamp, R. S. Ashton, and W. C. Crone. 2014. Micropattern width dependent sarcomere development in human ESC-derived cardiomyocytes. *Biomaterials* 35(15):4454–4464.

Segers, V. F. M. and R. T. Lee. 2008. Stem-cell therapy for cardiac disease. *Nature* 451(7181):937–942.

Segers, V. F. M. and R. T. Lee. 2011. Biomaterials to enhance stem cell function in the heart. *Circulation Research* 109(8):910–922.

Serena, E., E. Cimetta, S. Zatti, T. Zaglia, M. Zagallo, G. Keller, and N. Elvassore. 2012. Micro-arrayed human embryonic stem cells-derived cardiomyocytes for in vitro functional assay. *PLoS One* 7(11):e48483.

Severs, N. J. 2000. The cardiac muscle cell. *Bioessays* 22(2):188–199.

Shachar, M., N. Benishti, and S. Cohen. 2012. Effects of mechanical stimulation induced by compression and medium perfusion on cardiac tissue engineering. *Biotechnology Progress* 28(6):1551–1559.

Shamhart, P. E. and J. G. Meszaros. 2010. Non-fibrillar collagens: Key mediators of post-infarction cardiac remodeling? *Journal of Molecular and Cellular Cardiology* 48(3):530–537.

Shim, J., A. Grosberg, J. C. Nawroth, K. K. Parker, and K. Bertoldi. 2012. Modeling of cardiac muscle thin films: Pre-stretch, passive and active behavior. *Journal of Biomechanics* 45(5):832–841.

Shin, M., O. Ishii, T. Sueda, and J. P. Vacanti. 2004. Contractile cardiac grafts using a novel nanofibrous mesh. *Biomaterials* 25(17):3717–3723.

Shin, S. R., B. Aghaei-Ghareh-Bolagh, X. Gao, M. Nikkhah, S. M. Jung, A. Dolatshahi-Pirouz, S. B. Kim et al. 2014. Layer-by-layer assembly of 3D tissue constructs with functionalized graphene. *Advanced Functional Materials*, in press.

Shin, S. R., S. M. Jung, M. Zalabany, K. Kim, P. Zorlutuna, S. B. Kim, M. Nikkhah et al. 2013. Carbon-nanotube-embedded hydrogel sheets for engineering cardiac constructs and bioactuators. *ACS Nano* 7(3):2369–2380.

Simpson, D. L. and S. C. Dudley, Jr. 2013. Modulation of human mesenchymal stem cell function in a three-dimensional matrix promotes attenuation of adverse remodelling after myocardial infarction. *Journal of Tissue Engineering and Regenerative Medicine* 7(3):192–202.

Solan, A., S. Mitchell, M. Moses, and L. Niklason. 2003. Effect of pulse rate on collagen deposition in the tissue-engineered blood vessel. *Tissue Engineering* 9(4):579–586.

Soler-Botija, C., J. R. Bago, A. Llucia-Valldeperas, A. Valles-Lluch, C. Castells-Sala, C. Martinez-Ramos, T. Fernandez-Muinos et al. 2014. Engineered 3D bioimplants using elastomeric scaffold, self-assembling peptide hydrogel, and adipose tissue-derived progenitor cells for cardiac regeneration. *American Journal of Translational Research* 6(3):291–301.

Soliman, S., S. Pagliari, A. Rinaldi, G. Forte, R. Fiaccavento, F. Pagliari, O. Franzese et al. 2010. Multiscale three-dimensional scaffolds for soft tissue engineering via multi-modal electrospinning. *Acta Biomaterialia* 6(4):1227–1237.

Song, W., H. Lu, N. Kawazoe, and G. Chen. 2011. Gradient patterning and differentiation of mesenchymal stem cells on micropatterned polymer surface. *Journal of Bioactive and Compatible Polymers* 26(3):242–256.

Soonpaa, M. H., G. Y. Koh, M. G. Klug, and L. J. Field. 1994. Formation of nascent intercalated disks between grafted fetal cardiomyocytes and host myocardium. *Science* 264(5155):98–101.

Souders, C. A., S. L. K. Bowers, and T. A. Baudino. 2009. Cardiac fibroblast the renaissance cell. *Circulation Research* 105(12):1164–1176.

Tandon, N., C. Cannizzaro, P.-H. G. Chao, R. Maidhof, A. Marsano, H. T. H. Au, M. Radisic, and G. Vunjak-Novakovic. 2009. Electrical stimulation systems for cardiac tissue engineering. *Nature Protocols* 4(2):155–173.

Tandon, V., B. Zhang, M. Radisic, and S. K. Murthy. 2013. Generation of tissue constructs for cardiovascular regenerative medicine: From cell procurement to scaffold design. *Biotechnology Advances* 31(5):722–735.

Thavandiran, N., N. Dubois, A. Mikryukov, S. Masse, B. Beca, C. A. Simmons, V. S. Deshpande et al. 2013. Design and formulation of functional pluripotent stem cell-derived cardiac microtissues. *Proceedings of the National Academy of Sciences of the United States of America* 110(49):E4698–E4707.

Thery, M. 2010. Micropatterning as a tool to decipher cell morphogenesis and functions. *Journal of Cell Science* 123(24):4201–4213.

Tous, E., B. Purcell, J. L. Ifkovits, and J. A. Burdick. 2011. Injectable acellular hydrogels for cardiac repair. *Journal of Cardiovascular Translational Research* 4(5):528–542.

Venugopal, J., R. Rajeswari, M. Shayanti, R. Sridhar, S. Sundarrajan, R. Balamurugan, and S. Ramakrishna. 2013. Xylan polysaccharides fabricated into nanofibrous substrate for myocardial infarction. *Materials Science & Engineering C—Materials for Biological Applications* 33(3):1325–1331.

Vunjak-Novakovic, G., N. Tandon, A. Godier, R. Maidhof, A. Marsano, T. P. Martens, and M. Radisic. 2010. Challenges in cardiac tissue engineering. *Tissue Engineering Part B—Reviews* 16(2):169–187.

Wainwright, J. M., C. A. Czajka, U. B. Patel, D. O. Freytes, K. Tobita, T. W. Gilbert, and S. F. Badylak. 2010. Preparation of cardiac extracellular matrix from an intact porcine heart. *Tissue Engineering Part C—Methods* 16(3):525–532.

Walsh, R. G. 2005. Design and features of the acorn CorCap(TM) stop cardiac support device: The concept of passive mechanical diastolic support. *Heart Failure Reviews* 10(2):101–107.

Wang, F., Z. Li, M. Khan, K. Tamama, P. Kuppusamy, W. R. Wagner, C. K. Sen, and J. Guan. 2010a. Injectable, rapid gelling and highly flexible hydrogel composites as growth factor and cell carriers. *Acta Biomaterialia* 6(6):1978–1991.

Wang, H., W. Wang, L. Li, J. Zhu, W. Wang, D. Zhang, Z. Xie, H. Fuchs, Y. Lei, and L. Chi. 2014. Surface microfluidic patterning and transporting organic small molecules. *Small* 10(13):2549–2552.

Wang, H., J. Zhou, Z. Liu, and C. Wang. 2010b. Injectable cardiac tissue engineering for the treatment of myocardial infarction. *Journal of Cellular and Molecular Medicine* 14(5):1044–1055.

Wang, J., A. Chen, D. K. Lieu, I. Karakikes, G. Chen, W. Keung, C. W. Chan et al. 2013. Effect of engineered anisotropy on the susceptibility of human pluripotent stem cell-derived ventricular cardiomyocytes to arrhythmias. *Biomaterials* 34(35):8878–8886.

Wang, L. and R. L. Carrier. 2011. Biomimetic topography: Bioinspired cell culture substrates and scaffolds. In *Advances in Biomimetics*, ed. A. George. Rijeka, Croatia: InTech.

Whitesides, G. M., E. Ostuni, S. Takayama, X. Y. Jiang, and D. E. Ingber. 2001. Soft lithography in biology and biochemistry. *Annual Review of Biomedical Engineering* 3:335–373.

Xia, Y. N. and G. M. Whitesides. 1998. Soft lithography. *Annual Review of Materials Science* 28:153–184.

Yamada, S., T. J. Nelson, G. C. Kane, A. Martinez-Fernandez, R. J. Crespo-Diaz, Y. Ikeda, C. Perez-Terzic, and A. Terzic. 2013. Induced pluripotent stem cell intervention rescues ventricular wall motion disparity, achieving biological cardiac resynchronization post-infarction. *Journal of Physiology—London* 591(17):4335–4349.

Yang, M. and X. Zhang. 2007. Electrical assisted patterning of cardiac myocytes with controlled macroscopic anisotropy using a microfluidic dielectrophoresis chip. *Sensors and Actuators A—Physical* 135(1):73–79.

Yang, M.-C., S.-S. Wang, N.-K. Chou, N.-H. Chi, Y.-Y. Huang, Y.-L. Chang, M.-J. Shieh, and T.-W. Chung. 2009. The cardiomyogenic differentiation of rat mesenchymal stem cells on silk fibroin-polysaccharide cardiac patches in vitro. *Biomaterials* 30(22):3757–3765.

Yasukawa, A., Y. Takayama, T. Suzuki, and K. Mabuchi. 2013. Construction of an in vitro model system for the anatomical reentry phenomenon of cardiac tissues by using microfabrication techniques. *IEEJ Transactions on Electrical and Electronic Engineering* 8(3):308–309.

Yin, L. H., H. Bien, and E. Entcheva. 2004. Scaffold topography alters intracellular calcium dynamics in cultured cardiomyocyte networks. *American Journal of Physiology— Heart and Circulatory Physiology* 287(3):H1276–H1285.

Yoon, S. J., Y. H. Fang, C. H. Lim, B. S. Kim, H. S. Son, Y. Park, and K. Sun. 2009. Regeneration of ischemic heart using hyaluronic acid-based injectable hydrogel. *Journal of Biomedical Materials Research Part B—Applied Biomaterials* 91B(1):163–171.

Yu, H., C. Y. Tay, M. Pal, W. S. Leong, H. Li, H. Li, F. Wen, D. T. Leong, and L. P. Tan. 2013. A bio-inspired platform to modulate myogenic differentiation of human mesenchymal stem cells through focal adhesion regulation. *Advanced Healthcare Materials* 2(3):442–449.

Zhang, B., Y. Xiao, A. Hsieh, N. Thavandiran, and M. Radisic. 2011. Micro- and nanotechnology in cardiovascular tissue engineering. *Nanotechnology* 22(49):494003.

Zhang, D., I. Y. Shadrin, J. Lam, H.-Q. Xian, H. R. Snodgrass, and N. Bursac. 2013. Tissue-engineered cardiac patch for advanced functional maturation of human ESC-derived cardiomyocytes. *Biomaterials* 34(23):5813–5820.

Zhang, M., D. Methot, V. Poppa, Y. Fujio, K. Walsh, and C. E. Murry. 2001. Cardiomyocyte grafting for cardiac repair: Graft cell death and anti-death strategies. *Journal of Molecular and Cellular Cardiology* 33(5):907–921.

Zhang, T., L. Q. Wan, Z. Xiong, A. Marsano, R. Maidhof, M. Park, Y. Yan, and G. Vunjak-Novakovic. 2012. Channelled scaffolds for engineering myocardium with mechanical stimulation. *Journal of Tissue Engineering and Regenerative Medicine* 6(9):748–756.

Zieber, L., S. Or, E. Ruvinov, and S. Cohen. 2014. Microfabrication of channel arrays promotes vessel-like network formation in cardiac cell construct and vascularization in vivo. *Biofabrication* 6(2):024102.

Zimmermann, W. H., M. Didie, G. H. Wasmeier, U. Nixdorff, A. Hess, I. Melnychenko, O. Boy, W. L. Neuhuber, M. Weyand, and T. Eschenhagen. 2002a. Cardiac grafting of engineered heart tissue in syngenic rats. *Circulation* 106(13):I151–I157.

Zimmermann, W. H., I. Melnychenko, and T. Eschenhagen. 2004. Engineered heart tissue for regeneration of diseased hearts. *Biomaterials* 25(9):1639–1647.

Zimmermann, W. H., I. Melnychenko, G. Wasmeier, M. Didie, H. Naito, U. Nixdorff, A. Hess et al. 2006. Engineered heart tissue grafts improve systolic and diastolic function in infarcted rat hearts. *Nature Medicine* 12(4):452–458.

Zimmermann, W. H., K. Schneiderbanger, P. Schubert, M. Didie, F. Munzel, J. F. Heubach, S. Kostin, W. L. Neuhuber, and T. Eschenhagen. 2002b. Tissue engineering of a differentiated cardiac muscle construct. *Circulation Research* 90(2):223–230.

Zong, X. H., H. Bien, C. Y. Chung, L. H. Yin, D. F. Fang, B. S. Hsiao, B. Chu, and E. Entcheva. 2005. Electrospun fine-textured scaffolds for heart tissue constructs. *Biomaterials* 26(26):5330–5338.

Zorlutuna, P., N. Annabi, G. Camci-Unal, M. Nikkhah, J. M. Cha, J. W. Nichol, A. Manbachi, H. Bae, S. Chen, and A. Khademhosseini. 2012. Microfabricated biomaterials for engineering 3D tissues. *Advanced Materials* 24(14):1782–1804.

9 Polysaccharide-Based Biomaterials for Cell–Material Interface

Jorge Almodovar, David A. Castilla Casadiego, and Heleine V. Ramos Avilez

CONTENTS

9.1 INTRODUCTION

9.1.1 WHAT ARE POLYSACCHARIDES?

Polysaccharides are long carbohydrate molecules that contain repeated monosaccharide units joined together by means of glycosidic bonds.[1] Polysaccharides are present in a variety of living beings such as in algae as alginate, in plants as pectin and cellulose, in microbes as dextran and xanthan gum, and in animals as chitosan and heparin.[2] They may be classified according to their charge: cationic (e.g., chitosan), anionic (e.g., hyaluronic acid, heparin, and alginate), and nonionic (e.g., dextran).[1,3]

Polysaccharides may be classified by their function[4,5] and their composition.[6,7] For instance, structural polysaccharides (e.g., cellulose and chitin) provide protection, support, and shape to cells, tissues, organs, and organisms; storage polysaccharides (e.g., starch and dextran) serve as a means to store carbohydrates[4,5]; and hydrophilic polysaccharides (e.g., hyaluronan) prevent cell dehydration.[8] They can be classified according to their composition as homopolysaccharides—composed of a monosaccharide unit (e.g., cellulose, starch, and glycogen)—and heteropolysaccharides composed of two or more monosaccharide units.[6,7] An important class of heteropolysaccharides are glycosaminoglycans that are molecules of high molecular weight formed from two or more different types of monosaccharide units, linked by O-glycosidic bond such as hyaluronic acid, heparin, and chondroitin sulfate.[9] Glycosaminoglycans (GAGs) are a crucial component of native extracellular matrix (ECM).

9.1.2 WHAT IS THE ROLE OF POLYSACCHARIDES IN THE ECM?

The ECM provides structural support to cells, serves as a reservoir of signaling molecules, and provides biochemical, mechanical, and physical cues that dictate cellular behavior (Figure 9.1).[10,11] One of the major components of native ECM is glycosaminoglycans. GAGs carry out a variety of biological functions in the ECM, for example, hyaluronan interacts with diverse proteins or proteoglycans to organize the ECM and to maintain tissue homeostasis.[8] Its physical and mechanical properties help maintain tissue hydration and lubrication, and contribute to solute diffusion through the extracellular space.[8] Binding of hyaluronan (HA) with cell surface receptors activates various signaling pathways, which regulate cell function, tissue development, inflammation, wound healing, tumor progression, and metastasis.[8] Polysaccharides enable cellular interactions including cell fate, cell motility, and cell apoptosis.[12] They form noncovalent bioadhesion or react with functional molecules.[3] Heparin, a highly sulfated GAG, serves as a reservoir for multiple growth factors—such as the transforming growth factor β1 (TGFβ1) superfamily—protecting them from degradation and enhancing their biological activity.[13] Thus, polysaccharides are ideal candidates for the design of biomaterials to be used in regenerative medicine and tissue engineering.

9.1.3 WHY ARE THEY USEFUL AS BIOMATERIALS?

Polysaccharides are biocompatible, biodegradable, abundant in nature, and chemically modifiable, and they are involved in living systems processes.[1,3,14] They have the ability to mimic the ECM while offering temporary mechanical support.[12]

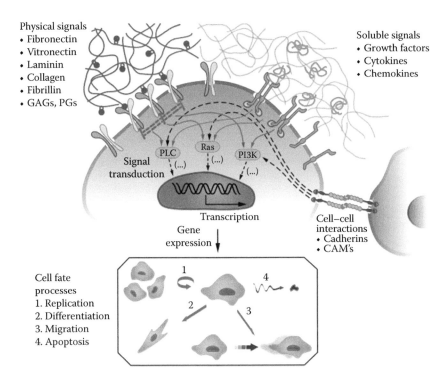

FIGURE 9.1 The surrounding extracellular matrix (ECM) is a complex environment that provides crucial signals, which control cellular behavior. The polysaccharides known as glycosaminoglycans (GAGs) are a main component of the ECM. GAGs provide structural support to cells, act as reservoir for growth factors, are involved in hydration, and more. (Reprinted by permission from Macmillan Publishers Ltd. *Nature Biotechnology*, Lutolf, M. P. and Hubbell, J. A., Synthetic biomaterials as instructive extracellular microenvironments for morphogenesis in tissue engineering, 23, 47–55, copyright 2005.)

Polysaccharide-based scaffolds have shown to be structures that support cellular attachment and growth while facilitating their organization and possibly differentiation toward a highly ordered biomimetic construct.[15–17] Also, they present biological properties that enhance tissue repair. For instance, dextran demonstrated to promote the neovascularization of third-degree burn wounds[18] and chitosan allowed neovasculature formation in damaged heart tissue by increasing vessel density and improving cellular microenvironment, thus preserving cardiac function.[19] Most natural polysaccharides have groups such as hydroxyl, carboxyl, and amino groups, which allow their chemical modifications in a simple manner.[1,3] In regenerative medicine and tissue engineering, a wide variety of polysaccharides have been used in the design of biomaterials including chitosan (CHI),[20] alginate (ALG),[21] hyaluronan (HA),[22] pullulan,[23] dextran (DEX),[24] and agarose.[25] Polysaccharide-based biomaterials have been designed using different methods and with different geometries such as thin films, nanofibers, and hydrogels.[2] This chapter will explore recent work on a number of polysaccharide-based biomaterials and their interactions with cells from different tissues.

9.2 POLYSACCHARIDE-BASED COATINGS

Polysaccharides are ideal biomacromolecules to generate surface coatings, particularly coatings based on the layer-by-layer (LbL) assembly. The LbL method, first established by Decher,[26] involves the sequential adsorption of molecules with complementary interactions (ionic, hydrogen bonding, hydrophobic/hydrophilic, etc.) onto a substrate. A number of polysaccharide-based LbL films have been designed—due to the polyelectrolyte nature of many polysaccharides—as matrix mimetic materials to investigate cell/material interactions.[9,27,28] The versatility of engineering polysaccharide-based coatings via the LbL method is that they can be applied to any charged substrate regardless of geometry, they are capable of acting as reservoirs for drugs or biomacromolecules, their mechanical and physical properties (e.g., stiffness, porosity) are easily tuned, and mild aqueous solutions are used during assembly. Polysaccharide-based polyelectolyte multilayered films constructed using the LbL method are often engineered using chitosan or chitin as the polycation and/or glycosaminoglycans such as HA, alginate, modified dextran, and cellulose as the polyanions. This section highlights recent achievements on polysaccharide-based LbL films used to investigate various cellular processes such as adhesion and differentiation.

9.2.1 CELLULAR ADHESION ON LbL FILMS

LbL films constructed using polysaccharides have demonstrated poor cellular adhesion to various mammalian cell types.[29–33] In order to improve cellular adhesion, various approaches have been proposed. Several investigators include a final layer of ECM proteins to improve adhesion.[29–31,33] Groth and colleagues observed an enhancement in osteoblast cell adhesion when CHI/HEP (chitosan/heparin) or PEI/HEP (polyethyleneimine/heparin) LbL films were terminated with a layer of serum or fibronectin.[29,31] Fibronectin also was used to enhance the adhesion of endothelial cells on films composed of PLL (poly-L-lysine) and sulfated dextran.[33] Kipper and colleagues have also investigated CHI/HEP films on a number of substrates.[30,34] They observed that the composition of the final layer greatly affects cell adhesion, where poor adhesion was observed on CHI-ending films as compared to HEP-ending films.[30] CHI/HEP films on flat surfaces such as tissue culture polystyrene and titanium required the addition of fibronectin to improve cell adhesion.[30] However, CHI/HEP films constructed on denaturalized bone does not require further addition of fibronectin for suitable cell adhesion.[34] This behavior highlights the influence of the underlying substrate on the control of cellular adhesion by polysaccharide-based LbL films.

Interestingly, when polysaccharides are incorporated onto synthetic polymeric LbL films an improvement in cell adhesion is observed. The team of Liefeith et al. prepared polypeptide films composed of PLL/PGA (poly glutamic acid) that contained a final layer of either heparin or chondroitin sulfate (CS). They noticed that PLL/PGA films with an outer layer of CS improved cell adhesion while when the outer layer was HEP the adhesion was drastically decreased.[35] The same group further investigated the rescuing effect of cell adhesion by the polypeptides in the presence of polysaccharides. Cellular adhesion decreases with increasing polysaccharide content, which enables them to generate patterned substrates of a PLL/CS film that is low cell adhesive with patterns of PGA to create cell adhesive regions.[36]

Another approach to enhance cellular adhesion on polysaccharide-based LbL films is their chemical modification with adhesive molecules. Alves et al. designed LbL films of a modified hyaluronan containing dopamine—an adhesive amino acid presented on mussel proteins.[37] The films prepared using chitosan and dopamine-hyaluronan exhibited an dramatic enhancement on osteoblastic and fibroblastic cell adhesion, proliferation, and viability as compared to chitosan/hyaluronan films.[37] Another commonly used adhesive molecule is the peptide sequence arginine–glycine–aspartic acid (RGD). LbL films have been constructed using chitosan and an elastin-like recombinamer containing RGD (ELR-RGD).[38] An osteoblast cell adhesion test revealed an enhanced adhesion on the ELR-RGD films compared to the chitosan ending films and to a scrambled RDG sequence (CHI/ELR-RDG).[38] In fact, these films have also been used to investigate cellular uptake of micro-capsules (CHI/ELR-RGD and CHI/ELR-RDG) prepared via the LbL method.[39] These capsules were loaded with a fluorescently labeled ovalbumin that shifts from red to green as it is being degraded (Figure 9.2). The internalization of these ovalbumin-containing capsules was followed via fluorescence microscopy, and it was observed that the cells are able to internalize both types of capsules, but the bioavailability of ovalbumin was higher on the

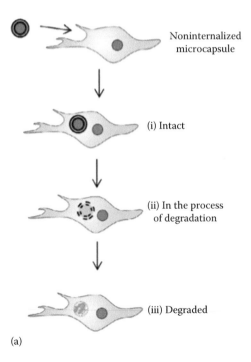

(a)

FIGURE 9.2 Scheme (a) depicts the various degradation stages that an ovalbumin undergoes. Representative microscopy images of (CHI/ELR-RGD) and (CHI/ELR-RDG) microcapsules loaded with ovalbumin incubated with hMSCs after 3 h (b and d) or 72 h (c and e) of incubation. The nucleus is stained with DAPI. *(Continued)*

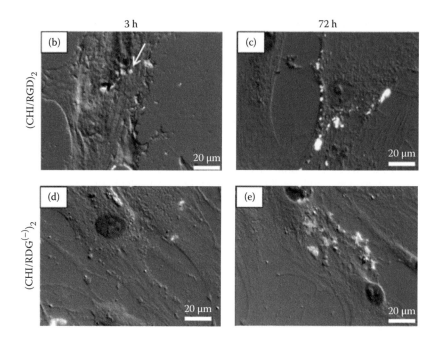

FIGURE 9.2 (*Continued*) In (b), the arrow points to a bright green spot (seen as bright gray). A shift from red/orange (dark gray spots on d and e) to green/yellow (as observed in c as bright gray spots) indicates a transition between fully intact ovalbumin to degraded ovalbumin. Refer to original manuscript for color version of the images. (Reprinted from *Acta Biomaterialia*, 10, Costa, R.R., Girotti, A., Santos, M., Javier Arias, F., Mano, J.F., and Carlos Rodriguez-Cabello J.; Cellular uptake of multilayered capsules produced with natural and genetically engineered biomimetic macromolecules, 2653–2662, Copyright 2014, with permission from Elsevier.)

CHI/ELR-RGD capsules.[39] Various cellular/material interactions are enhanced when chemically modified polysaccharides are used as a component of LbL films.

Other approaches include the chemical cross-linking of the films to improve cell adhesion. On hyaluronan containing films, Picart et al. employed the EDC (1-ethyl-3-[3-dimethylaminopropyl]carbodiimide hydrochloride) chemistry to increase the stiffness of the films. They observed an enhancement on myoblast cell attachment on the stiffer films.[32,40] They also employed the EDC chemistry on CHI/HA films and observed an enhancement in myoblast adhesion, although not as pronounced as PLL/HA films.[41] They observe formation of F-actin stress fibers and vinculin plaques only on the stiff films.[41] The flexibility of adjusting film stiffness to control cellular adhesion is desirable to investigate the decoupling between mechanical cues and biochemical cues to control cell adhesion. For example, on soft PLL/HA films (which are naturally cytophobic) an enhancement on myoblast cell adhesion is observed with the inclusion of the growth factor BMP-2.[42]

Combining LbL with microfluidics a gradient of EDC was generated yielding a gradient in stiffness from 600 to 200 kPa.[43] Cells attached on such stiffness gradient in a graded fashion, where cell number and cell area decreased with decreasing stiffness.[43] Another approach for cross-linking to enhance adhesion was proposed by

Labat et al., whereas chondroitin sulfate (CS) containing films were cross-linked using genipin.[44] Cellular adhesion and proliferation was enhanced on genipin cross-linked CS/PLL films; however, this enhancement was not dependent on genipin concentration.[44] Genipin cross-linking improves cellular adhesion on various polysaccharide-based films including CHI/HA and CHI/ALG systems.[45] Photo-cross-linking has also been used to modulate LbL film stiffness where a modified photoreactive hyaluronan was used to generate photo-cross-linkable films.[46] Cellular adhesion was enhanced on the photo-cross-linked films as compared to non-cross-linked films.[46]

9.2.2 Cellular Differentiation on LbL Films

LbL films of controlled stiffness are able to modulate cellular differentiation. Hyaluronan containing LbL films with various degree of cross-linking (using EDC chemistry) was evaluated by Picart et al. with regard to myoblast differentiation.[32] The differentiation of skeletal muscle cells was greatly enhanced on the stiffer (higher cross-linked) LbL films as noticed by the larger amount of well-defined myotubes formed.[32] A dependence of differentiation on degree of cross-linking was observed on osteoblasts seeded on genipin cross-linked chondroitin sulfate containing LbL films.[44] The early osteogenic marker ALP and calcium mineral deposition increased with higher degree of cross-linking.[44]

LbL films can be deposited to substrates with controllable stiffness and micropatterning. Polydimethylsiloxane (PDMS) substrate of tunable stiffness were coated with a CS/PLL LbL film to investigate osteoblast differentiation.[47] Osteblast differentiation was enhanced on the stiffer PDMS substrate with CS-ending films.[47] Microgrooved PDMS was also coated with PLL/HA films of tunable stiffness to control myoblast differentiation and aligment.[48] Suitable conditions for myoblast differentiation (i.e., well-aligned myotubes) were on stiff films with microgrooves of 5–10 μm.[48]

Polysaccharide-based LbL films also serve as reservoir for growth factors that are known to modulate cellular differentiation. This is due to the specific interactions between polysaccharides and growth factors that make LbL films an efficient delivery vehicle. Several growth factors have been successfully incorporated onto polysaccharide-based LbL coatings including BMP-2,[49] BMP-7,[50] FGF-2,[30,51] VEGF,[52] TGF-β3,[53] SDF-1α,[54] and aFGF.[55] However, more focus has been dedicated to BMPs, as they are the only clinically approved growth factors. Hammond et al. developed a chitosan containing tetralayer LbL system for the continuous delivery of BMP-2.[56–58] The tetralayer system successfully delivers optimal doses of BMP-2 modulating osteogenic differentiation and promoting bone tissue formation.[56–58]

Picart and colleagues have extensively used polysaccharide-based LbL films to deliver BMP-2[42,49,50,59–61] and recently BMP-7.[50] They observed that the amount of incorporated and released BMP-2 can be tuned with the degree of cross-linking and with the selection of polysaccharide (HA vs. HEP), and that osteogenic differentiation was enhanced when BMP-2 is delivered from the LbL films as opposed to in solution at optimal dosages.[49] Scaffolds currently used in the clinic were coated with LbL films to deliver BMP-2. Both ceramic granules[59] and titanium implants[61] were successfully coated with hyaluronan containing LbL films for delivery of BMP-2. *In vivo* results showed that bone regeneration was superior on implants coated

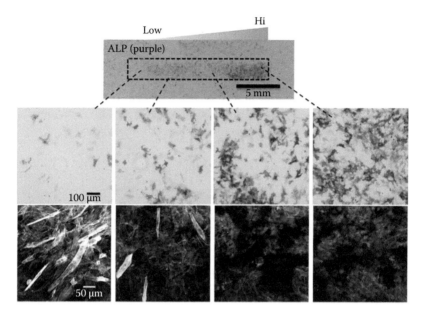

FIGURE 9.3 Differentiation of myoblasts on BMP-7 gradients. Overview image (top row) and representative images (center row) of ALP staining confirms osteogenic differentiation, while immunofluorescent imaging (bottom row) reveals a decrease of troponin T (myotube marker) positive cells with increasing BMP-7 concentration. Top and center row images: ALP (dark stain) and bottom images: ALP (black), troponin T (bright white), and actin (gray). (Reprinted from *Biomaterials*, 35, Almodóvar, J., Guillot, R., Monge, C., Vollaire, J., Selimović, S., Coll, J.-L., Khademhosseini, A., and Picart, C., Spatial patterning of BMP-2 and BMP-7 on biopolymeric films and the guidance of muscle cell fate, 3975–3985, Copyright 2014, with permission from Elsevier.)

with the LbL film containing BMP-2 as compared to BMP-2 delivery in solution.[59,61] Microfluidics was combined with polysaccharide-based LbL films to spatially arrange BMP-2 and BMP-7 on an ECM mimetic surface generating matrix-bound growth factor gradients.[50] Differentiation of myoblasts seeded on these BMP gradients was modulated, whereas on high BMP concentrations high ALP activity— indicating osteogenic differentiation—is observed, decreasing with decreasing BMP concentration (Figure 9.3). However, at a certain BMP concentration expression of myogenic markers is noticed, and it increases with decreasing BMP concentration to the point were fully developed myotubes are observed.[50]

9.2.3 LbL Films of Other Polysaccharides

Polysaccharide-based LbL films are typically constructed using the negatively charged glycosaminoglycans (heparin, chondroitin sulfate, hyaluronan, keratin sulfate, dermatan sulfate, etc.) and other natural polysaccharides that also behave as polyelectrolytes (alginate). Polysaccharides that are neutral are also used when they are chemically modified to become polyelectrolytes such as chitosan (derived from chitin), cellulose, and dextran. Polysaccharide chemistry permits for the easy modification of the saccharide units to become charged. Various research teams are focusing on other less common polysaccharides to generate LbL films such as xyloglucan,[62–64] levan,[65] and carrageenan.[66,67] These polysaccharides have exciting properties that can alleviate current limitations with the typically used polysaccharides. For instance, phosphonated levan/chitosan films had an increase in mammalian cell adhesion as compared to alginate/chitosan films.[65] Carrageenan is an interesting polysaccharide, as there exists various derivatives with increasing amounts of sulfur content. The team of Mano and colleagues, investigated LbL films composed of chitosan and κ-, ι-, λ-carrageenan (increasing sulfur content).[66] They evaluated osteoblast activity and observed that every film combination improved cellular proliferation compared to their uncoated polycaprolactone substrate. Mineralization was observed on all films but the ι-carrageenan (medium content of sulfur) exhibited a significant increase in mineralization compared to κ- and λ-carrageenan.[66]

9.3 POLYSACCHARIDE-BASED HYDROGELS

Hydrogels are 3D structures constituted from hydrophilic homopolymers, copolymers, or macromers cross-linked to design insoluble polymer matrices.[68,69] In regenerative medicine and tissue engineering, hydrogels are used as scaffolds offering structural integrity, cellular organization, and morphogenic guidance. They can be used for protein and growth factor delivery, to encapsulate and deliver cells, to act as drug storage reservoir, and as glues or barriers between tissue and material surfaces. Hydrogels are an adequate scaffolding material because of their similarities in mechanical properties and composition to the native extracellular matrix, thus they can be designed to mimic human tissues.[16,70,71] There are mainly two categories to classify hydrogels: physical and chemical. For physical hydrogels, cross-linking is given by physical entanglement and/or weak interactions such as ionic and/or hydrophobic interactions or crystallization. This cross-linking is reversible and does not present chemical reactions that may be unfavorable to the integrity of incorporated bioactive agents or cells. Nevertheless, they may interact with bodily functions physiologically and/or mechanically putting at risk their stability *in vivo*. For chemical hydrogels, cross-linking occurs by covalent bonding that allows controllable mechanical strength and higher physiological stability. These hydrogels can be manufactured via radical polymerization, chemical reaction of complementary groups, enzymatic cross-linking, or using high-energy radiation.[14,72–75] Polysaccharides are attractive candidates for the production of hydrogels to be used in regenerative medicine and tissue engineering due to their role in native ECM. Polysaccharide-based hydrogels have attracted the interest of many researchers who have investigated them for promoting cellular processes such as adhesion, proliferation, differentiation, and survival.

9.3.1 Cellular Adhesion on Polysaccharide-Based Hydrogels

Cell adhesion is the binding or contact between cells or between cell and extracellular matrix or other surfaces. Cells bind to other cells and to extracellular matrix through adhesion receptors or specialized proteins such as cadherins, immunoglobulin superfamily, and integrins. Cell adhesion proteins recognize and interact with either ECM proteins or adhesion receptors of neighboring cells. This biological process is crucial during tissue assembly, and it is essential in maintaining cellular structure; it also affects other cellular processes such as growth and differentiation.[76]

Polysaccharides used for the engineering of hydrogels include chitosan, agarose, hyaluronan, and dextran, among others. Some polysaccharides such as agarose lack the ability to enhance cell adhesion.[77] However, when agarose is combined with other polysaccharides, an enhancement in cell adhesion is observed.[25] A hydrogel synthesized from agarose and chitosan was proposed for improving the attachment and outgrowth of cortical neurons. Hydrogels were produced changing the concentration of chitosan from 0 to 3.0 wt% and keeping the concentration of agarose at 1 wt% in all experiments. It was observed that neurons on the surface extended down to a depth of 160 μm below the surface. Hydrogels with chitosan concentration of 0.33% expanded axons as a straight line without branching while with concentrations of 1.0%–3.0% axon morphology was vastly branched. Adequate concentration to improve neural cell adhesion was of 0.66–1.5 wt% (Figure 9.4).[25] The chitosan-agarose hydrogels displayed better support of neuron adhesion than the pure agarose hydrogel.[25] Nonspecific interaction between cells and chitosan is attributed to the electrostatic attractive force between the positive charges of the amine groups in chitosan chains and the negative charges of the phospholipid structure of the cell membranes.[78] The influence of charge and surface characteristics in the adhesion and extension of neurons was researched by Zuidema et al. who cultured cortical neurons on a methylcellulose and agarose hydrogel blend, which included dextran and chitosan.[24] The mixtures with highest content of dextran and chitosan enhanced neuron attachment. Blends having 3% methylcellulose/5.3% dextran and chitosan showed the most neurite extension. Particularly, the blend with 3% methylcellulose/4.6% dextran/1% chitosan demonstrated the highest neuronal attachment, and major neurite extension.[24] Therefore, chitosan improves neuron compatibility, possibly by decreasing the storage modulus and increasing the surface charge of the hydrogels. These results suggest that softer, more positively charged hydrogels enable superior neuron attachment and neurite extension.[24]

Chitosan blends have also been used as scaffolds to investigate cell/material interactions of different cell types such as corneal endothelial cells (CECs)[20] and cardiomyocytes.[79] Ultrathin chitosan–poly(ethylene glycol) (PEG) hydrogel films (CPHFs) were manufactured utilizing post-cross-linking via epoxy–amine chemistry, for corneal tissue applications.[20] The CPHFs displayed good mechanical, optical, and permeability properties; plus they supported the attachment of sheep CECs.[20] Furthermore, a chitosan chloride-glutathione (CHICl-GSH) hydrogel was produced to decrease oxidative stress injury in cardiomyocytes (CMs) caused by reactive oxygen species.[79] Results showed that the CHICl-GSH supported the adhesion and survival of CMs while removing reactive oxygen species that cause cell damage and apoptosis.[79]

Mammalian cells do not have specific interactions with alginates; however, the addition of cell-adhesive peptides,[21] gelatin,[80] or collagen type I with beta-tricalcium[81]

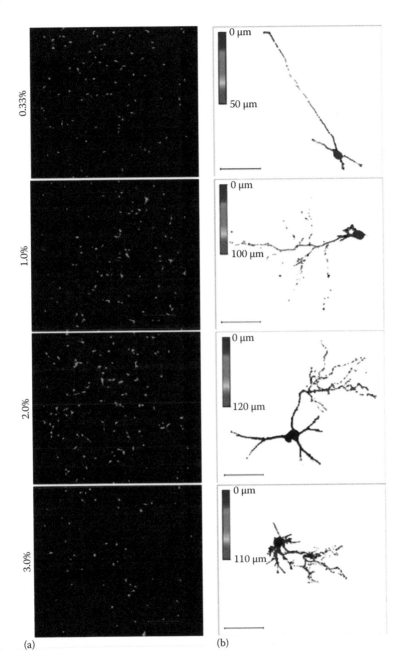

FIGURE 9.4   Confocal images of (a) neural network (scale bar = 300 μm) and (b) representative neuron morphology with color band, indicating the depth of the neurite location
(scale bar = 50 μm) on chitosan–agarose hydrogels. (Reprinted with permission from Cao,
Z., Gilbert, R.J., and He, W., Simple agarose-chitosan gel composite system for enhanced
neuronal growth in three dimensions, *Biomacromolecules*, 10, 2954–2959. Copyright 2009
American Chemical Society.)

FIGURE 9.5 SEM image showing chondrocytes attached to hydrogel pore walls (scale bar 100 µm). (Reprinted from *Acta Biomaterialia*, 10, Balakrishnan, B., Joshi, N., Jayakrishnan, A., and Banerjee, R., Self-crosslinked oxidized alginate/gelatin hydrogel as injectable, adhesive biomimetic scaffolds for cartilage regeneration, 3650–3663, Copyright 2014, with permission from Elsevier.)

confers mammalian cell adhesion to alginate hydrogels. For instance, Balakrishnan et al. proposed a hydrogel prepared by self-cross-linking of periodate oxidized alginate and gelatin in the presence of borax, to support the culture of articular chondrocytes.[21] Scanning electron microscopy images of primary chondrocytes, which are within the hydrogel, displayed cell adhesion to this matrix, as shown in Figure 9.5. This hydrogel did not cause inflammatory or oxidative stress responses, and there was a formation of hyaline cartilage as evidenced by expression of collagen type II and aggrecan.[21]

9.3.2 Cellular Differentiation

There are numerous factors that influence cell differentiation, such as the biochemical composition of the matrix and/or its physical properties such as elasticity or stiffness. Several investigations have shown that soft matrices direct differentiation of mesenchymal stem cells into neuronal-like cells, moderate elasticity guides myogenic differentiation, and a rigid matrix encourages osteogenic differentiation, as displayed in the Figure 9.6.[82,83] Leipzig et al. built a soft (Young's modulus less than 1 kPa) 3D scaffold using a streptavidin-modified methacrylamide chitosan (MAC) hydrogel containing recombinant biotin-IFN-γ, for the investigation of the differentiation of neural stem/progenitor cells (NSPCs) to neurons.[15] They found that immobilized recombinant biotin-IFN-γ encouraged neuronal differentiation to a similar degree as soluble recombinant biotin-IFN-γ.[15] These scaffolds stimulated the differentiation of NSPCs into neurons in a 3D matrix in the presence of only basic medium

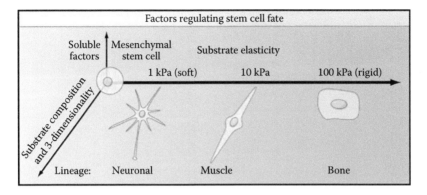

FIGURE 9.6 Stem cell differentiation is dependent on matrix stiffness, where neuronal differentiation is favored on soft scaffolds whereas osteogenic differentiation is favored on stiff scaffolds. (Reprinted from *Cell*, 126, Even-Ram, S., Artym, V., and Yamada, K.M., Matrix control of stem cell fate, 645–647, Copyright 2006, with permission from Elsevier.)

in 1 week.[15] Neural stem cells that can differentiate into the primary cell types found in the central nervous system (CNS) are of particular importance in the regeneration of lost or damaged nerve tissue. Polysaccharide-based hydrogels are used as scaffold material for neural tissue engineering because they can be designed to be mechanically similar to the human brain, and they can be an ideal environment for nerve tissue cell growth.[84]

Polysaccharide-based hydrogels are utilized to mimic variety of tissues in order to restore their native functions and to treat diseases that affect the quality of life and are life-threatening. Heart tissue suffers damages, which leads to irreversible cell loss and scar formation. For example, up to one billion cardiomyocytes can be impaired by ischemia after a major myocardial infarction, which results in deterioration in cardiac function, and ultimately heart failure.[85] A tissue engineering strategy to repair damaged myocardium based on an injectable chitosan hydrogel was researched as a carrier for brown adipose–derived stem cells (BADSCs) into ischemic hearts.[86] The results obtained *in vitro* showed that chitosan improved cardiac differentiation of BADSCs. Furthermore, it enhanced the survival of engrafted BADSCs and increased the differentiation rate of BADSCs into cardiomyocytes *in vivo*, prevented adverse matrix remodeling, increased angiogenesis, and preserved heart function.[86] Alginate and high-molecular-weight hyaluronan (HA) hydrogels were engineered as a scaffold for osteogenesis.[16] Pre-osteoblasts seeded on these scaffolds were used for assessing their biocompatibility and bioactivity. Alginate hydrogels showed higher alkaline phosphatase (ALP) activity levels and calcium content compared to HA hydrogels.[16] Furthermore, culture with alginate increased osteocalcin mRNA levels, while HA hydrogels decrease alkaline phosphatase, bone sialoprotein, and osteocalcin expression.[16] These results suggest that, for applications in bone tissue engineering, alginate hydrogels are more convenient than high molecular weight HA hydrogels.[16] Wood-derived nanofibrillar cellulose (NFC) and hyaluronan-gelatin (HG) hydrogels were evaluated as scaffolds for liver progenitor cells.[22] These hydrogels induced formation of 3D multicellular spheroids with

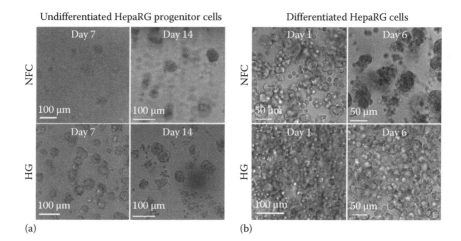

FIGURE 9.7 Growth of HepaRG cells (liver progenitor cells) upon seeding in nanofibrillar cellulose (NFC) and hyaluronan-gelatin (HG) hydrogels. The HepaRG cells formed spherical aggregates, spheroids, in the hydrogels when they were embedded (a) as hepatic undifferentiated progenitors at density of 1 million/mL (low-density) and (b) as differentiated cells at density of 9 million/mL (high-density). The spheroids grew more in size when the cells were seeded at low-density compared to cultures seeded at high-density. (From Malinen, M.M. et al., *Biomaterials*, 35, 5110, 2014.)

apicobasal polarity and functional bile canaliculi-like structures, structural hallmarks of the liver tissue (Figure 9.7).[22] The spheroids showed expression of hepatocyte markers, metabolic activity, and vectorial molecular transport toward bile duct compartment. Improved cell culture models of hepatocytes are needed in the drug discovery, drug development, and chemical testing.[22]

9.3.3 CELLULAR SURVIVAL, VIABILITY, AND PROLIFERATION

Cellular survival, viability, and proliferation are biological processes that indicate that (1) cells are alive, (2) they are self-renewing, and (3) they are increasing in number as a result of cell division and growth. For these processes to be carried out, it is necessary that the cells are in a nontoxic and biocompatible environment similar to their native environment. Hydrogels with a large variety of polysaccharides have been used as scaffolding for different types of cells. Hydrogels containing a biodegradable dextran (DEX) chain grafted with a hydrophobic poly(ε-caprolactone)-2-hydroxylethyl methacrylate (PCL-HEMA) chain and a thermoresponsive poly(N-isopropylacrylamide) (PNIPAAm) chain were manufactured to be used for endothelial cell growth.[87] The results displayed that when the concentration of dextran in the hydrogel was between 0.0032 and 50 g/L the hydrogel was not toxic to the cells, and there was no reduction in cell viability.[87] Hydrogel scaffolds from blends of chitosan with hydroxypropylcellulose (CHI–HPC), collagen (CHI–COL), or elastin (CHI–ELA) cross-linked with genipin were evaluated as supports for human corneal epithelial cells to be used during epithelium transplantation.[88] The results *in vitro* demonstrated that (CHI–COL) composite allowed the regular stratified growth of the epithelium cells, good surface coverage, and increased number of the cell layers. These carriers can be considered as eligible for grafting in humans.[88] Methacrylated gellan gum (MeGG) hydrogels were synthesized for tissue engineering applications by Coutinho et al.[17] They combined physical cross-linking methods (temperature and the addition of cations) with chemical cross-linking approaches (through photo-cross-linking) to obtain hydrogels with values of Young modulus between 0.15 and 148 kPa. Fibroblasts were encapsulated in these MeGG networks, and their viability was assessed. The results *in vitro* showed biocompatibility of these materials confirmed by high cell survival.[17] These hydrogels due to highly tunable mechanical and degradation properties may be applicable for a wide range of tissue engineering approaches (see Figure 9.3).[17] The response of human adipose-derived stem cell culture (ASCs) encapsulated within 3D scaffolds incorporating decellularized adipose tissue (DAT) as a bioactive matrix within photo-cross-linkable methacrylated glycol chitosan (MGC) or methacrylated chondroitin sulfate (MCS) delivery vehicles was characterized.[89] Higher viability and adipogenic differentiation was observed in the MCS composites containing 5 wt% DAT.[89]

9.3.4 *IN VIVO* RESULTS

A myocardial infarction (MI) is a damage that happens to a part of cardiac tissue due to the shortage of oxygen occurred by the blockage of one of the coronary arteries that supplies blood to the heart. In the ischemic myocardium, there is a hostile environment that presents inadequate angiogenesis, inflammation, and generation of reactive oxygen species (ROS). This leads to a low rate of survival of transplanted cells when the damage is treated using cellular therapy. Injectable chitosan hydrogels have shown to modulate the unfriendly microenvironment of MI replacing and repairing the extracellular matrix. Chitosan hydrogels allow to overcome some limitations of cell therapy applied to the ischaemic cardiac tissue.[79,90–92] A chitosan hydrogel mixed with embryonic stem cells (ESCs) for injection inside the ischemic myocardium of rat infarction models was manufactured.[19] It was found that this scaffold enhanced implanted ESC retention and survival. Chitosan allowed neovasculature formation in damaged heart tissue by increasing vessel density and improving the cellular microenvironment, thereby helping to preserve cardiac function.[19] Liu et al. also used a chitosan hydrogel for treating myocardial infarction using adipose-derived mesenchymal stem cells for engraftment. The results of this study support that chitosan hydrogels are suitable scaffolds for improving the myocardial infarction microenvironment, facilitated engraftment, and survival of transplanted stem cells as well as homing of endogenous stem cells.[91]

Vascularization is essential to successful transplantation of tissue constructs and for the recovery of ischemic and wounded tissues. A healthy circulatory system connection is necessary in the development and maintenance of functional tissues and organs.[18] Sun et al. produced dextran-based hydrogels for treating third-degree burn wounds on mice without to use growth factors, cytokines, or cells. They found that hydrogels promoted the neovascularization and skin regeneration with hair follicles, sebaceous glands, and thickness similar to a normal mouse skin.[93] This dextran hydrogels are soft and flexible, and they help to regenerate epithelial tissue after a third-degree burn, which destroy epidermis and all skin layers under the subcutaneous tissue destroying epithelial appendages, nerve endings, blood vessels, and all the specialized cells of skin.[93–95]

9.3.5 NOVEL POLYSACCHARIDES

Morelli and Chiellini functionalized and prepared hydrogels from ulvan, which is extracted from green seaweeds like *Ulva armoricana*.[96] Ulvan is an anionic, water soluble, sulfated, and semi-crystalline polysaccharide. It is mainly composed of rhamnose, glucuronic acid, iduronic acid, xylose, and sulfate.[68,96] They functionalized ulvan with methacrylate groups using either methacrylic anhydride or glycidyl methacrylate. Hydrogels showed antioxidant activity, thus they could be used as a matrix for cell encapsulation. Also, these hydrogels can be utilized as scaffolds due their softness.[68,96] Cytotoxicity of these hydrogels has not been evaluated but Alves et al. assessed the filmogenic properties of ulvan membranes and its usefulness as a wound dressing or in drug delivery. Membranes showed ability to uptake water up to ~1800% of its initial dry weight and a mechanical performance of 1.76 MPa related with cross-linking.[69] The results of the use of a model drug showed an initial steady release of the drug of approximately 49% followed by slower and sustained release up to 14 days. These membranes are good candidates to be used as wound dressing.[69]

Novel hydrophilic materials were presented by Reis and colleagues who produced bioactive-glass-reinforced gellan-gums spongy-like hydrogels (GG-BAG) for use as the scaffolding in bone-tissue engineering.[97] The composite scaffold showed lower mechanical properties than desired for application in bone tissue, but the reinforcement with bioactive-glass particles improved the microstructure and the mechanical properties of the material, which depended on the composition and was enhanced with the amount of bioactive glass. By incorporating the bioactive-glass particles, the composite material acquired the ability to form an apatite layer when soaked in simulated body fluid.[97] Also, human-adipose-derived stem cells were able to adhere and spread within the gellan-gum, spongy-like hydrogels reinforced with the bioactive glass, and remained viable.[97]

9.4 POLYSACCHARIDE-BASED FIBERS

Nanofibers have been widely used for biomedical applications due to their resemblance to native ECM—both in scale and geometry—and because of the various cues they can provide to cells. The design of a nanofibrous scaffold to be used in tissue engineering and regenerative medicine is based on the selection of a suitable material that is biodegradable, biocompatible, and nontoxic. An ideal scaffold would promote a better surface for cell attachment, proliferation, and differentiation. Polysaccharides are attractive candidates for nanofiber production. Recently, researchers have developed polysaccharide-based nanofibers using chitosan,[98–100] alginate,[101–103] cellulose,[104–106] dextran,[107,108] chitin,[109,110] hyaluronic acid,[111] and heparin.[112] Often, these nanofibers are prepared in combination with a synthetic polymer that provides structural support, using the polysaccharide to enhance biological activity as tested both *in vitro*[103,104,108] and *in vivo*.[112] Cellular interactions with polysaccharide-based nanofibers have been evaluated with numerous mammalian cell types including brain tumor cells,[104] human mesenchymal stem cells (hMSC),[113] human neural stem cells (hNSCs),[114] mouse embryonic fibroblasts,[105] vascular endothelial cells,[112] chondrocytes from articular cartilage from rabbits,[106] human fetal osteoblastic cells,[115] and MSCs from rat bone marrow.[116] The most commonly used technique for the manufacturing of polysaccharide-based nanofibers is electrospinning.[113,117,118] However, other techniques such as wet spinning,[113] force spinning (FS),[118] and an electrospinning/electrospraying hybrid technique[119] have also been used. This section highlights recent achievements on polysaccharide-based nanofibers used to investigate various cellular processes such as adhesion and differentiation.

9.4.1 CELLULAR ADHESION AND CYTOTOXICITY ON POLYSACCHARIDE-BASED NANOFIBERS

Recent research has shown that the addition of polysaccharides to nanofibers of synthetic polymers significantly improves cellular adhesion. For instance, Shalumon et al. fabricated CHI/PCL nanofiber scaffolds via electrospinning and evaluated the bioactivity, cytocompatibility, cell adhesion, and cytotoxicity using fibroblasts, osteoblasts, and adipocytes.[113] The cytocompatibility studies showed an increase in cell viability for all cell types, demonstrating that the CHI/PCL nanofibers were not toxic.[113] They observed that the addition of chitosan improved cellular attachment of each of the cell types evaluated.[113] Fibroblast adhesion has been also investigated on a nanofibrous polyurethane/dextran scaffolds prepared via electrospinning, where dextran improved adhesion.[107] Inclusion of cellulose on starch nanofibers also enhanced chondrocyte adhesion as compared to starch nanofibers alone.[106] The addition of alginate to polyoxyethylene fibers demonstrated an enhancement on fibroblast adhesion.[101]

Chitosan has been a popular polysaccharide in the preparation of nanofibrous scaffolds due to its attractive properties and similarity to ECM components. Electrospun nanofibers composed of polyethersulfone, cellulose acetate, and chitosan were evaluated by Du et al.[114] They demonstrated that the nanofiber mats resulting from this mixture did not have a cytotoxic effect on neuronal-like cells, and inclusion of chitosan improved cell proliferation.[114] Polyethylene oxide electrospun nanofibers containing silica and various percentages of chitosan was evaluated with regard to osteoblast adhesion.[117] Osteoblast adhesion increased with increasing chitosan concentration.[117] The addition of chitosan to PCL electrospun nanofibers also enhanced the adhesion of neuron-like cells.[100] Not only chitosan improves cellular adhesion, but also it imparts nanofibers with antimicrobial properties. A chitosan/pullulan nanofibrous mat loaded with tannic acid was developed as a wound dressing.[118] This mat not only demonstrated high fibroblast adhesion but also exhibited antimicrobial activity against *Escherichia coli*.[118]

Cell/material interactions of nanofibers prepared from other natural polymers can be enhanced using polysaccharides. For instance, gelatin nanofibers were prepared via electrospinning and stabilized by cross-linking using a modified dextran aldehyde.[108]

Cross-linking with dextran aldehyde rendered the gelatin fibers insoluble in water and thus suitable for cell culture. The cross-linked gelatin fibers were not toxic to fibroblasts, and they were able to attach and spread.[108] Often the combination of two or more polysaccharides, to generate nanofibers, yields scaffolds with improved cellular activity. For instance, a combination of chitosan/alginate yields cross-linked fibers with enhanced cellular attachment as compared to scaffolds of only alginate.[102] In a similar manner, an electrospinning–electrospraying hybrid technique was used to prepare cellulose acetate/chitosan nanofibers in a layer-by-layer fashion.[119] They observed that the chitosan ending fiber mats promoted adhesion and proliferation of lung fibroblasts over the cellulose acetate ending mats.[119] Polysaccharide-based nanofibers have also been prepared in combination with nanostructured materials. For instance, cellulose acetate nanofibers containing carbon nanotubes were generated via electrospinning.[105] The incorporated nanotubes did not impart any toxicity and, in fact, promoted fibroblast attachment.[105] Organic rectorite has also been incorporated on nanofibers of carboxymethyl chitin via electrospinning.[110] The addition of the organic rectorite improves the thermal properties of the fiber mat without inducing any cellular toxicity.[110]

9.4.2 CELLULAR PROLIFERATION ON POLYSACCHARIDE-BASED NANOFIBERS

Cellular proliferation of multiple cell types also appears to be enhanced by polysaccharides in nanofibers. The addition of alginate to PCL nanofibers successfully enhanced the proliferation of pre-osteoblastic cells.[103] The presence of alginate significantly enhanced the metabolic activity of pre-osteoblasts cells during a culture period of 7 days, indicating an increase in cellular proliferation (Figure 9.8).[103] Similarly, the addition of cellulose to starch nanofibers also improved cell proliferation.[106] Cellulose has also been demonstrated to enhance the proliferation of both human mesenchymal cells and brain tumor stem cells on gelatin nanofibers.[104] Heparin also enhances cellular attachment and growth of both fibroblasts and endothelial cells cultured on silk fibroin nanofibers.[112] Cell proliferation is enhanced on gelatin nanofibers cross-linked with dextran aldehyde,[108] and on polyoxyethylene electrospun fibers containing alginate.[101]

Inclusion of chitosan on PCL nanofibers also enhances cell proliferation.[120] In an interesting combination, chitosan, calcium phosphate cement, and polyglactin yields a suitable nanofibrous scaffold that promotes the proliferation of mesenchymal stem cells.[121] Chitosan/collagen electrospun scaffolds have been prepared to investigate the effect of percentage of chitosan in the proliferation of smooth muscle cells and endothelial cells.[98] It was observed that scaffolds containing 20% and 50% chitosan enhanced the proliferation of both cell types.[98] Chitosan also enhanced proliferation and metabolic activity of cells seeded on electrospun alginate nanofibers.[102] On a comparative study of chitosan, cellulose acetate, and polyethersulfone electrospun nanofibers, the chitosan nanofibers were superior on promoting the proliferation of human neuronal stem cells.[114] Other researchers investigated whether the type of morphology of the nanofibers had any effect on the proliferation capacity.[100] Comparing aligned versus randomly oriented electrospun chitosan/PCL nanofibers, it was observed that aligned fibers regulated the growth of neuronal-like cells.[100]

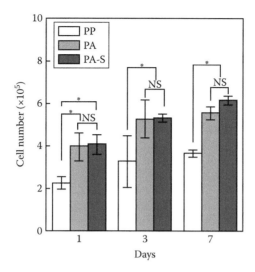

FIGURE 9.8   Results of MTT assay—which is indicative of cellular proliferation—after 1, 3, and 7 days of culture (PP = PCL, PA = PCL + Alginate and PA-S = PCL struts + Alginate). *P < 0.05 indicates a significant difference and NS, a nonsignificant difference. (Reprinted from *Carbohydrate Polymers*, 114, Kim, M.S. and Kim, G., Three-dimensional electrospun polycaprolactone (PCL)/alginate hybrid composite scaffolds, 213–221, Copyright 2014, with permission from Elsevier.)

Polysaccharide-based nanofibers also serve as a reservoir for multiple growth factors.[116,122,123] FGF-2 released from electrospun chitosan nanofibers retained its biological activity, promoting MSC proliferation even after 30 days of incubation.[123] Chitosan nanofibers prepared via wet spinning served as reservoir for both BMP-2 and BMP-7 for simultaneous or sequential delivery.[116] The BMPs were either incorporated in the surfaces of the fibers or within the fibers. BMP incorporated on the surface of the fibers exhibited higher MSC proliferation compared to within the fibers. Moreover, BMP-7 had a higher proliferative effect than BMP-2, and the simultaneous delivery of both BMP-2 and BMP-7 showed the highest proliferation rates.[116]

9.4.3   CELLULAR DIFFERENTIATION ON POLYSACCHARIDE-BASED NANOFIBERS

Polysaccharide-based or polysaccharide-containing nanofibers provide a suitable scaffold for cellular differentiation. A comparative study of electrospun chitosan, silk fibroin, or a mixture of both demonstrated an enhanced osteogenic differentiation of MSCs on chitosan fibers, whereas proliferation was enhanced on the silk fibroin nanofibers.[120] On nanofibers of both polymers, the proliferative effect of silk fibroin was maintained as well as the osteoinductive effect chitosan had.[120] This trend is also applicable to the differentiation of human fetal osteoblastic cells, where proliferation is enhanced by silk fibroin and differentiation by chitosan.[115] To further improve the osteoinductive capabilities of chitosan electrospun nanofibers, hydroxyapatite can be included in the fiber mat.[99] Osteoblast-like cells seeded

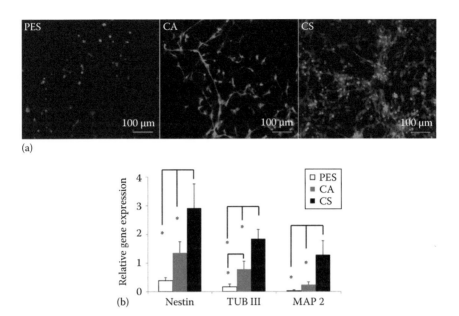

(a)

(b)   Nestin   TUB III   MAP 2

FIGURE 9.9 Neuronal differentiation of hNSCs cultured on chitosan (CS), cellulose acetate (CA), and polyethersulfone (PES) scaffolds. (a) Immunofluorescent images of cells cultured on PES, CA, and CS for 7 days in differentiation media. The phenotypic differentiation of hNSCs was assessed by immunocytochemistry for TUJ1, a class III β-tubulin protein that marks early neurons (bright gray). (b) RT-PCR for neural differentiation markers, nestin (a neural stem cell marker), β-tubulin (early neuron marker), and MAP2 (a mature neuronal marker) by day 7. The reference condition was laminin coated, tissue culture plastic; all data were normalized to this condition. Asterisks denote statistical significance ($P < 0.05$) by comparison with the other fiber substrates, as indicated. Bars represent mean ± standard deviation (n = 3). (Reprinted from *Carbohydrate Polymers*, 99, Du, J., Tan, E., Kim, H.J., Zhang, A., Bhattacharya, R., and Yarema, K.J., Comparative evaluation of chitosan, cellulose acetate, and polyethersulfone nanofiber scaffolds for neural differentiation, 483–490, Copyright 2014, with permission from Elsevier.)

on hydroxyapatite-containing chitosan fibers exhibited higher alkaline phosphatase expression compared to chitosan only films.[99]

The differentiation potential of neuronal-like cells and neuronal stem cells on electrospun nanofibers of chitosan, cellulose acetate (CA), and polyethersulfone (PES) were evaluated.[114] As shown in Figure 9.9, neuronal differentiation was significantly enhanced on both polysaccharide-based nanofibers over the PES fibers, with chitosan having the highest differentiation potential.[114] An increase on TUJ1 (early neuronal marker) protein expression and increase in gene expression of neuronal markers (nestin, β-tubulin, and MAP2) indicate the positive effect of chitosan on neuronal differentiation (Figure 9.9).[114] A similar trend was observed for neuronal-like cells, where a higher degree of differentiated cells was observed on the chitosan scaffolds, followed by the cellulose acetate scaffold, and the least amount of differentiated cells were on the PES fibers.[114]

Not only chitosan has demonstrated an enhancement on cell differentiation, but other polysaccharides such as alginate,[103] cellulose,[104] and methacrylated hyaluronan also promote cellular differentiation. Alginate-containing PCL nanofibers showed an increase in alkaline phosphatase and mineralization of pre-osteoblastic cells as compared to PCL only scaffolds.[103] Cellulose-containing gelatin nanofibers also improved the osteogenic and adipogenic differentiation of human MSCs.[104] Lastly, a mechanically tunable electrospun nanofibrous scaffold prepared from methacrylated hyaluronan influenced the differentiation of human MSCs.[111] An enhancement in chondrogenesis was observed on the softer scaffolds compared to the stiffer nanofibrous scaffold.[111]

9.5 CONCLUSIONS

Polysaccharide-based biomaterials hold great promise in the development of novel solutions for tissue engineering and regenerative medicine. Countless polyssacharides are available for the design of novel biomaterials. Being a crucial component of the ECM, polysaccharides are ideal in the design of scaffolds to be used to investigate cell–material interactions. Due to the versatility of polysaccharide, one can design thin coatings via the layer-by-layer method, hydrogels of controllable chemical and mechanical properties, and nanofibrous scaffolds of various size and configurations. Polysaccharide-based biomaterials have been shown to enhance cellular processes such as adhesion, proliferation, and differentiation. Moreover, polysaccharides have natural properties that make them suitable for biomaterials such as antimicrobial activity, growth factor binding capabilities, and lubrication. Polysaccharides have been shown to enhance the biological activity of synthetic biomaterials. Due to their versatile chemistry, they can be finely tuned to any desired application, yielding materials that are temperature sensitive and/or mechanically modifiable, for example. Certainly, popular polysaccharides such as chitosan, alginate, hyaluronan, dextran, heparin, etc., continue to be explored as candidates for biomaterials. However, new biomaterials have begun to emerge with less common polysaccharide such as ulvan, suggesting that there is much more to explore in the field of polysaccharide-based biomaterials.

ABBREVIATIONS

ALG	Alginate
ALP	Alkaline phosphatase
ASCs	Adipose-derived stem cell culture
BADSCs	Brown adipose–derived stem cells
BMP-2	Bone morphogenetic protein 2
BMP-7	Bone morphogenetic protein 7
CA	Cellulose acetate
CAMs	Cell adhesion molecules
CECs	Corneal endothelial cells
CHI	Chitosan
CHI–Col	Chitosan with collagen
CHI–Ela	Chitosan with elastin

CHI–HPC	Chitosan with hydroxypropylcellulose
CHICl-GSH	Chitosan chloride-glutathione
CMs	Cardiomyocytes
CNS	Central nervous system
CPHFs	Chitosan–poly(ethylene glycol) hydrogel films
CS	Chondroitin sulfate
DAT	Decellularized adipose tissue
DEX	Dextran
ECM	Extracellular matrix
EDC	1-Ethyl-3-(3-dimethylaminopropyl) carbodiimide
ELR	Elastin-like recombinamer
ESCs	Embryonic stem cells
FE-SEM	Field emission scanning electron microscopy
FGF-2	Fibroblast growth factor 2
FS	Force spinning
GAGs	Glycosaminoglycans
GG-BAG	Gellan-gums spongy-like hydrogels
HA	Hyaluronan
HEP	Heparin
HG	Hyaluronan-gelatin
hMSC	Human mesenchymal stem cells
hNSCs	Human neural stem cells
IFN	Interferon
LbL	Layer-by-layer
MAC	Methacrylamide chitosan
MAP2	Microtubule-associated protein 2
MCS	Methacrylated chondroitin sulfate
MeGG	Methacrylated gellan gum
MGC	Methacrylated glycol chitosan
MI	Myocardial infarction
MSC	Mesenchymal stem cell
MTS	(3-(4,5-Dimethylthiazol-2-yl)-5-(3-carboxymethoxyphenyl)-2-(4-sulfophenyl)-2H-tetrazolium)
MTT	Methylthiazol tetrazolium
NFC	Nanofibrillar cellulose
NSPCs	Neural stem/progenitor cells
PA	PCL/alginate-5 wt% fibrous scaffolds
PA-S	PCL/alginate-5 wt% fibrous scaffolds with micro-sized PCL struts
PCL	Polycaprolactone
PCL-HEMA	Hydrophobic poly (ε-caprolactone)-2-hydroxylethyl methacrylate
PDMS	Polydimethylsiloxane
PES	Polyethersulfone
PEG	Poly(ethylene glycol)
PEI	Poly(ethylene imine)
PGs	Proteoglycans
PGA	Poly(glycolic acid)

PLL	Poly-L-lysine
PNIPAAm	Poly(N-isopropylacrylamide)
PP	Pure PCL
RGD	Arginine-glycine-aspartic acid
ROS	Reactive oxygen species
RT-PCR	Reverse transcription polymerase chain reaction
SDF-1α	Stromal-derived factor alpha 1
SEM	Scanning electron microscopy
TGF-β1	Transforming growth factor beta
TGF-β3	Transforming growth factor beta 3
TUJ1	Neuron-specific class III beta-tubulin
VEGF	Vascular endothelial growth factor

REFERENCES

1. Silva, A.; Juenet, M.; Letourneur, D. Polysaccharide-based strategies for heart tissue engineering. *Carbohydrate Polymers* 2015, *116*, 267–277.
2. Lee, K. Y.; Jeong, L.; Kang, Y. O.; Lee, S. J.; Park, W. H. Electrospinning of polysaccharides for regenerative medicine. *Advanced Drug Delivery Reviews* 2009, *61*, 1020–1032.
3. Nitta, S. K.; Numata, K. Biopolymer-based nanoparticles for drug/gene delivery and tissue engineering. *International Journal of Molecular Sciences* 2013, *14*, 1629–1654.
4. Koolman, J.; Rohm, K.-H. *Bioquímica: Texto Y Atlas*, 3rd edn.; Medica Panamericana S.A.: Madrid, Spain, 2004; pp. 44–45.
5. Melo Ruiz, V.; Cuamatzi Tapia, O. *Bioquímica de Los Procesos Metabólicos*; Editorial Reverte: Mexico D.F., Mexico, 2007; pp. 61–75.
6. Pena, A. "Los carbohidratos, almacenes de energía solar" *en* Pena, A, *Bioquimica*, Ed., 2nd edn., Mexico D.F., Mexico: Editorial Limusa S.A de C.V; Grupo Noriega, 2004; pp. 143–148.
7. Garrido Pertierra, A.; Teijon Riviera, J. M. *Fundamentos de Bioquímica Estructural*, 2nd edn.; Editorial Tébar: Madrid, Spain, 2006; pp. 325–341.
8. Dicker, K. T.; Gurski, L. A.; Pradhan-bhatt, S.; Witt, R. L.; Farach-carson, M. C.; Jia, X. Hyaluronan : A simple polysaccharide with diverse biological functions Q. *Acta Biomaterialia* 2014, *10*, 1558–1570.
9. Boddohi, S.; Kipper, M. J. Engineering nanoassemblies of polysaccharides. *Advanced Materials* 2010, *22*, 2998–3016.
10. Lu, P.; Weaver, V. M.; Werb, Z. The extracellular matrix: A dynamic niche in cancer progression. *The Journal of Cell Biology* 2012, *196*, 395–406.
11. Lutolf, M. P.; Hubbell, J. A. Synthetic biomaterials as instructive extracellular microenvironments for morphogenesis in tissue engineering. *Nature Biotechnology* 2005, *23*, 47–55.
12. Rederstorff, E.; Weiss, P.; Sourice, S.; Pilet, P.; Xie, F.; Sinquin, C.; Colliec-Jouault, S.; Guicheux, J.; Laïb, S. An in vitro study of two GAG-like marine polysaccharides incorporated into injectable hydrogels for bone and cartilage tissue engineering. *Acta Biomaterialia* 2011, *7*, 2119–2130.
13. Gallagher, J. T.; Lyon, M.; Steward, W. P. Structure and function of heparan sulphate proteoglycans. *Biochemical Journal* 1986, *236*, 313–325.
14. Roldo, M.; Fatouros, D. G. Chitosan derivative based hydrogels: Applications in drug delivery and tissue engineering. *Studies in Mechanobiology, Tissue Engineering and Biomaterials* 2011, *8*, 351–376.
15. Leipzig, N. D.; Wylie, R. G.; Kim, H.; Shoichet, M. S. Differentiation of neural stem cells in three-dimensional growth factor-immobilized chitosan hydrogel scaffolds. *Biomaterials* 2011, *32*, 57–64.

16. Rubert, M.; Alonso-Sande, M.; Monjo, M.; Ramis, J. M. Evaluation of alginate and hyaluronic acid for their use in bone tissue engineering. *Biointerphases* 2012, *7*, 1–11.

17. Coutinho, D. F.; Sant, S. V; Shin, H.; Oliveira, J. T.; Gomes, M. E.; Neves, N. M.; Khademhosseini, A.; Reis, R. L. Modified gellan gum hydrogels with tunable physical and mechanical properties. *Biomaterials* 2010, *31*, 7494–7502.

18. Sun, G.; Shen, Y.-I.; Kusuma, S.; Fox-Talbot, K.; Steenbergen, C. J.; Gerecht, S. Functional neovascularization of biodegradable dextran hydrogels with multiple angiogenic growth factors. *Biomaterials* 2011, *32*, 95–106.

19. Lu, W.-N.; Lü, S.-H.; Wang, H.-B.; Li, D.-X.; Duan, C.-M.; Liu, Z.-Q.; Hao, T. et al. Functional improvement of infarcted heart by co-injection of embryonic stem cells with temperature-responsive chitosan hydrogel. *Tissue Engineering Part A* 2009, *15*, 1437–1447.

20. Ozcelik, B.; Brown, K. D.; Blencowe, A.; Daniell, M.; Stevens, G. W.; Qiao, G. G. Ultrathin chitosan-poly(ethylene glycol) hydrogel films for corneal tissue engineering. *Acta Biomaterialia* 2013, *9*, 6594–6605.

21. Balakrishnan, B.; Joshi, N.; Jayakrishnan, A.; Banerjee, R. Self-crosslinked oxidized alginate/gelatin hydrogel as injectable, adhesive biomimetic scaffolds for cartilage regeneration. *Acta Biomaterialia* 2014, *10*, 3650–3663.

22. Malinen, M. M.; Kanninen, L. K.; Corlu, A.; Isoniemi, H. M.; Lou, Y.-R.; Yliperttula, M. L.; Urtti, A. O. Differentiation of liver progenitor cell line to functional organotypic cultures in 3D nanofibrillar cellulose and hyaluronan-gelatin hydrogels. *Biomaterials* 2014, *35*, 5110–5121.

23. Xu, F.; Weng, B.; Gilkerson, R.; Materon, L. Development of tannic acid chitosan pullulan composite nanofibers from aqueous solution for potential applications as wound dressing. *Carbohydrate Polymers* 2015, *15*, 16–24.

24. Zuidema, J. M.; Pap, M. M.; Jaroch, D. B.; Morrison, F. A.; Gilbert, R. J. Fabrication and characterization of tunable polysaccharide hydrogel blends for neural repair. *Acta Biomaterialia* 2011, *7*, 1634–1643.

25. Cao, Z.; Gilbert, R. J.; He, W. Simple agarose-chitosan gel composite system for enhanced neuronal growth in three dimensions. *Biomacromolecules* 2009, *10*, 2954–2959.

26. Decher, G. Fuzzy nanoassemblies: Toward layered polymeric multicomposites. *Science* 1997, *277*, 1232–1237.

27. Picart, C.; Crouzier, T.; Boudou, T. Polysaccharide-based polyelectrolyte multilayers. *Current Opinion in Colloid & Interface Science* 2010, *15*, 417–426.

28. Boudou, T.; Crouzier, T.; Ren, K.; Blin, G.; Picart, C. Multiple functionalities of polyelectrolyte multilayer films: New biomedical applications. *Advanced Materials* 2010, *22*, 441–467.

29. Kirchhof, K.; Hristova, K.; Krasteva, N.; Altankov, G.; Groth, T. Multilayer coatings on biomaterials for control of MG-63 osteoblast adhesion and growth. *Journal of Materials Science: Materials in Medicine* 2009, *20*, 897–907.

30. Almodóvar, J.; Bacon, S.; Gogolski, J.; Kisiday, J. D.; Kipper, M. J. Polysaccharide-based polyelectrolyte multilayer surface coatings can enhance mesenchymal stem cell response to adsorbed growth factors. *Biomacromolecules* 2010, *11*, 2629–2639.

31. Niepel, M. S.; Peschel, D.; Sisquella, X.; Planell, J. A.; Groth, T. pH-dependent modulation of fibroblast adhesion on multilayers composed of poly(ethylene imine) and heparin. *Biomaterials* 2009, *30*, 4939–4947.

32. Ren, K.; Crouzier, T.; Roy, C.; Picart, C. Polyelectrolyte multilayer films of controlled stiffness modulate myoblast cells differentiation. *Advanced Functional Materials* 2008, *18*, 1378–1389.

33. Wittmer, C. R.; Phelps, J. A.; Saltzman, W. M.; Van Tassel, P. R. Fibronectin terminated multilayer films: Protein adsorption and cell attachment studies. *Biomaterials* 2007, *28*, 851–860.

34. Almodóvar, J.; Mower, J.; Banerjee, A.; Sarkar, A. K.; Ehrhart, N. P.; Kipper, M. J. Chitosan-heparin polyelectrolyte multilayers on cortical bone: Periosteum-mimetic, cytophilic, antibacterial coatings. *Biotechnology and Bioengineering* 2013, *110*, 609–618.

35. Grohmann, S.; Rothe, H.; Frant, M.; Liefeith, K. Colloidal force spectroscopy and cell biological investigations on biomimetic polyelectrolyte multilayer coatings composed of chondroitin sulfate and heparin. *Biomacromolecules* 2011, *12*, 1987–1997.

36. Grohmann, S.; Rothe, H.; Liefeith, K. Investigations on the secondary structure of polypeptide chains in polyelectrolyte multilayers and their effect on the adhesion and spreading of osteoblasts. *Biointerphases* 2012, *7*, 62.

37. Neto, A. I.; Cibrão, A. C.; Correia, C. R.; Carvalho, R. R.; Luz, G. M.; Ferrer, G. G.; Botelho, G.; Picart, C.; Alves, N. M.; Mano, J. F. Nanostructured polymeric coatings based on chitosan and dopamine-modified hyaluronic acid for biomedical applications. *Small (Weinheim an der Bergstrasse, Germany)* 2014, *10*, 2459–2469.

38. Costa, R. R.; Custódio, C. A.; Arias, F. J.; Rodríguez-Cabello, J. C.; Mano, J. F. Layer-by-layer assembly of chitosan and recombinant biopolymers into biomimetic coatings with multiple stimuli-responsive properties. *Small (Weinheim an der Bergstrasse, Germany)* 2011, *7*, 2640–2649.

39. Costa, R.R., Girotti, A., Santos, M., Javier Arias, F., Mano, J.F., and Carlos Rodriguez-Cabello J. Cellular uptake of multilayered capsules produced with natural and genetically engineered biomimetic macromolecules. *Acta Biomaterialia* 2014, *10*, 2653–2662.

40. Richert, L.; Boulmedais, F.; Lavalle, P.; Mutterer, J.; Ferreux, E.; Decher, G.; Schaaf, P.; Voegel, J.-C.; Picart, C. Improvement of stability and cell adhesion properties of polyelectrolyte multilayer films by chemical cross-linking. *Biomacromolecules* 2004, *5*, 284–294.

41. Boudou, T.; Crouzier, T.; Nicolas, C.; Ren, K.; Picart, C. Polyelectrolyte multilayer nanofilms used as thin materials for cell mechano-sensitivity studies. *Macromolecular Bioscience* 2011, *11*, 77–89.

42. Crouzier, T.; Fourel, L.; Boudou, T.; Albigès-Rizo, C.; Picart, C. Presentation of BMP-2 from a soft biopolymeric film unveils its activity on cell adhesion and migration. *Advanced Materials (Deerfield Beach, Fla.)* 2011, *23*, H111–H118.

43. Almodóvar, J.; Crouzier, T.; Selimović, Š.; Boudou, T.; Khademhosseini, A.; Picart, C.; Almodo, J.; Selimovic, S. Gradients of physical and biochemical cues on polyelectrolyte multilayer films generated via microfluidics. *Lab on a Chip* 2013, *13*, 1562–1570.

44. Gaudière, F.; Morin-Grognet, S.; Bidault, L.; Lembré, P.; Pauthe, E.; Vannier, J.-P.; Atmani, H.; Ladam, G.; Labat, B. Genipin-cross-linked layer-by-layer assemblies: Biocompatible microenvironments to direct bone cell fate. *Biomacromolecules* 2014, *15*, 1602–1611.

45. Hillberg, A. L.; Holmes, C. A.; Tabrizian, M. Effect of genipin cross-linking on the cellular adhesion properties of layer-by-layer assembled polyelectrolyte films. *Biomaterials* 2009, *30*, 4463–4470.

46. Vázquez, C. P.; Boudou, T.; Dulong, V.; Nicolas, C.; Picart, C.; Glinel, K. Variation of polyelectrolyte film stiffness by photo-cross-linking: A new way to control cell adhesion. *Langmuir: The ACS Journal of Surfaces and Colloids* 2009, *25*, 3556–3563.

47. Gaudière, F.; Masson, I.; Morin-Grognet, S.; Thoumire, O.; Vannier, J.-P.; Atmani, H.; Ladam, G.; Labat, B. Mechano-chemical control of cell behaviour by elastomer templates coated with biomimetic layer-by-layer nanofilms. *Soft Matter* 2012, *8*, 8327.

48. Monge, C.; Ren, K.; Berton, K.; Guillot, R.; Peyrade, D.; Picart, C. Engineering muscle tissues on microstructured polyelectrolyte multilayer films. *Tissue Engineering Part A* 2012, *18*, 1664–1676.

49. Crouzier, T.; Ren, K.; Nicolas, C.; Roy, C.; Picart, C. Layer-by-layer films as a biomimetic reservoir for rhBMP-2 delivery: Controlled differentiation of myoblasts to osteoblasts. *Small* 2009, 598–608.

50. Almodóvar, J.; Guillot, R.; Monge, C.; Vollaire, J.; Selimović, S.; Coll, J.-L.; Khademhosseini, A.; Picart, C. Spatial patterning of BMP-2 and BMP-7 on biopolymeric films and the guidance of muscle cell fate. *Biomaterials* 2014, *35*, 3975–3985.

51. Macdonald, M. L.; Rodriguez, N. M.; Shah, N. J.; Hammond, P. T. Characterization of tunable FGF-2 releasing polyelectrolyte multilayers. *Biomacromolecules* 2010, *11*, 2053–2059.

52. Wang, H. G.; Yin, T. Y.; Ge, S. P.; Zhang, Q.; Dong, Q. L.; Lei, D. X.; Sun, D. M.; Wang, G. X. Biofunctionalization of titanium surface with multilayer films modified by heparin-VEGF-fibronectin complex to improve endothelial cell proliferation and blood compatibility. *Journal of Biomedical Materials Research Part A* 2013, *101*, 413–420.

53. Park, J. S.; Park, K.; Woo, D. G.; Yang, H. N.; Chung, H.-M.; Park, K.-H. PLGA microsphere construct coated with TGF-Beta 3 loaded nanoparticles for neocartilage formation. *Biomacromolecules* 2008, *9*, 2162–2169.

54. Dalonneau, F.; Liu, X. Q. X.; Sadir, R.; Almodovar, J.; Mertani, H. C.; Bruckert, F.; Albiges-Rizo, C.; Weidenhaupt, M.; Lortat-Jacob, H.; Picart, C. The effect of delivering the chemokine SDF-1α in a matrix-bound manner on myogenesis. *Biomaterials* 2014, *35*, 4525–4535.

55. Mao, Z.; Ma, L.; Zhou, J.; Gao, C.; Shen, J. Bioactive thin film of acidic fibroblast growth factor fabricated by layer-by-layer assembly. *Bioconjugate Chemistry* 2005, *16*, 1316–1322.

56. Macdonald, M. L.; Samuel, R. E.; Shah, N. J.; Padera, R. F.; Beben, Y. M.; Hammond, P. T. Tissue integration of growth factor-eluting layer-by-layer polyelectrolyte multilayer coated implants. *Biomaterials* 2011, *32*, 1446–1453.

57. Shah, N. J.; Hyder, M. N.; Moskowitz, J. S.; Quadir, M. A.; Morton, S. W.; Seeherman, H. J.; Padera, R. F.; Spector, M.; Hammond, P. T. Surface-mediated bone tissue morphogenesis from tunable nanolayered implant coatings. *Science Translational Medicine* 2013, *5*, 191ra83.

58. Min, J.; Braatz, R. D.; Hammond, P. T. Tunable staged release of therapeutics from layer-by-layer coatings with clay interlayer barrier. *Biomaterials* 2014, *35*, 2507–2517.

59. Crouzier, T.; Sailhan, F.; Becquart, P.; Guillot, R.; Logeart-avramoglou, D.; Picart, C. Osteoinductive TCP/HAP porous ceramics loaded with BMP-2: Polyelectrolyte multilayer film coating versus direct adsorption. *Biomaterials* 2011, *32*, 7543–7554.

60. Gilde, F.; Maniti, O.; Guillot, R.; Mano, J. F.; Logeart-Avramoglou, D.; Sailhan, F.; Picart, C. Secondary structure of rhBMP-2 in a protective biopolymeric carrier material. *Biomacromolecules* 2012, *13*, 3620–3626.

61. Guillot, R.; Gilde, F.; Becquart, P.; Sailhan, F.; Lapeyrere, A.; Logeart-Avramoglou, D.; Picart, C. The stability of BMP loaded polyelectrolyte multilayer coatings on titanium. *Biomaterials* 2013, *34*, 5737–5746.

62. Villares, A.; Moreau, C.; Capron, I.; Cathala, B. Impact of ionic strength on chitin nanocrystal-xyloglucan multilayer film growth. *Biopolymers* 2014, *101*, 924–930.

63. Cerclier, C. V; Guyomard-Lack, A.; Cousin, F.; Jean, B.; Bonnin, E.; Cathala, B.; Moreau, C. Xyloglucan-cellulose nanocrystal multilayered films: Effect of film architecture on enzymatic hydrolysis. *Biomacromolecules* 2013, *14*, 3599–3609.

64. Villares, A.; Moreau, C.; Capron, I.; Cathala, B. Chitin Nanocrystal-xyloglucan multilayer thin films. *Biomacromolecules* 2014, *15*, 188–194.

65. Costa, R. R.; Neto, A. I.; Calgeris, I.; Correia, C. R.; Pinho, A. C. M.; Fonseca, J.; Öner, E. T.; Mano, J. F. Adhesive nanostructured multilayer films using a bacterial exopolysaccharide for biomedical applications. *Journal of Materials Chemistry B* 2013, *1*, 2367.

66. Oliveira, S. M.; Silva, T. H.; Reis, R. L.; Mano, J. F. Nanocoatings containing sulfated polysaccharides prepared by layer-by-layer assembly as models to study cell–material interactions. *Journal of Materials Chemistry B* 2013, *1*, 4406.

67. Briones, A. V.; Sato, T.; Bigol, U. G. Antibacterial activity of polyethylenimine/carrageenan multilayer against pathogenic bacteria. *Advances in Chemical Engineering and Science* 2014, *04*, 233–241.

68. Alves, A.; Caridade, S. G.; Mano, J. F.; Sousa, R. A.; Reis, R. L. Extraction and physicochemical characterization of a versatile biodegradable polysaccharide obtained from green algae. *Carbohydrate Research* 2010, *345*, 2194–2200.

69. Alves, A.; Pinho, E. D.; Neves, N. M.; Sousa, R. A.; Reis, R. L. Processing ulvan into 2D Structures: Cross-linked ulvan membranes as new biomaterials for drug delivery applications. *International Journal of Pharmaceutics* 2012, *426*, 76–81.

70. Slaughter, B. V; Khurshid, S. S.; Fisher, O. Z.; Khademhosseini, A.; Peppas, N. A. Hydrogels in regenerative medicine. *Advanced Materials (Deerfield Beach, Fla.)* 2009, *21*, 3307–3329.

71. Silva, R.; Fabry, B.; Boccaccini, A. R. Fibrous protein-based hydrogels for cell encapsulation. *Biomaterials* 2014, *35*, 6727–6738.

72. Teixeira, L. S. M.; Feijen, J.; van Blitterswijk, C. A.; Dijkstra, P. J.; Karperien, M. Enzyme-catalyzed crosslinkable hydrogels: Emerging strategies for tissue engineering. *Biomaterials* 2012, *33*, 1281–1290.

73. Zhao, X.; Huebsch, N.; Mooney, D. J.; Suo, Z. Stress-relaxation behavior in gels with ionic and covalent crosslinks. *Journal of Applied Physics* 2010, *107*, 63509.

74. Gupta, S.; Pramanik, A. K.; Kailath, A.; Mishra, T.; Guha, A.; Nayar, S.; Sinha, A. Composition dependent structural modulations in transparent poly(vinyl alcohol) hydrogels. *Colloids and Surfaces B: Biointerfaces* 2009, *74*, 186–190.

75. Delair, T. In situ forming polysaccharide-based 3D-hydrogels for cell delivery in regenerative medicine. *Carbohydrate Polymers* 2012, *87*, 1013–1019.

76. Lodish, H.; Berk, A.; Matsudaira, P.; Kaiser, C. A.; Krieger, M.; Scott, M. P.; Zipursky, L.; Darnell, J. *Molecular Cell Biology*, 5th edn., W. H. Freeman: New York, 2003.

77. Lin, P.-W.; Wu, C.-C.; Chen, C.-H.; Ho, H.-O.; Chen, Y.-C.; Sheu, M.-T. Characterization of cortical neuron outgrowth in two- and three-dimensional culture systems. *Journal of Biomedical Materials Research, Part B: Applied Biomaterials* 2005, *75*, 146–157.

78. Dillon, G. P.; Yu, X.; Bellamkonda, R. V. The polarity and magnitude of ambient charge influences three-dimensional neurite extension from DRGs. *Journal of Biomedical Materials Research* 2000, *51*, 510–519.

79. Li, J.; Shu, Y.; Hao, T.; Wang, Y.; Qian, Y.; Duan, C.; Sun, H.; Lin, Q.; Wang, C. A. Chitosan-glutathione based injectable hydrogel for suppression of oxidative stress damage in cardiomyocytes. *Biomaterials* 2013, *34*, 9071–9081.

80. Liu, Y.; Sakai, S.; Taya, M. Impact of the composition of alginate and gelatin derivatives in bioconjugated hydrogels on the fabrication of cell sheets and spherical tissues with living cell sheaths. *Acta Biomaterialia* 2013, *9*, 6616–6623.

81. Lawson, M. A.; Barralet, J. E.; Wang, L.; Shelton, R. M.; Triffitt, J. T. Adhesion and growth of bone marrow stromal cells on modified alginate hydrogels. *Tissue Engineering* 2004, *10*, 1480–1491.

82. Even-Ram, S.; Artym, V.; Yamada, K. M. Matrix control of stem cell fate. *Cell* 2006, *126*, 645–647.

83. Engler, A. J.; Sen, S.; Sweeney, H. L.; Discher, D. E. Matrix elasticity directs stem cell lineage specification. *Cell* 2006, *126*, 677–689.

84. Jain, A.; Kim, Y.-T.; McKeon, R. J.; Bellamkonda, R. V. In situ gelling hydrogels for conformal repair of spinal cord defects, and local delivery of BDNF after spinal cord injury. *Biomaterials* 2006, *27*, 497–504.

85. Mummery, C. L.; Davis, R. P.; Krieger, J. E. Challenges in using stem cells for cardiac repair. *Science Translational Medicine* 2010, *2*, 27ps17.

86. Wang, H.; Shi, J.; Wang, Y.; Yin, Y.; Wang, L.; Liu, J.; Liu, Z.; Duan, C.; Zhu, P.; Wang, C. Promotion of cardiac differentiation of brown adipose derived stem cells by chitosan hydrogel for repair after myocardial infarction. *Biomaterials* 2014, *35*, 3986–3998.

87. Wu, D.-Q.; Qiu, F.; Wang, T.; Jiang, X.-J.; Zhang, X.-Z.; Zhuo, R.-X. Toward the development of partially biodegradable and injectable thermoresponsive hydrogels for potential biomedical applications. *ACS Applied Materials & Interfaces* 2009, *1*, 319–327.

88. Grolik, M.; Szczubiałka, K.; Wowra, B.; Dobrowolski, D.; Orzechowska-Wylęgała, B.; Wylęgała, E.; Nowakowska, M. Hydrogel membranes based on genipin-crosslinked chitosan blends for corneal epithelium tissue engineering. *Journal of Materials Science: Materials in Medicine* 2012, *23*, 1991–2000.

89. Cheung, H. K.; Han, T. T. Y.; Marecak, D. M.; Watkins, J. F.; Amsden, B. G.; Flynn, L. E. Composite hydrogel scaffolds incorporating decellularized adipose tissue for soft tissue engineering with adipose-derived stem cells. *Biomaterials* 2014, *35*, 1914–1923.

90. Dobric, M.; Ostojic, M.; Giga, V.; Djordjevic-dikic, A.; Stepanovic, J. Glycogen phosphorylase BB in myocardial infarction. *Clinica Chimica Acta* 2015, *438*, 107–111.

91. Liu, Z.; Wang, H.; Wang, Y.; Lin, Q.; Yao, A.; Cao, F.; Li, D.; Zhou, J.; Duan, C.; Du, Z.; Wang, Y.; Wang, C. The influence of chitosan hydrogel on stem cell engraftment, survival and homing in the ischemic myocardial microenvironment. *Biomaterials* 2012, *33*, 3093–3106.

92. Menasché, P. Embryonic stem cells for severe heart failure: Why and how? *Journal of Cardiovascular Translational Research* 2012, *5*, 555–565.

93. Sun, G.; Zhang, X.; Shen, Y.-I.; Sebastian, R.; Dickinson, L. E.; Fox-Talbot, K.; Reinblatt, M.; Steenbergen, C.; Harmon, J. W.; Gerecht, S. Dextran hydrogel scaffolds enhance angiogenic responses and promote complete skin regeneration during burn wound healing. *Proceedings of the National Academy of Sciences of the United States of America* 2011, *108*, 20976–20981.

94. Bozkurt, M.; Kapi, E.; Acar, Y.; Bayram, Y. Spontaneous healing and treatment alternatives in burns: Review. *Archives of Clinical and Experimental Surgery (ACES)* 2013, *2*(3), 186–196.

95. Boucard, N.; Viton, C.; Agay, D.; Mari, E.; Roger, T.; Chancerelle, Y.; Domard, A. The use of physical hydrogels of chitosan for skin regeneration following third-degree burns. *Biomaterials* 2007, *28*, 3478–3488.

96. Morelli, A.; Chiellini, F. Ulvan as a new type of biomaterial from renewable resources: Functionalization and hydrogel preparation. *Macromolecular Chemistry and Physics* 2010, *211*, 821–832.

97. Gantar, A.; Silva, L. P. da; Oliveira, J. M.; Marques, A. P.; Correlo, V. M.; Novak, S.; Reis, R. L. Nanoparticulate bioactive-glass-reinforced gellan-gum hydrogels for bone-tissue engineering. *Materials Science and Engineering C: Materials for Biological Applications* 2014, *43*, 27–36.

98. Chen, Z. G.; Wang, P. W.; Wei, B.; Mo, X. M.; Cui, F. Z. Electrospun collagen-chitosan nanofiber: A biomimetic extracellular matrix for endothelial cell and smooth muscle cell. *Acta Biomaterialia* 2010, *6*, 372–382.

99. Frohbergh, M. E.; Katsman, A.; Botta, G. P.; Lazarovici, P.; Schauer, C. L.; Wegst, U. G. K.; Lelkes, P. I. Electrospun hydroxyapatite-containing chitosan nanofibers crosslinked with genipin for bone tissue engineering. *Biomaterials* 2012, *33*, 9167–9178.

100. Cooper, A.; Bhattarai, N.; Zhang, M. Fabrication and cellular compatibility of aligned chitosan–PCL fibers for nerve tissue regeneration. *Carbohydrate Polymers* 2011, *85*, 149–156.

101. Ma, G.; Fang, D.; Liu, Y.; Zhu, X.; Nie, J. Electrospun sodium alginate/poly(ethylene oxide) core–shell nanofibers scaffolds potential for tissue engineering applications. *Carbohydrate Polymers* 2012, *87*, 737–743.

102. Jeong, S. I.; Krebs, M. D.; Bonino, C. A.; Samorezov, J. E.; Khan, S. A.; Alsberg, E. Electrospun chitosan-alginate nanofibers with in situ polyelectrolyte complexation for use as tissue engineering scaffolds. *Tissue Engineering Part A* 2011, *17*, 59–70.

103. Kim, M. S.; Kim, G. Three-dimensional electrospun polycaprolactone (PCL)/alginate hybrid composite scaffolds. *Carbohydrate Polymers* 2014, *114*, 213–221.
104. Xing, Q.; Zhao, F.; Chen, S.; McNamara, J.; Decoster, M. A.; Lvov, Y. M. Porous biocompatible three-dimensional scaffolds of cellulose microfiber/gelatin composites for cell culture. *Acta Biomaterialia* 2010, *6*, 2132–2139.
105. Luo, Y.; Wang, S.; Shen, M.; Qi, R.; Fang, Y.; Guo, R.; Cai, H.; Cao, X.; Tomás, H.; Zhu, M.; Shi, X. Carbon nanotube-incorporated multilayered cellulose acetate nanofibers for tissue engineering applications. *Carbohydrate Polymers* 2013, *91*, 419–427.
106. Nasri-Nasrabadi, B.; Mehrasa, M.; Rafienia, M.; Bonakdar, S.; Behzad, T.; Gavanji, S. Porous starch/cellulose nanofibers composite prepared by salt leaching technique for tissue engineering. *Carbohydrate Polymers* 2014, *108*, 232–238.
107. Unnithan, A. R.; Barakat, N. A. M.; Pichiah, P. B. T.; Gnanasekaran, G.; Nirmala, R.; Cha, Y.-S.; Jung, C.-H.; El-Newehy, M.; Kim, H. Y. Wound-dressing materials with antibacterial activity from electrospun polyurethane-dextran nanofiber mats containing ciprofloxacin HCl. *Carbohydrate Polymers* 2012, *90*, 1786–1793.
108. Jalaja, K.; Kumar, P. R. A.; Dey, T.; Kundu, S. C.; James, N. R. Modified dextran cross-linked electrospun gelatin nanofibres for biomedical applications. *Carbohydrate Polymers* 2014, *114*, 467–475.
109. Yoo, C. R.; Yeo, I.-S.; Park, K. E.; Park, J. H.; Lee, S. J.; Park, W. H.; Min, B.-M. Effect of chitin/silk fibroin nanofibrous bicomponent structures on interaction with human epidermal keratinocytes. *International Journal of Biological Macromolecules* 2008, *42*, 324–334.
110. Xin, S.; Li, Y.; Li, W.; Du, J.; Huang, R.; Du, Y.; Deng, H. Carboxymethyl chitin/organic rectorite composites based nanofibrous mats and their cell compatibility. *Carbohydrate Polymers* 2012, *90*, 1069–1074.
111. Kim, I. L.; Khetan, S.; Baker, B. M.; Chen, C. S.; Burdick, J. A. Fibrous hyaluronic acid hydrogels that direct MSC chondrogenesis through mechanical and adhesive cues. *Biomaterials* 2013, *34*, 5571–5580.
112. Wang, S.; Zhang, Y.; Wang, H.; Dong, Z. Preparation, characterization and biocompatibility of electrospinning heparin-modified silk fibroin nanofibers. *International Journal of Biological Macromolecules* 2011, *48*, 345–353.
113. Shalumon, K. T.; Anulekha, K. H.; Chennazhi, K. P.; Tamura, H.; Nair, S. V; Jayakumar, R. Fabrication of chitosan/poly(caprolactone) nanofibrous scaffold for bone and skin tissue engineering. *International Journal of Biological Macromolecules* 2011, *48*, 571–576.
114. Du, J.; Tan, E.; Kim, H. J.; Zhang, A.; Bhattacharya, R.; Yarema, K. J. Comparative evaluation of chitosan, cellulose acetate, and polyethersulfone nanofiber scaffolds for neural differentiation. *Carbohydrate Polymers* 2014, *99*, 483–490.
115. Chen, J.-P.; Chen, S.-H.; Lai, G.-J. Preparation and characterization of biomimetic silk fibroin/chitosan composite nanofibers by electrospinning for osteoblasts culture. *Nanoscale Research Letters* 2012, *7*, 170.
116. Yilgor, P.; Tuzlakoglu, K.; Reis, R. L.; Hasirci, N.; Hasirci, V. Incorporation of a sequential BMP-2/BMP-7 delivery system into chitosan-based scaffolds for bone tissue engineering. *Biomaterials* 2009, *30*, 3551–3559.
117. Toskas, G.; Cherif, C.; Hund, R.-D.; Laourine, E.; Mahltig, B.; Fahmi, A.; Heinemann, C.; Hanke, T. Chitosan(PEO)/silica hybrid nanofibers as a potential biomaterial for bone regeneration. *Carbohydrate Polymers* 2013, *94*, 713–722.
118. Xu, F.; Weng, B.; Gilkerson, R.; Materon, L. A.; Lozano, K. Development of tannic acid/chitosan/pullulan composite nanofibers from aqueous solution for potential applications as wound dressing. *Carbohydrate Polymers* 2015, *115*, 16–24.
119. Li, W.; Li, X.; Wang, T.; Li, X.; Pan, S.; Deng, H. Nanofibrous mats layer-by-layer assembled via electrospun cellulose acetate and electrosprayed chitosan for cell culture. *European Polymer Journal* 2012, *48*, 1846–1853.

120. Lai, G.-J.; Shalumon, K. T.; Chen, S.-H.; Chen, J.-P. Composite chitosan/silk fibroin nanofibers for modulation of osteogenic differentiation and proliferation of human mesenchymal stem cells. *Carbohydrate Polymers* 2014, *111*, 288–297.

121. Zhao, L.; Burguera, E. F.; Xu, H. H. K.; Amin, N.; Ryou, H.; Arola, D. D. Fatigue and human umbilical cord stem cell seeding characteristics of calcium phosphate-chitosan-biodegradable fiber scaffolds. *Biomaterials* 2010, *31*, 840–847.

122. Almodóvar, J.; Kipper, M. J. Coating electrospun chitosan nanofibers with polyelectrolyte multilayers using the polysaccharides heparin and *N,N,N*-trimethyl chitosan. *Macromolecular Bioscience* 2011, *11*, 72–76.

123. Zomer, F.; Almodóvar, J.; Erickson, K.; Popat, K. C.; Migliaresi, C.; Kipper, M. J. Preservation of FGF-2 bioactivity using heparin-based nanoparticles, and their delivery from electrospun chitosan fibers. *Acta Biomaterialia* 2012, *8*, 1551–1559.

10 Macrophage-Mediated Foreign Body Responses

*Vladimir Riabov, Alexandru Gudima,
and Julia Kzhyshkowska*

CONTENTS

10.1 MACROPHAGE ORIGIN AND SUBPOPULATIONS IN TISSUES IN HOMEOSTASIS

Currently, most of the available information about origins of macrophages comes from murine studies utilizing genetic fate-mapping techniques. F4/80+ cells first appear in mouse yolk sac on the days 8–9 of embryo development, and then populate embryonic liver at day 10, reaching the peak at days 12–14 of development [1]. Yolk sac macrophages differentiate from restricted progenitors that give rise to macrophages and red blood cells, whereas fetal liver macrophages differentiate from hematopoietic stem cells (HSCs) [2]. Genetic fate-mapping experiments revealed that resident macrophages in many adult organs at least partially originate from yolk sac and fetal liver macrophages. For example, it is believed that microglia (resident macrophages of brain and spinal cord) almost exclusively originates from yolk sac macrophages [2]. Macrophages in other organs such as liver, heart, and skin partially originate from yolk sac. However, embryonic HSCs (in case of liver) or embryonic and adult HSCs (in case of skin and heart) also give rise to resident macrophage populations in these organs. In general, most of the organs in adult animals are composed of embryonically derived and adult-derived macrophage subpopulations (Figure 10.1). Moreover, recent studies conclude that populations of resident macrophages in most of the organs exist autonomously and are not completely replaced by circulating monocytes [2].

245

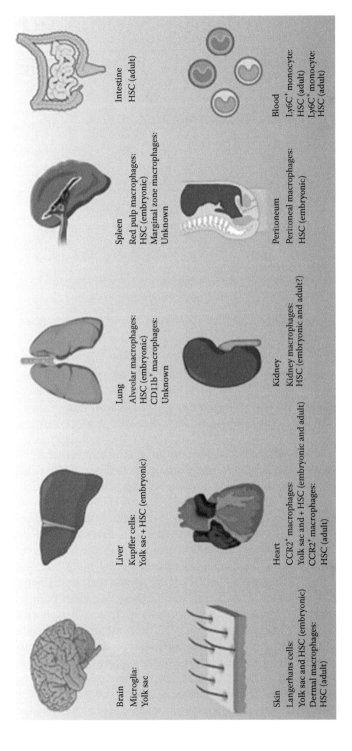

FIGURE 10.1 Contribution of macrophage lineages to populations of adult resident macrophages. HSC-derived populations include embryonic populations and no definitive evidence of yolk sac origin (embryonic), whereas HSCs (adult) have passed through a FLT3+ stage and are continually replaced by circulating adult monocytes. (Adapted from Epelman, S. et al., *Immunity*, 41, 21, 2014.)

In the perinatal period, macrophages originate from bone marrow HSCs and differentiate into monocyte-derived macrophages and dendritic cells from common monocyte-macrophage dendritic cell progenitor (MDP). Specifically, monocytes originate from Ly6c$^+$ monocytic progenitor downstream of MDP, which gives rise to classical Ly6c^{high} monocytes and nonclassical Ly6c^{low} monocytes. The latter subpopulation was found to differentiate from Ly6c^{high} monocytes presumably in circulation [3]. Functionally, in homeostatic conditions, Ly6c^{high} monocytes extravasate and patrol tissues delivering antigens into draining lymph nodes without differentiation into mature macrophages. However, in case of infection, Ly6c^{high} monocytes rapidly infiltrate inflamed tissue and differentiate into macrophages. In contrast, Ly6c^{low} monocytes are known to stay in the intravascular space to monitor and remove damaged endothelial cells. It is worth to note that Ly6c^{high} monocytes reveal functional similarity with human CD14+ CD16− "inflammatory" monocytes, whereas Ly6c^{low} cells resemble CD14^{low} CD16+ "resident" monocytes [3].

The distribution and function of macrophages in homeostasis is relatively well-characterized in mice. However, less is known about macrophage subpopulations in human. In adult organism, resident populations of macrophages are present in most of the organs in homeostatic conditions and display functional and phenotypic heterogeneity. Moreover, multiple macrophage subpopulations that differ in expression of surface markers, spatial distribution, and functional specialization are usually observed within the same organ.

For example, in mouse bone marrow stromal stellate macrophages are located in the center of hematopoietic islands. These cells express F4/80 and associate with myeloblasts and erythroblasts to engulf erythroid nuclei. In addition, they participate in hemoglobin uptake and iron recycling through scavenger receptor (SR) CD163 and may control release of myeloid cells into the circulation. Another resident macrophage population in bone marrow is osteoclasts that are represented by large F4/80− multinucleated cells controlling bone resorption and osteoblast differentiation. Apart from these two macrophage subpopulations, other distinct types of F4/80+ cells are found in bone marrow at the sites of muscle attachment [1]. Similarly, resident spleen macrophages are represented by several heterogeneous populations. Among them red pulp macrophages are known to engulf aged erythrocytes participating in iron homeostasis, whereas white pulp macrophages mediate uptake of apoptotic lymphocytes in germinal centers. Other splenic macrophages known as metallophilic and marginal zone macrophages specialize on engulfment of bacteria and viruses [4]. Liver resident macrophages known as Kupffer cells also participate in the clearance of senescent red blood cells. Besides this, Kupffer cells participate in liver regeneration through induction of hepatocyte proliferation, maintain immune tolerance by suppressing T cell activation, and contribute to liver injury during alcoholic hepatotoxicity, nonalcoholic fatty liver, ischemia-reperfusion injury, and other diseases [5,6]. In intestine, resident macrophages are abundantly present in *lamina propria* and play an important role in the maintenance of intestinal tolerance by production of interleukin IL10 and expansion of regulatory T cells [7]. Resident alveolar macrophages in the lung express high levels of CD206, SRs-AI/II and MARCO, and mediate clearance of inhaled particles and pathogens. The recognition of foreign particles such as silica and asbestos by alveolar macrophages may result in inflammation and fibrotic lung diseases [1]. It is important to note that resident macrophages in many organs originate during prenatal development from yolk sac and fetal liver precursors, and maintain their numbers in the organs by local proliferation. Recent data suggest that circulating monocyte-derived macrophages do not significantly contribute to tissue macrophage compartments in homeostasis. However, their influx and differentiation into macrophages is increased during inflammatory diseases [3]. Overall, organ-specific populations of resident macrophages differ in gene expression, phenotype, and functions reflecting their ubiquitous role in homeostatic regulation. It is anticipated that distinct microenvironment and ligand repertoire in different organs is responsible for such phenotypic diversity.

10.2   MACROPHAGE ACTIVATION AND POLARIZATION: PLASTICITY OF MACROPHAGE PHENOTYPE

Macrophages are primary initiators of inflammation and immune response to various pathogens. At the same time, these cells are major regulators of wound healing and resolution of inflammation. To manage efficiently these opposite physiological processes, macrophages developed extreme phenotypic plasticity, which allows them to react promptly to constantly changing microenvironment. Due to broad repertoire of recognition receptors and transcription factors, macrophages are able to respond to versatile environmental stimuli by switching on and off specific transcriptional modules. It was recently described that each molecular stimuli potentially induces unique macrophage transcriptional program, the hypothesis known as spectral model of macrophage activation [8].

Initially, differential macrophage activation in response to cytokine stimulation was described by Gordon and colleagues [9]. As opposed to classically activated pro-inflammatory macrophages, IL4-stimulated alternatively activated macrophages showed reduced inflammatory cytokine expression and enhanced endocytic capacity with prominent expression of macrophage mannose receptor (CD206) [9]. Later, it was found out that other stimuli such as IL13, IL10, TGFβ, and glucocorticoids are also able to induce anti-inflammatory macrophage phenotype (also termed M2 phenotype) that reveal certain similarities with IL4-differentiated macrophages [10].

Among stimuli that induce classical pro-inflammatory (M1) macrophage activation are Th1-derived cytokine IFNγ and bacterial products such as lipopolysaccharide (LPS). Although all of these stimuli induce pro-inflammatory macrophage phenotype, the resulting gene expression profiles significantly differ. This difference may reflect activation of stimulus-specific signal transduction pathways and transcription factors. For example, IFNγ induces activation of STAT1, IRF-1, and IRF-8 transcription factors through heterodimeric IFNγ receptor. This results in elevated expression of cytokine receptors, cell activation markers, and adhesion molecules. LPS is known to signal primarily through TLR4 and activates NFκB, AP-1, STAT1, and EGR transcription factors, resulting in potent production of pro-inflammatory cytokines IL12, tumor necrosis factor TNFα, IL6, and IL1β, as well as upregulated expression of MHC and co-stimulatory molecules [11]. Functionally, M1 macrophages are characterized by elevated antimicrobial and tumoricidal activity and are able to activate Th1 responses. At the same time, M1 macrophages and their pro-inflammatory molecular products are involved in several pathologies, includ-ing diabetes, rheumatoid arthritis, systemic lupus erythematosus, obesity-associated inflammation, and insulin resistance [12].

Alternatively, activated (M2) macrophages utilize different signaling pathways, resulting in the activation of anti-inflammatory and healing programs. For example, IL4 signals through IL4 receptor consisting of IL4Rα1 paired with common γc chain or IL13Rα1 (binds both IL4 and IL13). The binding of IL4 induces signaling through STAT6, c-Myc and IRF-4 transcription factors, resulting in elevated expression of CD206, IL1Ra, CCL18, arginase 1 (in mice), and low production of pro-inflammatory factors such as TNFα and IL12 [11]. IL10 and glucocorticoids are usually included in M2-polarizing factors since they mediate anti-inflammatory effects. However, they induce transcriptional programs significantly different from IL4 stimulation. IL10 signals through heterodimeric IL10 receptor activating STAT3 transcription factor and inducing formation of p50 NFκB homodimers. IL10-induced genes include CD206, SR MARCO, IL10, TGFβ1, CCL18, and others. Glucocorticoids signal through intracellular glucocorticoid receptor alpha (GCRα), resulting in expression of CD206, CD163, thrombospondin-1, IL10, and IL1R2 [11,12]. M2 macrophages are involved in wound healing, tissue remodeling, resolution of inflammation, and antiparasitic immunity. At the same time, M2 macrophages contribute to several pathologies, including allergy and cancer.

Nowadays, it is postulated that M1/M2 nomenclature does not accurately repre-sent macrophage phenotypes since gene expression studies reveal that each stimulus induces unique transcriptional program in macrophages (this concept is known as a spectral model of macrophage activation) [8,13]. It is also important to note that clear M1 and M2 phenotypes are rarely observed in vivo, whereas mixed M1/M2 signatures prevail reflecting macrophage heterogeneity in complex molecular microenvironment.

One of the most important aspects of macrophage physiology is their unique pheno-typic plasticity caused by immediate need to react to rapidly changing microenviron-ment. To fulfill this need, macrophages are able to revert their phenotype in response to newly appeared stimulus. For example, IL4-differentiated anti-inflammatory mac-rophages maintain their ability to react to exogenous danger stimuli such as LPS by production of TNFα and IL1β. Similarly, IFNγ-polarized macrophages respond to IL4 stimulation by production of anti-inflammatory cytokine IL1Ra and decreased bacterial killing demonstrating critical role in regulation of immune homeostasis [14]. This and other studies suggest that macrophages do not differentiate into stable sub-sets upon cytokine stimulation but rather transiently change their expression profiles in order to execute stimulus-directed function [15,16]. The extreme plasticity of mac-rophage phenotype also suggests that chromatin in these cells is permanently opened for plethora of transcription factors with synergic and opposing effects.

10.3 FOREIGN BODIES, BIOMATERIALS, AND THEIR RECOGNITION BY MACROPHAGES

A foreign body is defined as any object or substance that is introduced from outside in any organ or tissue in which it does not belong under normal circumstances [17]. Foreign bodies can be introduced in the organism both intentionally (e.g., in the form of food) or by accident (e.g., inhalation of asbestos and coal particles by construction workers or miners). Foreign bodies can enter the organism through a variety of ways. The most common way is through the natural orifices of the human body: the mouth, nostrils, ear canals, eyes, urethra, anus, and vagina. Foreign body ingestion frequently occurs in children between 6 months and 6 years of age; they account for up to 80% of the cases [18]. The prevalent ingested objects are coins; they are found in 66% of cases. In adults or children above 11 years old, 60% of the foreign bodies retained in the gastrointestinal tract are food boluses. Although in the United States alone more than 1500 patients annually die from foreign body ingestion, in more than 80% of the cases, there are no complications and the foreign bodies pass spontaneously, with only less than 20% of cases requiring a medical intervention for the removal of the foreign body [19].

Another way of entering the organism requires penetration of the skin, which can occur during an accidental injury or during medical procedures. All surgeries involve the insertion of foreign bodies into the organism. Sometimes, they remain in the body in the form of implants to support, enhance, or replace a biological structure or organ. Occasionally, surgical instruments such as sponges, needles, or towels are forgotten inside the patient's body during a surgery. This can lead to serious injuries like formation of an abscess, sepsis, or even death. Although there are a lot of measures taken to prevent these mistakes (counting of all surgical instruments before and after surgery, careful inspection of the body cavity, x-rays films, etc.), recent studies estimate the incidence of retained surgical instruments from 1–5,500 to 1–18,760 operations [20].

Because foreign bodies may be harmful to the organism, our body developed special mechanisms of protection. For example, the cilia in the respiratory epithelium or the peristalsis in the gastrointestinal tract will push the foreign bodies outward; tears will protect the eyes and drain small irritating particles through the nasolacrimal duct into the nasal cavity. Additionally, the immune's system surveillance sometimes also takes part in the elimination of the foreign bodies. Macrophages, which are present in virtually all tissues, are the cells responsible for this. They can either phagocyte the foreign body or initiate an inflammatory reaction against it. If the foreign body cannot be removed, the inflammatory reaction can become chronic and lead to fibrosis.

One of the important aspects of macrophage biology is their primary role in the initiation of foreign body responses (FBR), which result in inflammation and fibrotic encapsulation of introduced foreign object. When a foreign body penetrates the organism, macrophages are able to recognize it either through pattern-recognition receptors or through opsonic receptors. Pattern-recognition receptors are usually activated by bacteria, fungi, or parasites. Opsonic receptors are involved in opsonization, which is a process that helps phagocytes identify a foreign body or invading agent. Without opsonization, the identification and elimination of the foreign body would be much less efficient.

Most of the biomaterials generated for the purposes of implantation are recognized by macrophages as foreign bodies and thus may induce adverse reactions such as chronic inflammation and excessive fibrosis. The introduction of biomaterials (such as surgical implants) into the body is accompanied by local tissue trauma followed

TABLE 10.1

Macrophage Receptors with Known or Proposed Function in Foreign Body Recognition

Receptor	Ligand
FcγRI (CD64)	IgG1, IgG3, IgG4
FcγRIIa (CD32a)	IgG3, IgG1, IgG2
FcγRIIc (CD32c)	IgG
FcγRIIIa (CD16a)	IgG
CR1 (CD35)	Mannan-binding lectin, C1q, C4b, C3b
CR3 (αMβ2, CD11b/CD18, Mac-1)	iC3b
CR4 (αXβ2, CD11c/CD18)	iC3b
α5β1	Fibronectin, vitronectin
αvβ3	Vitronectin
αvβ5	Vitronectin
TLR1/2	Gram-positive bacteria, HSP60, HSP70, HMGB1
TLR4	LPS, HSP60, HSP70, HMGB1, fibronectin EDA, fibrinogen, polycationic and polyanionic biomaterials, divalent cations (Ni^{2+} and Co^{2+})
SR-AI/II	TiO_2 particles
MARCO	TiO_2 particles, silica

by the release of damage-associated molecular patterns (DAMPs), activation of complement, and blood coagulation systems. As a result of protein cascades activation, chemoattractants such as complement factors C3a and C5a are locally released, and fibrin deposition is initiated on the implant surface. These events induce recruitment of innate immune cells and formation of protein layer on the biomaterial surface. Adsorbed proteins include components of complement system, extracellular matrix (ECM) molecules (e.g., fibronectin, fibrinogen, fibrin, vitronectin), albumins, immunoglobulins (Ig), and DAMPs [21,22]. These proteins play the role of opsonins and can be recognized by pattern-recognition receptors (PRRs) and opsonic receptors on macrophages and other immune cells initiating inflammatory response. Among proteins involved in the recognition of adsorbed proteins, three groups of receptors are characterized. These include integrins, toll-like receptors (TLRs), and scavenger receptors (SRs) that are abundantly expressed by tissue macrophages. A list of macrophage receptors and their ligands with known or proposed function in foreign body recognition can be found in Table 10.1.

10.3.1 INTEGRINS

The family of integrin receptors consists of 24 transmembrane proteins that share common αβ heterodimeric structure (Figure 10.2). Integrins are involved in cell adhesion through interactions with ECM proteins including collagen, fibronectin, laminin, vitronectin, and others. They are also known to induce intracellular signaling regulating actin cytoskeleton assembly, cell migration, proliferation, and apoptosis. The ligand specificity of integrins is determined by combination of extracellular

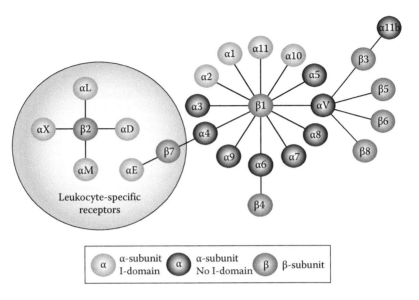

FIGURE 10.2 The integrin superfamily. The integrins can be subdivided according to their β chains. However, some α chains can combine with several β chains. Twenty-four different integrins are present in humans. (Adapted from Niu, G. and Chen, X., *Theranostics*, 1, 30, 2011.)

domains of α and β chains. Integrins are evolutionary conserved proteins and are expressed by multiple cell types. Macrophages express several members of integrin receptor family including β1 integrins α1β1, α2β1, α3β1, α4β1, α5β1, β2 integrins αLβ2, αMβ2, and αXβ2, as well as αvβ3, and αvβ5 [23,24].

Proteins adsorbed on the surface of biomaterials play a major role in the implant recognition and initiation of FBR [25]. Some of these proteins denature on the hydrophobic biomaterials exposing binding sites that can be recognized by several types of macrophage integrins including αMβ2 (Mac-1, CD11b/CD18) and RGD binding integrins αvβ3, αvβ5, and α5β1 [22,24]. For example, surface-adsorbed proteins of blood coagulation cascade including high molecular weight kininogen (HMWK), fibrinogen, factor X, and complement component C3b can ligate αMβ2 integrin (CD11b/CD18), which is abundantly expressed by macrophages [22].

Nonspecific adsorption of Ig and C3 component of complement on the surface of biomaterials launches classical and alternative complement pathways. Moreover, lack of negative regulators of complement cascade activation on the biomaterial surface results in uncontrolled complement activation and release of phagocyte chemoattractant C3a. In turn, recruited monocytes recognize adsorbed opsonins (e.g., fibronectin and vitronectin) using β1 and β2 integrins that may initiate formation of foreign body giant cells.

Integrins are not only involved in the initial adhesion to biomaterials but also mediate inflammatory response upon contact with particulate biomaterials [24]. For example, in human macrophages, it was observed that CD11b/CD18 integrin recognizes titanium alloy particles followed by signaling through transcription factors NFκB and NF-IL6 inducing expression of pro-inflammatory cytokines TNFα and IL6 [26]. Interestingly, the presence of particle-adsorbed LPS seems to be important in

CD11b/CD18-dependent release of pro-inflammatory cytokines, indicating that this receptor may work in coordination with CD14 and TLR4 [24]. CD11b/CD18 and RGD-binding integrins are also involved in the formation of fibrous capsule around foreign material since their specific targeting significantly reduces capsule thickness [24].

Overall, targeting of leukocyte-expressed integrins is a potentially perspective approach to block FBR at initial stages. However, since integrin function is redundant in the way that multiple receptors are able to recognize the same adsorbed protein ligand, only extensive local targeting of multiple integrins may be beneficial to prevent adhesion and activation of recruited macrophages.

10.3.2 TOLL-LIKE RECEPTORS

TLRs are the family of cell surface and intracellular transmembrane PRRs containing leucine-rich repeats (LRRs) in the extracellular domain that recognize wide range of exogenous and endogenous structurally conserved molecules named pathogen-associated molecular patterns (PAMPs). Upon ligand binding, TLRs initiate intracellular signaling through adaptor molecules MyD88 and TRIF. This results in activation of transcription factor NFκB and several interferon regulatory factors (IRFs) followed by expression of pro-inflammatory cytokines and type I interferons (Figure 10.3). In recent years, it has become evident that TLRs recognize wide range of endogenous ligands known as damage-associated molecular patterns (DAMPs). These molecules such as heat shock proteins, certain components of ECM, self DNA, RNA, high-mobility group box 1 (HMGB1), and others are released from cells following tissue injury and become available for TLR recognition. Macrophages express most of the known TLRs, indicating their involvement in biomaterial recognition. Moreover, TLR4- and TLR9-expressing macrophages were found at the interfacial membrane of aseptically loosened hip replacement implants [27].

Since surgical implantation of biomaterials is accompanied by wound and tissue injury, DAMPs can be locally released and become associated with biomaterial surface [22]. Thus, as in the case of integrins TLRs rather recognize surface-associated ligands than biomaterial itself. For example, in the study by Greenfield et al. mouse macrophages were able to recognize only bacterial debris-coated titanium particles through TLR2 and TLR4 engagement but not endotoxin-free particles [28]. However, several types of biomaterials such as polycationic compounds polyethyleneimine, polylysine, cationic dextran, and cationic gelatin can be directly recognized by TLRs such as TLR4. Similarly, polyanionic compound alginate as well as hydroxyapatite is recognized by both TLR2 and TLR4 [22]. Moreover, TLR4 is known to recognize directly divalent metal cations such as Ni^{2+} and Co^{2+}, resulting in inflammatory complications in metal-on-metal joint replacements [29,30]. Overall, a number of studies demonstrate particular importance of TLR1/2 and TLR4-mediated pathways in particle-induced osteolysis that involve MyD88 and NFκB pro-inflammatory signaling [30]. For example, titanium particle-induced osteolysis and TNFα production were significantly inhibited in TLR2$^{-/-}$ and TLR4$^{-/-}$ mice [28]. It was observed that exogenous and endogenous PAMPs and DAMPs such as adsorbed endotoxin and macrophage-released HSP60 are responsible for TLR4-induced inflammation and osteolysis [31]. However, TLR1/2 ligands during biomaterial recognition are insufficiently characterized.

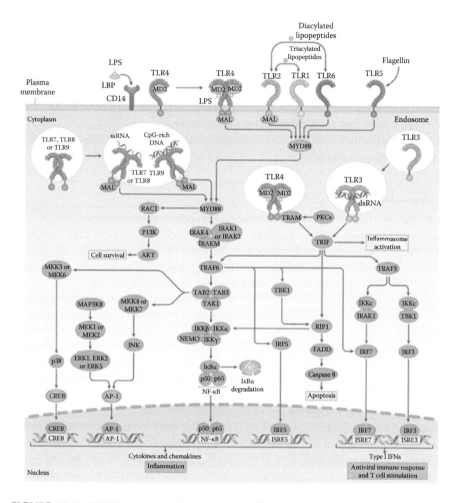

FIGURE 10.3 Toll-like receptor family and signaling cascades. (Adapted from Gay, N.J. et al., *Nat. Rev. Immunol.*, 14, 546, 2014.)

10.3.2 SCAVENGER RECEPTORS

SRs are structurally heterogeneous family of transmembrane proteins that participate in the clearance of modified low-density lipoproteins (LDL) from circulation. Besides modified LDL, SRs are able to recognize and bind a broad range of oxidized proteins, lipoproteins, and lipids, altogether named "neo-self" antigens. In addition, ligands of SR include apoptotic bodies and pathogen-associated structures, indicating that these proteins may function as PRRs. Currently, the SR family is subdivided into eight classes (A–H) according to the structure (Figure 10.4). Most of them are expressed by myeloid cells and specifically by macrophages [32]. One of the main characteristics of SR is their high redundancy in ligand repertoire. SRs are known to participate in signaling cascades. However, since most of the SRs have very short

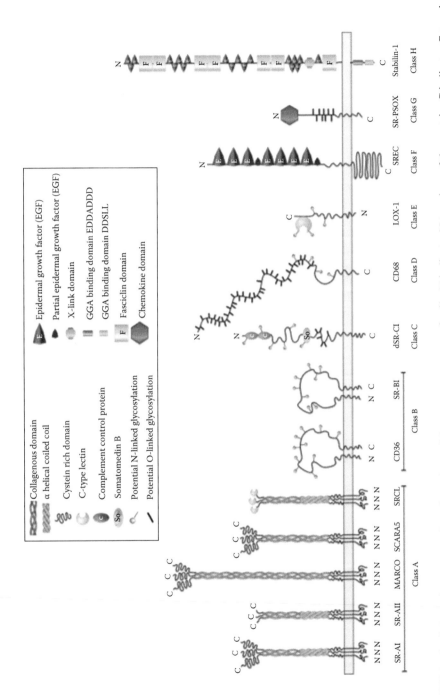

FIGURE 10.4 Schematic representation of members of the scavenger receptor family. N indicates N-terminus of the protein; C indicates C-terminus of the protein. Plasma membrane is presented as a yellow bar. (Adapted from Kzhyshkowska, J. et al., *Immunobiology*, 217, 492, 2012.)

cytoplasmic tails that lack signaling motifs, the exact mechanisms of SR-mediated signaling are obscure. It is suggested that SR form multimeric protein complexes (signalosomes) with other receptors such as TLRs to mediate intracellular signaling. In this case, the nature of co-receptor defines whether SR will induce inflammatory or anti-inflammatory signaling [33].

Participation of SR in FBRs to biomaterials is less investigated compared to integrins and TLRs. Most of the published studies involving SR focused specifically on the reactions of alveolar macrophages during the recognition of inhaled environmental particles such as titanium dioxide (TiO_2) and silica. The consequences of such recognition in terms of macrophage activation are not completely clarified and downstream signaling pathways are obscure. Up to date, only SR of class A including SR-AI/II and MARCO (SR-AII) were found to be involved in the recognition of foreign materials. For example, MARCO participates in engulfment of titanium dioxide, iron, and silica particles in alveolar macrophages [22]. Similarly, SR-AI/II is involved in the recognition of inhaled TiO_2 particles in mice [34]. SR may bind negatively charged particles directly using SRCR domain (in case of MARCO) or collagenous domain (in case of SR-AI/II) [35]. It is not known whether protein opsonization of particles plays a role in SR-mediated recognition. Noteworthy, MARCO-mediated uptake of silica particles results in cytotoxicity and macrophage apoptosis, whereas TiO_2 uptake by alveolar macrophages does not induce cytotoxic effect [35]. Some of the studies using knockout mouse models suggest that the expression of SR-AI/II and MARCO is beneficial during exposure to titanium wear debris since SR-AI/II$^{-/-}$ and MARCO$^{-/-}$ mice develop exacerbated lung inflammation after challenge with TiO_2 particles [34]. Thus, overexpression of certain SRs on macrophages during FBR may aid to reduce inflammation and needs additional investigation. The summary of known mechanisms of macrophage-mediated biomaterials recognition is presented in Figure 10.5.

10.4 THERAPEUTIC CONTROL OF MACROPHAGE PHENOTYPE DURING IMPLANT-INDUCED COMPLICATIONS

As discussed in the previous paragraph, nearly all implanted materials induce FBR. The nature and surface topography of biomaterials define the severity of FBR that may also depend on individual reactions of patients with implanted devices. The recognition of biomaterials by macrophage surface receptors may result in macrophage activation, recruitment of other immune cells, and acute inflammation followed by chronic inflammation, formation of foreign body giant cells, and fibrous encapsulation of implanted material [25]. In addition, the presence of implant-associated infection may strongly amplify inflammatory response and induce implant failure [36]. One of the perspective strategies to increase tissue integration of implant and prevent excessive implant-associated inflammation is therapeutic manipulation of macrophage phenotype.

Multiple studies demonstrate that interaction of macrophages with various biomaterials including titanium, polyethylene terephthalate, polymethylmethacrylate, and others induce expression and production of pro-inflammatory (M1) cytokines and chemokines such as TNFα, MCP-1 (CCL2), MCP-3 (CCL7), IL1β, IL6, MIP-1α

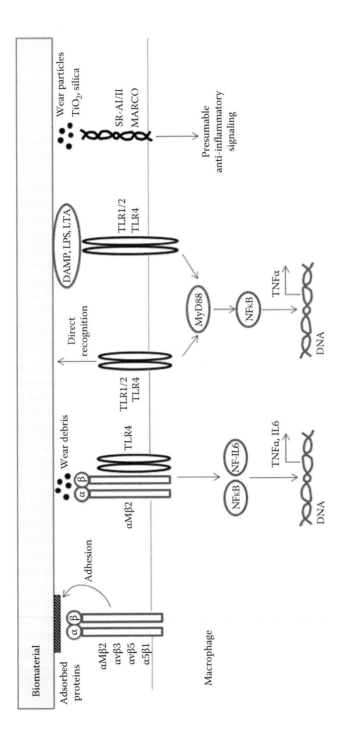

FIGURE 10.5 Biomaterial recognition by macrophage receptors. Protein adsorption on the surface of implanted biomaterials induces macrophage adhesion through integrin receptors such as αMβ2 (CD11b/CD18), αvβ3, αvβ5, and α5β1. In addition, ligation of CD11b/CD18 by wear particles may induce pro-inflammatory signaling in cooperation with TLR4. TLRs including TLR1/2 and TLR4 are able to directly recognize certain biomaterials or respond to material-adsorbed damage-associated molecular patterns (DAMP) and bacterial products (LPS, LTA) inducing expression of pro-inflammatory cytokines. In contrast, expression of scavenger receptors (SR) such as macrophage receptor with collagenous structure (MARCO) and SR-AI/II may dampen macrophage inflammatory response. TLR, toll-like receptor; LPS, lipopolysaccharide; and LTA, lipoteichoic acid.

(CCL3), and high M1/M2 index. Wear debris produced by implanted materials during exploitation is one of the primary causes of inflammatory response [37–41]. It is important to note that the magnitude of cytokine responses may vary strongly between individuals, indicating that personalized approaches are required for the treatment of such complications [41].

Since M2 cytokines such as IL4 are known to alleviate excessive inflammation, it was suggested that macrophage re-polarization toward M2 pro-healing phenotype may prevent implant-related inflammatory complications [42,43]. For example, it was observed that pro-inflammatory response to titanium particles was suppressed in IL4-treated human macrophages [44]. Moreover, IL4 decreased production of pro-inflammatory cytokine TNFα and increased production of anti-inflammatory cytokine IL1Ra in M1-polarized mouse macrophages stimulated by polymethylmethacrylate [45]. In addition, local administration of IL4 reduced inflammation and osteolysis caused by polyethylene particles in mouse calvarial model [46]. Although being perspective approach, in vivo studies involving local M2 re-polarization of macrophages are sporadic and are restricted to application of IL4 as an anti-inflammatory agent. It is anticipated that more potent cytokines and their combinations may be generated in order to dampen implant-induced inflammation.

Another potentially perspective strategy for M2 macrophage re-polarization is surface modifications of biomaterials [47]. Different surface properties such as roughness, porosity, and micropatterning were shown to affect macrophage polarization [47]. It was previously described that changing macrophage morphology using micropatterned surfaces affects their polarization and may be used to maintain M2 phenotype [48]. At the same time, several studies involving nano- and microstructured biomaterials revealed that surface patterning results in only moderate changes of macrophage phenotype. Moreover, such modifications induce expression of both pro- and anti-inflammatory factors, suggesting that additional studies are necessary to clarify optimal pattern and size of surface modifications that will result in preferential induction of healing macrophage phenotype [47,49,50].

The third approach for modulation of implant-associated macrophage phenotype is an application of anti-inflammatory implant coatings. These thin coatings based on degradable biomaterials serve to create biocompatible surface, resulting in reduced FBR. In general, two types of coatings can be used to improve outcome of cell interaction with implants. The first type of coating based on natural polymers (e.g., collagen, hyaluronan, alginate) aims to increase interaction of cells with implanted biomaterials [25]. In contrast, the second type of coating based on highly hydrophilic synthetic polymers prevents protein adsorption and cell activation on the surface of biomaterials [51]. Both types of coatings have their specific advantages and disadvantages. For example, natural polymers can be rapidly degraded and lose protective effect over time, whereas synthetic polymers form passive layer on the surface of implants and do not promote active interaction of biomaterials with host cells, which may potentially delay implant integration. One of the widely used coatings with high biocompatibility is hydrogel. Hydrogels are

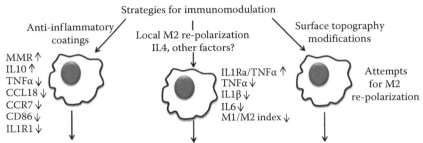

Strategies for immunomodulation

Anti-inflammatory coatings / Local M2 re-polarization IL4, other factors? \ Surface topography modifications

MMR ↑
IL10 ↑
TNFα ↓
CCL18 ↓
CCR7 ↓
CD86 ↓
IL1R1 ↓

IL1Ra/TNFα ↑
TNFα ↓
IL1β ↓
IL6 ↓
M1/M2 index ↓

Attempts for M2 re-polarization

Decreased fibrotic encapsulation and osteolysis, improved healing and implant integration

FIGURE 10.6 Current strategies for therapeutic control of macrophage phenotype during implant-induced complications.

macromolecular structures composed of network of cross-linked polymer chains. They are well hydrated and have high porosity, resulting in efficient transfer of nutrients and bioactive molecules [52]. Several studies demonstrated the ability of hydrogels to modulate macrophage phenotype and promote healing. For example, zwitterionic poly(carboxybetaine methacrylate) (PCBMA)-based hydrogels were resistant to fibrotic incapsulation in mice and induced M2-like macrophage phenotype with increased expression of MMR (CD206), Arg1, IL10, and SR-BI/II [53]. In rats, implantation of polypropylene meshes coated with ECM-based hydrogels also increased M2/M1 index of locally recruited macrophages [54]. Lastly, coatings composed of polyelectrolyte multilayers (PEM) offer a good opportunity for local immunomodulation. These structures that consist of multiple layers of polyanions and polycations match most of the implant surfaces. In addition, these coatings can be loaded with anti-inflammatory and antimicrobial drugs and allow their controlled local release [55,56]. Several studies showed the ability of PEM-coated biomaterials to reduce FBR and increase IL10 levels [57,58]. A summary of current strategies for the modulation of macrophage phenotype during implant-induced complications is presented in Figure 10.6.

Overall, multiple studies demonstrate that the manipulation of macrophage phenotype using local cytokine administration and implant surface modifications (including topography changes and application of coatings) are feasible approaches to reduce FBR and increase implant functionality. However, improvements such as optimization of potent anti-inflammatory cytokine combinations and prolongation of controlled substance release from implant coatings are necessary in order to overcome inflammatory complications. It is also anticipated that combination of different approaches can be beneficial to reach this aim.

ACKNOWLEDGMENTS

This research was supported by Tomsk State University Competitiveness Improvement Program (J.K.) and European Union's Seventh Framework Programme for research, technological development, and demonstration under grant agreement no. 602694 (J.K.).

REFERENCES

1. Gordon, S., Pluddemann, A., Martinez Estrada, F. (2014) Macrophage heterogeneity in tissues: Phenotypic diversity and functions. *Immunological Reviews* **262**, 36–55.
2. Epelman, S., Lavine, K. J., Randolph, G. J. (2014) Origin and functions of tissue macrophages. *Immunity* **41**, 21–35.
3. Ginhoux, F., Jung, S. (2014) Monocytes and macrophages: Developmental pathways and tissue homeostasis. *Nature Reviews Immunology* **14**, 392–404.
4. Davies, L. C., Jenkins, S. J., Allen, J. E., Taylor, P. R. (2013) Tissue-resident macrophages. *Nature Immunology* **14**, 986–995.
5. You, Q., Cheng, L., Kedl, R. M., Ju, C. (2008) Mechanism of T cell tolerance induction by murine hepatic Kupffer cells. *Hepatology* **48**, 978–990.
6. Bilzer, M., Roggel, F., Gerbes, A. L. (2006) Role of Kupffer cells in host defense and liver disease. *Liver International: Official Journal of the International Association for the Study of the Liver* **26**, 1175–1186.
7. Hadis, U., Wahl, B., Schulz, O., Hardtke-Wolenski, M., Schippers, A., Wagner, N., Muller, W., Sparwasser, T., Forster, R., Pabst, O. (2011) Intestinal tolerance requires gut homing and expansion of FoxP3+ regulatory T cells in the lamina propria. *Immunity* **34**, 237–246.
8. Xue, J., Schmidt, S. V., Sander, J., Draffehn, A., Krebs, W., Quester, I., De Nardo, D. et al. (2014) Transcriptome-based network analysis reveals a spectrum model of human macrophage activation. *Immunity* **40**, 274–288.
9. Stein, M., Keshav, S., Harris, N., Gordon, S. (1992) Interleukin 4 potently enhances murine macrophage mannose receptor activity: A marker of alternative immunologic macrophage activation. *Journal of Experimental Medicine* **176**, 287–292.
10. Goerdt, S., Orfanos, C. E. (1999) Other functions, other genes: Alternative activation of antigen-presenting cells. *Immunity* **10**, 137–142.
11. Martinez, F. O., Gordon, S. (2014) The M1 and M2 paradigm of macrophage activation: Time for reassessment. *F1000Prime Reports* **6**, 13.
12. Sica, A., Mantovani, A. (2012) Macrophage plasticity and polarization: In vivo veritas. *Journal of Clinical Investigation* **122**, 787–795.
13. Murray, P. J., Allen, J. E., Biswas, S. K., Fisher, E. A., Gilroy, D. W., Goerdt, S., Gordon, S. et al. (2014) Macrophage activation and polarization: Nomenclature and experimental guidelines. *Immunity* **41**, 14–20.
14. Gratchev, A., Kzhyshkowska, J., Kothe, K., Muller-Molinet, I., Kannookadan, S., Utikal, J., Goerdt, S. (2006) Mphi1 and Mphi2 can be re-polarized by Th2 or Th1 cytokines, respectively, and respond to exogenous danger signals. *Immunobiology* **211**, 473–486.
15. Stout, R. D., Jiang, C., Matta, B., Tietzel, I., Watkins, S. K., Suttles, J. (2005) Macrophages sequentially change their functional phenotype in response to changes in microenvironmental influences. *Journal of Immunology* **175**, 342–349.
16. Stout, R. D., Suttles, J. (2004) Functional plasticity of macrophages: Reversible adaptation to changing microenvironments. *Journal of Leukocyte Biology* **76**, 509–513.
17. Mosby (2009) *Mosby's Medical Dictionary*, 8th edn. Elsevier: Philadelphia, PA.
18. Ambe, P., Weber, S. A., Schauer, M., Knoefel, W. T. (2012) Swallowed foreign bodies in adults. *Deutsches Arzteblatt International* **109**, 869–875.
19. Sugawa, C., Ono, H., Taleb, M., Lucas, C. E. (2014) Endoscopic management of foreign bodies in the upper gastrointestinal tract: A review. *World Journal of Gastrointestinal Endoscopy* **6**, 475–481.
20. Hariharan, D., Lobo, D. N. (2013) Retained surgical sponges, needles and instruments. *Annals of the Royal College of Surgeons of England* **95**, 87–92.
21. Wilson, C. J., Clegg, R. E., Leavesley, D. I., Pearcy, M. J. (2005) Mediation of biomaterial-cell interactions by adsorbed proteins: A review. *Tissue Engineering* **11**, 1–18.

22. Love, R. J., Jones, K. S. (2013) The recognition of biomaterials: Pattern recognition of medical polymers and their adsorbed biomolecules. *Journal of Biomedical Materials Research Part A* **101**, 2740–2752.

23. Ammon, C., Meyer, S. P., Schwarzfischer, L., Krause, S. W., Andreesen, R., Kreutz, M. (2000) Comparative analysis of integrin expression on monocyte-derived macrophages and monocyte-derived dendritic cells. *Immunology* **100**, 364–369.

24. Zaveri, T. D., Lewis, J. S., Dolgova, N. V., Clare-Salzler, M. J., Keselowsky, B. G. (2014) Integrin-directed modulation of macrophage responses to biomaterials. *Biomaterials* **35**, 3504–3515.

25. Anderson, J. M., Rodriguez, A., Chang, D. T. (2008) Foreign body reaction to biomaterials. *Seminars in Immunology* **20**, 86–100.

26. Nakashima, Y., Sun, D. H., Trindade, M. C., Maloney, W. J., Goodman, S. B., Schurman, D. J., Smith, R. L. (1999) Signaling pathways for tumor necrosis factor-alpha and interleukin-6 expression in human macrophages exposed to titanium-alloy particulate debris in vitro. *Journal of Bone and Joint Surgery, America* **81**, 603–615.

27. Takagi, M., Tamaki, Y., Hasegawa, H., Takakubo, Y., Konttinen, L., Tiainen, V. M., Lappalainen, R., Konttinen, Y. T., Salo, J. (2007) Toll-like receptors in the interface membrane around loosening total hip replacement implants. *Journal of Biomedical Materials Research Part A* **81**, 1017–1026.

28. Greenfield, E. M., Beidelschies, M. A., Tatro, J. M., Goldberg, V. M., Hise, A. G. (2010) Bacterial pathogen-associated molecular patterns stimulate biological activity of orthopaedic wear particles by activating cognate Toll-like receptors. *Journal of Biological Chemistry* **285**, 32378–32384.

29. Raghavan, B., Martin, S. F., Esser, P. R., Goebeler, M., Schmidt, M. (2012) Metal allergens nickel and cobalt facilitate TLR4 homodimerization independently of MD2. *EMBO Reports* **13**, 1109–1115.

30. Lin, T. H., Tamaki, Y., Pajarinen, J., Waters, H. A., Woo, D. K., Yao, Z., Goodman, S. B. (2014) Chronic inflammation in biomaterial-induced periprosthetic osteolysis: NF-kappaB as a therapeutic target. *Acta Biomaterialia* **10**, 1–10.

31. Gu, Q., Shi, Q., Yang, H. (2012) The role of TLR and chemokine in wear particle-induced aseptic loosening. *Journal of Biomedicine and Biotechnology* **2012**, 596870.

32. Kzhyshkowska, J., Neyen, C., Gordon, S. (2012) Role of macrophage scavenger receptors in atherosclerosis. *Immunobiology* **217**, 492–502.

33. Canton, J., Neculai, D., Grinstein, S. (2013) Scavenger receptors in homeostasis and immunity. *Nature Reviews Immunology* **13**, 621–634.

34. Arredouani, M. S., Yang, Z., Imrich, A., Ning, Y., Qin, G., Kobzik, L. (2006) The macrophage scavenger receptor SR-AI/II and lung defense against pneumococci and particles. *American Journal of Respiratory Cell and Molecular Biology* **35**, 474–478.

35. Thakur, S. A., Hamilton, R. F., Jr., Holian, A. (2008) Role of scavenger receptor a family in lung inflammation from exposure to environmental particles. *Journal of Immunotoxicology* **5**, 151–157.

36. Arciola, C. R., Campoccia, D., Speziale, P., Montanaro, L., Costerton, J. W. (2012) Biofilm formation in Staphylococcus implant infections. A review of molecular mechanisms and implications for biofilm-resistant materials. *Biomaterials* **33**, 5967–5982.

37. Grotenhuis, N., Bayon, Y., Lange, J. F., Van Osch, G. J., Bastiaansen-Jenniskens, Y. M. (2013) A culture model to analyze the acute biomaterial-dependent reaction of human primary macrophages. *Biochemical and Biophysical Research Communications* **433**, 115–120.

38. Pearl, J. I., Ma, T., Irani, A. R., Huang, Z., Robinson, W. H., Smith, R. L., Goodman, S. B. (2011) Role of the Toll-like receptor pathway in the recognition of orthopedic implant wear-debris particles. *Biomaterials* **32**, 5535–5542.

39. Purdue, P. E., Koulouvaris, P., Nestor, B. J., Sculco, T. P. (2006) The central role of wear debris in periprosthetic osteolysis. *HSS Journal* **2**, 102–113.

40. Nakashima, Y., Sun, D. H., Trindade, M. C., Chun, L. E., Song, Y., Goodman, S. B., Schurman, D. J., Maloney, W. J., Smith, R. L. (1999) Induction of macrophage C-C chemokine expression by titanium alloy and bone cement particles. *Journal of Bone and Joint Surgery, British* **81**, 155–162.

41. Stankevich, K. S., Gudima, A., Filimonov, V. D., Kluter, H., Mamontova, E. M., Tverdokhlebov, S. I., Kzhyshkowska, J. (2015) Surface modification of biomaterials based on high-molecular polylactic acid and their effect on inflammatory reactions of primary human monocyte-derived macrophages: Perspective for personalized therapy. *Materials Science and Engineering Part C: Materials for Biological Applications* **51**, 117–126.

42. Yagil-Kelmer, E., Kazmier, P., Rahaman, M. N., Bal, B. S., Tessman, R. K., Estes, D. M. (2004) Comparison of the response of primary human blood monocytes and the U937 human monocytic cell line to two different sizes of alumina ceramic particles. *Journal of Orthopaedic Research* **22**, 832–838.

43. Goodman, S. B., Gibon, E., Pajarinen, J., Lin, T. H., Keeney, M., Ren, P. G., Nich, C. et al. (2014) Novel biological strategies for treatment of wear particle-induced periprosthetic osteolysis of orthopaedic implants for joint replacement. *Journal of the Royal Society Interface* **11**, 20130962.

44. Pajarinen, J., Kouri, V. P., Jamsen, E., Li, T. F., Mandelin, J., Konttinen, Y. T. (2013) The response of macrophages to titanium particles is determined by macrophage polarization. *Acta Biomaterialia* **9**, 9229–9240.

45. Rao, A. J., Gibon, E., Ma, T., Yao, Z., Smith, R. L., Goodman, S. B. (2012) Revision joint replacement, wear particles, and macrophage polarization. *Acta Biomaterialia* **8**, 2815–2823.

46. Rao, A. J., Nich, C., Dhulipala, L. S., Gibon, E., Valladares, R., Zwingenberger, S., Smith, R. L., Goodman, S. B. (2013) Local effect of IL-4 delivery on polyethylene particle induced osteolysis in the murine calvarium. *Journal of Biomedical Materials Research Part A* **101**, 1926–1934.

47. Rostam, H. M., Singh, S., Vrana, N. E., Alexander, M. R., Ghaemmaghami, A. M. (2015) Impact of surface chemistry and topography on the function of antigen presenting cells. *Biomaterials Science* **3**, 424–441.

48. McWhorter, F. Y., Wang, T., Nguyen, P., Chung, T., Liu, W. F. (2013) Modulation of macrophage phenotype by cell shape. *Proceedings of the National Academy of Sciences of the United States of America* **110**, 17253–17258.

49. Paul, N. E., Skazik, C., Harwardt, M., Bartneck, M., Denecke, B., Klee, D., Salber, J., Zwadlo-Klarwasser, G. (2008) Topographical control of human macrophages by a regularly microstructured polyvinylidene fluoride surface. *Biomaterials* **29**, 4056–4064.

50. Chen, S., Jones, J. A., Xu, Y., Low, H. Y., Anderson, J. M., Leong, K. W. (2010) Characterization of topographical effects on macrophage behavior in a foreign body response model. *Biomaterials* **31**, 3479–3491.

51. Wisniewski, N., Reichert, M. (2000) Methods for reducing biosensor membrane biofouling. *Colloids and Surfaces B: Biointerfaces* **18**, 197–219.

52. Peppas, N. A., Huang, Y., Torres-Lugo, M., Ward, J. H., Zhang, J. (2000) Physicochemical foundations and structural design of hydrogels in medicine and biology. *Annual Review of Biomedical Engineering* **2**, 9–29.

53. Zhang, L., Cao, Z., Bai, T., Carr, L., Ella-Menye, J. R., Irvin, C., Ratner, B. D., Jiang, S. (2013) Zwitterionic hydrogels implanted in mice resist the foreign-body reaction. *Nature Biotechnology* **31**, 553–556.

54. Wolf, M. T., Dearth, C. L., Ranallo, C. A., LoPresti, S. T., Carey, L. E., Daly, K. A., Brown, B. N., Badylak, S. F. (2014) Macrophage polarization in response to ECM coated polypropylene mesh. *Biomaterials* **35**, 6838–6849.

55. Wong, S. Y., Moskowitz, J. S., Veselinovic, J., Rosario, R. A., Timachova, K., Blaisse, M. R., Fuller, R. C., Klibanov, A. M., Hammond, P. T. (2010) Dual functional polyelectrolyte multilayer coatings for implants: Permanent microbicidal base with controlled release of therapeutic agents. *Journal of the American Chemical Society* **132**, 17840–17848.

56. Vrana, N. E., Erdemli, O., Francius, G., Fahs, A., Rabineau, M., Debry, C., Tezcaner, A., Keskin, D., Lavalle, P. (2014) Double entrapment of growth factors by nanoparticles loaded into polyelectrolyte multilayer films. *Journal of Materials Chemistry B* **2**, 999–1008.

57. Schultz, P., Vautier, D., Richert, L., Jessel, N., Haikel, Y., Schaaf, P., Voegel, J.-C., Ogier, J., Debry, C. (2005) Polyelectrolyte multilayers functionalized by a synthetic analogue of an anti-inflammatory peptide, α-MSH, for coating a tracheal prosthesis. *Biomaterials* **26**, 2621–2630.

58. Ma, M., Liu, W. F., Hill, P. S., Bratlie, K. M., Siegwart, D. J., Chin, J., Park, M., Guerreiro, J., Anderson, D. G. (2011) Development of cationic polymer coatings to regulate foreign-body responses. *Advanced Materials* **23**, H189–H194.

59. Niu, G., Chen, X. (2011) Why integrin as a primary target for imaging and therapy. *Theranostics* **1**, 30–47.

60. Gay, N. J., Symmons, M. F., Gangloff, M., Bryant, C. E. (2014) Assembly and localization of Toll-like receptor signalling complexes. *Nature Reviews Immunology* **14**, 546–558.

Index